Springer Series in Optical Sciences Volume 20
Edited by David L. MacAdam

Springer Series in Optical Sciences

Edited by David L. MacAdam

Yu. I. Ostrovsky · M. M. Butusov
G. V. Ostrovskaya

Interferometry by Holography

With 184 Figures

Springer-Verlag Berlin Heidelberg GmbH 1980

Prof. Dr. Yu. I. Ostrovsky

A. F. Ioffe Physico-Technical Institute,
Academy of Sciences of the USSR,
SU-194021 Leningrad

Prof. Dr. M. M. Butusov

Leningrad Polytechnical Institute,
SU-195251 Leningrad

Prof. Dr. G. V. Ostrovskaya

A. F. Ioffe Physico-Technical Institute,
Academy of Sciences of the USSR,
SU- 194021 Leningrad

Authorized Translation from the Russian:

Ю. И. ОСТРОВСКИЙ, М. М. БУТУСОВ, Г. В. ОСТРОВСКАЯ
ГОЛОГРАФИЧЕСКАЯ ИНТЕРФЕРОМЕТРИЯ

Yu. I. Ostrovsky, M. M. Butusov, G. V. Ostrovskaya
Golograverscheskaya Interferometriya

DOI 10.1007/978-3-540-39008-4

Library of Congress Cataloging in Publication Data: Ostrovskiĭ, Yuriĭ Isaevich. Interferometry by holography. (Springer series in optical sciences ; v. 20) Translation of Gologaficheskaîa interferometriîa. Bibliography: p. Includes index. 1. Holographic interferometry. I. Butusov, Mikhail Mikhaĭlovich, joint author. II. Ostrovskaîa, Galîa Vsevolodovna, joint author. TA1555.O8713 621.36'75 79-27443

Originally published by Springer-Verlag Berlin Heidelberg New York in 1980.

MyCopy version of the original edition 1980

2153/3130-543210

www.springer.com/mycopy

Preface

This book is an introduction to holographic interferometry — a field of holography having a great number of important and practically useful applications. It is intended for specialists working in the field of optics and holography, and also for students of the relevant specialities.

At present, a greater and greater number of mechanical engineers, turbine designers, testers of diverse equipment, biologists, crystallographers, and so on have to do with holographic interferometry. To allow these specialists, who are comparatively far from optics, to master the subject too, the main content of the book is preceded by an introductory chapter treating the fundamental concepts of the interference of light, optical interferometry, holography and holographic interferometry.

The following chapters deal with the fundamentals of the theory of holographic interferometry and of experimental equipment. The authors have set themselves the task of sharing their more than ten year of experience of work in the field of holographic interferometry with their readers. In this connection, the questions which they dealt with directly are considered in somewhat greater detail, as a rule, than those with which they have become acquainted only from publications on the subject.

A sufficiently detailed (although far from complete) bibliography gives any interested reader an opportunity to improve his knowledge of this field.

Of course, the present book cannot cover all of the numerous branches of science and engineering in which holographic interferometry is used at present. Particularly, the control of optical surfaces and the investigation of stresses in transparent models are not considered in it. This subject is treated in detail in Volume 11 of this Series, "Matrix Theory of Photoelasticity" by P.S. Theocaris and E.E. Gdoutos. Also omitted are problems associated with speckle-interferometry, which formally does not relate to holographic interferometry, but is its very close neighbor.

The main work of writing the separate chapters of the book was distributed as follows: Yu.I. Ostrovsky and G.V. Ostrovskaya – Chaps.1,3 and Sect. 4.3, M.M. Butusov – Chap. 2 and Sects. 4.1,2,4, and Yu.I. Ostrovsky – Chap. 5.

The authors are greatly indebted to I.S. Klimenko who attentively read the manuscript and made a number of remarks that undoubtedly helped us to improve it.

Authors wish to thank Dr. Helmuth K.V. Lotsch for his initiative and constructive help in the edition of the English translation of this book.

Special thanks are due to Dr. David L.MacAdam, the editor of this series, whose help made this book more readable.

Leningrad, November 1979

Yu.I. Ostrovsky
M.M. Butusov
G.V. Ostrovskaya

Contents

1. General Principles

1.1 Interference of Light

Interference is the most striking manifestation of the wave nature of light. This phenomenon consists of the spatial redistribution of the volume density of energy in the region of superpositon of two or more waves.

Interference phenomena were first described in the 17th century in the works of Boyle, Grimaldi, and Hooke, who observed colored fringes when light was reflected from thin plates and films. Hooke tried to explain this phenomenon from the wave view point, although he did not have sufficiently clear notions on the nature of light waves.

The first most detailed experimental investigation of this phenomenon belongs to NEWTON [1.1], who studied the pattern of the fringes formed in the gap when he placed a convex lens onto a flat glass plate (Newton's fringes). Although NEWTON adhered to corpuscular notions on the nature of light, when explaining these phenomena he was forced to assume that the light particles interacting with transparent bodies induce oscillations in them.

The fundamental principles of the interference of light as a phenomenon resulting from summation of light oscillations were first formulated by YOUNG and appreciably developed by FRESNEL. These two scientists clearly can be considered to be the founders of wave optics [1.2,3].

Figure 1.1 shows schematically Young's classical double-aperture experiment which he did in 1802. The light from an extended source I illuminates a pinhole S_0 in an opaque screen A. A second screen B with two pinholes S_1 and S_2 is placed in the cone of rays diffracted at the pinhole S_0. A viewing screen C is placed where the waves diffracted at the pinholes S_1 and S_2 overlap, and interference fringes are observed on it. If one of the pinholes is closed, the fringes vanish, and the viewing screen C is illuminated uniformly. The appearance of the fringes cannot be explained by summation of the illuminations created on the viewing screen by each of the pinholes. According to YOUNG and FRESNEL, it is due to the summation of the light waves with account taken of their phases, the latter being determined by the distances from the pinholes S_1 and S_2 to the relevant points of the screen. The spatial frequency of the fringes (their number per unit length) increases with increase of the angle α subtended by the segment S_1S_2 from points on the plane of the screen C. It also depends on the wavelength of the light. Therefore, when a source that has a continuous spectrum (an incandescent lamp, the sun) is used, the fringes observed, except for the central one, will be colored, and the number of visible fringes will be comparatively small (about ten). Narrowing of the spectrum of the source is attended by increase of the number of fringes that have a contrast sufficient for observation. The contrast of the fringes also depends on the size of the pinhole S_0 – the fringes gradually vanish when its size is increased.

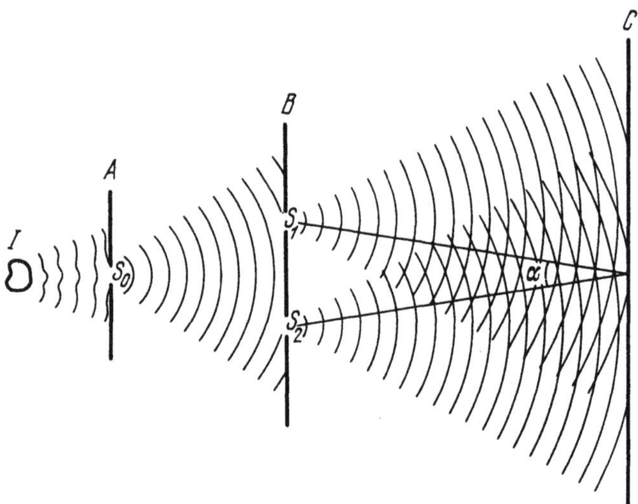

Fig.1.1. Young's double-aperture experiment

YOUNG and FRESNEL also formulated the indispensable conditions for interference: fringes (an interference pattern) are formed only when both waves are parts of the same primary wave emerging from the pinhole S_0.

Young's and Fresnel's wave theory explained all the diverse interference phenomena observed, but the nature of light waves remained unclear until Maxwell developed the foundation of the electromagnetic theory of light.

1.1.1 Fundamental Relationships

Although a number of interference phenomena are known that can be interpreted consistently only in terms of quantum optics [1.4], in holography, and in holographic interferometry, manifestations of the quantum nature of light are very few and slight. Therefore, in the present book, we shall use the notions of the classical electromagnetic theory of light according to which a monochromatic light wave can be represented in the form

$$\vec{E}(x,y,z,t) = \vec{a}(x,y,z)\cos[\omega t + \varphi(x,y,z)] \quad , \tag{1.1}$$

where $\vec{E}(x,y,z,t)$ is the electric field intensity vector, $\vec{a}(x,y,z)$ is the amplitude vector at the given point of space, ω is the angular frequency of oscillations, related to the ordinary frequency ν, the period of oscillations T and the wavelength λ by the expression

$$\omega = 2\pi\nu = \frac{2\pi}{T} = \frac{2\pi v}{\lambda} \quad , \tag{1.2}$$

where v is the speed of light in the given medium. The quantity $\varphi(x,y,z)$ is the initial phase of the oscillations at the given point of space.

A light wave is also characterized by the wave vector $\vec{k}$, whose direction coincides with the direction of propagation of the wave (for an isotropic medium this direction is perpendicular to the surface of the wavefront). The magnitude of the wave vector $\vec{k}$ is

$$k = \frac{2\pi}{\lambda} = \frac{n\omega}{c} = \frac{\omega}{v} \quad , \tag{1.3}$$

where n is the refractive index of the medium, equal to the ratio of the phase speeds in a vacuum (c) and in the given medium (v).

Let us consider the explicit form of the function $\varphi(x,y,z)$ for the important cases of a plane and a spherical light wave. We shall limit ourselves to the case of homogeneous waves, i.e., waves for which surfaces of a constant phase are simultaneously surfaces of constant amplitude.

A wave is called plane if at any moment of time the surfaces of equal phase are planes. Let us consider a plane wavefront S (Fig.1.2). If at the initial moment $t = 0$ the phase of the oscillations at the point $O(x = 0, y = 0, z = 0)$ is δ, then at the point P, which is the base of the perpendicular dropped from the point O onto the plane S, the phase of the oscillations is equal to

$$\varphi(P) = \delta - \frac{2\pi l}{\lambda} \quad , \tag{1.4}$$

where l is the length of OP.

Let us consider the arbitrary point Q in the plane S. The length l is the projection of the position vector $\vec{r}$ of the point Q onto the direction of a normal to the surface S, i.e.,

$$l = \vec{r} \cdot \vec{n} \quad , \tag{1.5}$$

where $\vec{n} = \vec{k}/k$ is a unit vector of a normal to the wavefront.

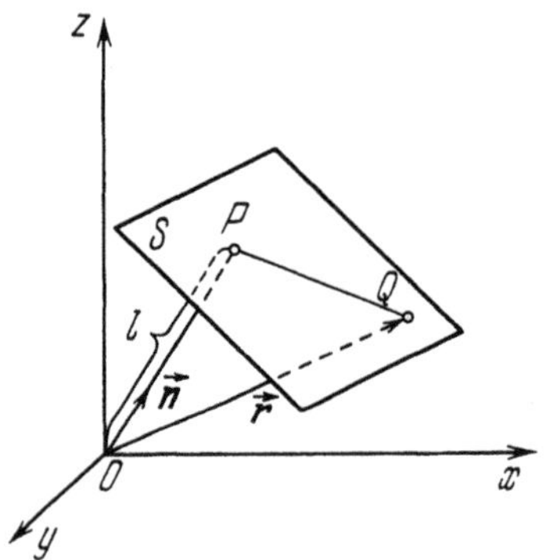

Fig.1.2. Derivation of the equation of a plane wave

The condition that $\vec{r}\cdot\vec{n} = l$ is observed for all of the points of the plane S; therefore, the phase of the wave at these points is constant and equals

$$\varphi(\vec{r}) = \delta - \frac{2\pi}{\lambda}(\vec{r}\cdot\vec{n}) = \delta - \vec{r}\cdot\vec{k} \quad . \tag{1.6}$$

Thus, (1.1) for a plane wave can be written in the form

$$\vec{E}(\vec{r},t) = \vec{a}(\vec{r}\cdot\vec{k})\cos(\omega t + \delta - \vec{r}\cdot\vec{k}) \quad . \tag{1.7}$$

If a plane wave propagates in the direction of one of the coordinate aces (for example z), then $\vec{r}\cdot\vec{k} = zk$, and

$$\vec{E}(z,t) = \vec{a}(z)\cos(\omega t + \delta - zk) \quad . \tag{1.8}$$

For a spherical wave that emerges from a point source at the origin of coordinates, the phase of the wave is constant for all points at the distance r from the center (point 0). If at the initial moment the phase at the point 0 is δ, then at all the points of the sphere of radius r it is

$$\varphi(r) = \delta - \frac{2\pi r}{\lambda} = \delta - kr \quad . \tag{1.9}$$

Thus, (1.1) for a spherical wave has the form

$$\vec{E}(r,t) = \vec{a}(r)\cos(\omega t + \delta - kr) \quad . \tag{1.10}$$

The action of a light wave on a detector of radiation is determined, as a rule, by its <u>intensity</u> — the value of the <u>energy flux density</u> u averaged in time,

$$u = < vw > \quad , \tag{1.11}$$

where the brackets < > signify averaging in time, and w is the density of the energy of the light field, equal (in the cgs system) to

$$w = \frac{\varepsilon}{4\pi}\vec{E}\cdot\vec{E} \quad . \tag{1.12}$$

Here ε is the permittivity (dielectric constant) of the medium. Hence,

$$u = \frac{\varepsilon v}{4\pi} < \vec{E}\cdot\vec{E} > \quad . \tag{1.13}$$

We find the average value of $\vec{E}\cdot\vec{E}$ during the time τ

$$< \vec{E}\cdot\vec{E} > = \frac{1}{\tau}\int_{-\tau/2}^{\tau/2} E^2(x,y,z)dt \quad , \tag{1.14}$$

where E(x,y,z,t) is the magnitude of the electric field for $\vec{E}(x,y,z,t)$

$$E(x,y,z,t) = a(x,y,z)\cos[\omega t + \varphi(x,y,z)] \quad , \tag{1.15}$$

where a(x,y,z) is the magnitude of the amplitude vector $\vec{a}(x,y,z)$.

Using this value of E(x,y,z,t) in (1.14), we have

$$< \vec{E}\cdot\vec{E} > = \frac{a^2(x,y,z)}{\tau} \int_{-\tau/2}^{\tau/2} \cos^2[\omega t + \varphi(x,y,z)]dt =$$

$$= \frac{a^2(x,y,z)}{2\tau} \int_{-\tau/2}^{\tau/2} \{1 + \cos 2[\omega t + \varphi(x,y,z)]\}dt \quad . \tag{1.16}$$

Because the integration time $\tau >> T$,

$\int_{-\tau/2}^{\tau/2} \cos\{2[\omega t + \varphi(x,t,z)]\}dt \approx 0$, and we have

$$< \vec{E}\cdot\vec{E} > = \frac{[a(x,y,z)]^2}{2} \quad . \tag{1.17}$$

Introducing this value into (1.13), we get

$$u = \frac{\varepsilon V}{8\pi} [a(x,y,z)]^2 \quad . \tag{1.18}$$

Thus, the action of a light wave on a detector is determined by the square of its amplitude; all of the information concerning the phase of the wave is lost. In this connection, all detectors of optical radiation are said to be quadratic.

We shall omit the constant factor $\varepsilon V/8\pi$, as a rule, and define the intensity as

$$I = [a(x,y,z)]^2 \quad . \tag{1.19}$$

It is often convenient to represent a wave in the complex form

$$\underline{\vec{E}}(x,y,z,t) = \vec{a}(x,y,z)e^{-i[\omega t+\varphi(x,y,z)]} \quad . \tag{1.20}$$

The quantity $\vec{E}(x,y,z,t)$ determined by (1.1) is

$$\vec{E}(x,y,z,t) = \mathrm{Re}\{\underline{\vec{E}}(x,y,z,t)\} \quad . \tag{1.21}$$

Similarly, for the magnitude of the electric field intensity vector, we have

$$E(x,y,z,t) = \mathrm{Re}\{\underline{\vec{E}}(x,y,z,t)\} \quad . \tag{1.22}$$

When expressions are written in the complex form, mathematical operations with trigonometric functions are replaced by simpler operations with exponential functions. It must, however, be remembered that after completing all mathematical operations with the complex quantitites the real part of the result obtained must be used.

The complex quantity $\underline{\vec{E}}(x,y,z,t)$ can be represented in the form of two multipliers, one of which depends only on the time, and the other only on the corrdinates of the given point

$$\underline{\vec{E}}(x,y,t) = \vec{a}(x,y,z)e^{-i\varphi(x,y,t)}e^{-i\omega t} = \vec{A}(x,y,z)e^{-i\omega t} \quad . \tag{1.23}$$

The quantity

$$\vec{A}(x,y,z) = \vec{a}(x,y,z)e^{-i\varphi(x,y,z)} \quad , \tag{1.24}$$

not depending on time, is called the complex vector of the amplitude of a wave, in contrast to $\vec{a}(x,y,z)$, which is called simply the vector of the amplitude of a wave. The magnitudes of the vectors $\vec{A}$ and $\vec{a}$ are called complex amplitude A and the amplitude a of a wave, respectively. When a wave is written in the complex form, its intensity is

$$I = \vec{E}\cdot\vec{E}^* = EE^* = \vec{A}\cdot\vec{A}^* = AA^* \quad , \tag{1.25}$$

where the asterisk (*) stands for a complex-conjugate quantity.

1.1.2 Addition of Wave Fields

If several light waves simultaneously propagate in space, then, in accordance with the principle of superposition, the resultant field can be written in the form

$$\vec{E}(x,y,z,t) = \sum_{i=1}^{n} \vec{E}_i(x,y,z,t) \quad . \tag{1.26}$$

If the waves being added have the same frequency, then, using the complex form of writing, we have

$$\underline{\vec{E}}(x,y,z,t) = \vec{A}e^{-i\omega t} = e^{-i\omega t}\sum_{i=1}^{n} \vec{A}_i \quad . \tag{1.27}$$

Hence

$$\vec{A} = \sum_{i=1}^{n} \vec{A}_i \quad . \tag{1.28}$$

Thus, instead of summing the electric field intensity vectors, we can sum the complex vectors of the amplitudes, omitting the time factor $\exp(-i\omega t)$. It is quite obvious that this simplification is impossible when waves that have different frequencies are summed.

In most of this book, we shall limit ourselves to treatment of the case when the intensity vectors of the light fields being added are parallel. Here we can pass over from summation of the complex vectors of the amplitudes to summation of the complex amplitudes, i.e.,

$$A = \sum_{i=1}^{n} A_i \quad . \tag{1.29}$$

1.1.3 Interference of Two Plane Monochromatic Waves of Identical Frequency

Let us consider the superposition of two plane waves of identical frequency characterized by the wave vectors $\vec{k}_1$ and $\vec{k}_2$ in the plane of the drawing (Fig.1.3) as an example of the addition of wave fields. We shall write the complex vectors of the amplitudes of these waves in accordance with (1.6, 24),

$$\left.\begin{aligned} \vec{A}_1 &= \vec{a}_i(\vec{r})e^{-i(\delta_1-\vec{r}\cdot\vec{k}_1)} \quad , \\ \vec{A}_2 &= \vec{a}_2(\vec{r})e^{-i(\delta_2-\vec{r}\cdot\vec{k}_2)} \quad . \end{aligned}\right\} \tag{1.30}$$

Let us assume that the electric field intensity vectors for both waves are perpendicular to the plane of the drawing. This permits us to pass over to a consideration of scalar quantities. The complex amplitude of the resultant wave, in accordance with (1.29), is

$$A(\vec{r}) = A_1(\vec{r}) + A_2(\vec{r}) = a_1e^{-i(\delta_1-\vec{r}\cdot\vec{k}_1)} + a_2e^{-i(\delta_2-\vec{r}\cdot\vec{k}_2)} \quad . \tag{1.31}$$

For homogeneous plane waves propagating in a nonabsorbing medium, their amplitudes do not depend on the coordinates, therefore a_1 and a_2 in (1.31) are independent of r.

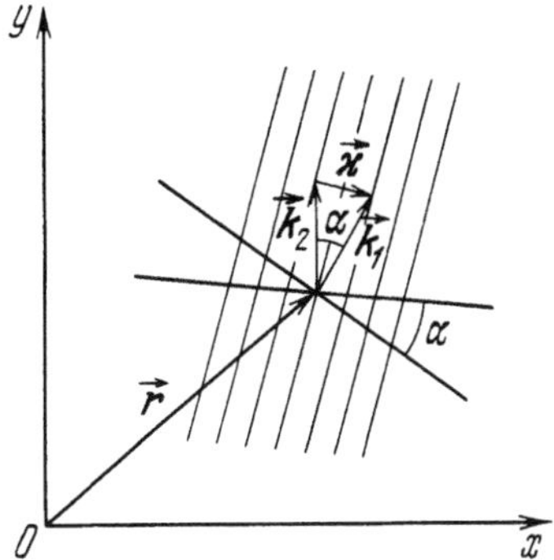

Fig.1.3. Addition of plane waves of the same frequency

We find the resultant intensity in accordance with (1.25),

$$I(r) = A(\vec{r})A^*(\vec{r}) =$$

$$= a_1^2 + a_2^2 + a_1a_2e^{-i[(\delta_1-\delta_2)-\vec{r}\cdot(\vec{k}_1-\vec{k}_2)]} +$$

$$+ a_1a_2e^{i[(\delta_1-\delta_2)-\vec{r}\cdot(\vec{k}_1-\vec{k}_2)]} = a_1^2 + a_2^2 +$$

$$+ 2a_1a_2\cos[\vec{r}\cdot(\vec{k}_1 - \vec{k}_2) - (\delta_1 - \delta_2)] \quad . \tag{1.32}$$

It follows from (1.32) that the total intensity changes periodically, reaching maxima (antinodes) at points for which

$$\vec{r}\cdot(\vec{k}_1 - \vec{k}_2) - (\delta_1 - \delta_2) = 2m\pi \quad , \tag{1.33}$$

where m is an integer.

The condition

$$\vec{r}\cdot(\vec{k}_1 - \vec{k}_2) - (\delta_1 - \delta_2) = (2m + 1)\pi \tag{1.34}$$

corresponds to minima of the intensity (nodes).

If the difference $(\delta_1 - \delta_2)$ is constant in time, then the position of the lines of the nodes and antinodes determined by (1.33, 34) remains unchanged in space, i.e., a stable interference pattern is observed.

The phase of the light oscillations emitted by any real source changes chaotically with time. Consequently, if the sources of the light that forms the plane waves being considered are independent, then the quantity $(\delta_1 - \delta_2)$ changes chaotically; this leads to blurring of the interference pattern. The average value of $\cos[\vec{r}(\vec{k}_1 - \vec{k}_2) - (\delta_1 - \delta_2)]$ during the time of measurements (if it is sufficiently great) will equal zero, and (1.32) will give us simple addition of the intensities of two waves,

$$I(r) = a_1^2 + a_2^2 = I_1 + I_2 \quad . \tag{1.35}$$

It is evident that if the phases of light waves emitted by independent sources change slowly, which occurs in highly stable lasers, we can also observe an interference pattern from two independent sources, especially when the intensity is measured fast enough.

When conventional (non-laser) sources of light are used to obtain an interference pattern, the phases δ_1 and δ_2 must not be independent. This is assured if the interfering waves are parts of the same primary wave, as in the arrangement used in Young's experiment, in which the pinholes S_1 and S_2 separate two portions of the primary wave that emerges from the pinhole S_0 (see Fig.1.1). The primary wave is divided similarly in a number of other interference devices, by which the interfering waves are formed from a single primary wave as in Young's experiment, by wavefront division (Fresnel biprism, Lloyd mirror, Fresnel mirrors, etc.), or by amplitude division (Jamin, Michelson, Mach-Zehnder interferometers, etc.). In such cases, notwithstanding that the phases δ_1 and δ_2 change chaotically with time, their difference $\delta_1 - \delta_2$ remains constant; this is what produces the stability of the interference pattern. This will be treated in greater detail in Sect. 1.1.9.

1.1.4 Orientation and Frequency of the Interference Structure

Let us consider the condition for the formation of antinodes (1.33), assuming that $\delta_1 - \delta_2 = 0$,

$$\vec{r}(\vec{k}_1 - \vec{k}_2) = 2m\pi \quad . \tag{1.36}$$

Equation (1.36) is an equation of a family of planes perpendicular to the vector $\vec{\kappa} = \vec{k}_1 - \vec{k}_2$.

Because $|\vec{k}_1| = |\vec{k}_2| = 2\pi/\lambda$, the vector κ (see Fig.1.3) is the base of an isosceles triangle whose sides are the vectors $\vec{k}_1$ and $\vec{k}_2$. Thus, the planes of the antinodes are parallel to the bisector of the internal angle α between the vectors $\vec{k}_1$ and $\vec{k}_2$. The distance from these planes to the origin of coordinates O is (see, e.g. [1.5]):

$$a_m = \frac{2m\pi}{|\vec{\kappa}|} , \tag{1.37}$$

whereas the distances between adjacent planes, labelled m and m + 1, are

$$d = a_{m+1} - a_m = \frac{2\pi}{|\vec{\kappa}|} . \tag{1.38}$$

It follows from Fig.1.3 that

$$|\vec{\kappa}| = 2|\vec{k}_1|\sin\frac{\alpha}{2} = \frac{4\pi}{\lambda}\sin\frac{\alpha}{2} . \tag{1.39}$$

Therefore

$$d = \frac{\lambda}{2\sin(\alpha/2)} . \tag{1.40}$$

The spatial frequency of the structure is

$$\nu = \frac{1}{d} = \frac{2\sin(\alpha/2)}{\lambda} . \tag{1.41}$$

1.1.5 Contrast of Interference Pattern

The contrast or visibility of a periodic structure is defined as the quantity

$$p = \frac{I_{max} - I_{min}}{I_{max} + I_{min}} , \tag{1.42}$$

where I_{max} and I_{min} are the intensities of the light wave in the antinodes and nodes of the interference pattern. It follows from (1.32) that

$$I_{max} = a_1^2 + a_2^2 + 2a_1a_2 \quad , \tag{1.43}$$

$$I_{min} = a_1^2 + a_2^2 - 2a_1a_2 \quad , \tag{1.44}$$

whence

$$p = \frac{2a_1a_2}{a_1^2 + a_2^2} = \frac{2\sqrt{I_1I_2}}{I_1 + I_2} = \frac{2\sqrt{\alpha}}{\alpha + 1} \quad . \tag{1.45}$$

Here $\alpha = a_1^2/a_2^2 = I_1/I_2$ is the ratio of the intensities of the interfering waves.

The contrast of an interference pattern is maximum and equals unity when $\alpha = 1$, i.e., when the amplitudes of the waves being added are equal.

1.1.6 Interference of Plane Monochromatic Waves of Different Frequency [1.6]

Let us now consider how the interference structure will change if the monochromatic plane waves described above differ in frequency. Unlike the preceding case, here we cannot sum the complex amplitudes, but must also take into account the difference between the time factors of the waves being added. Let us write the scalar equations of these waves in the form

$$\begin{aligned} \underline{E}_1 &= a_1 e^{-i(\omega_1 t + \delta_1 - \vec{r}\cdot\vec{k}_1)} \quad , \\ \underline{E}_2 &= a_2 e^{-i(\omega_2 t + \delta_2 - \vec{r}\cdot\vec{k}_2)} \quad . \end{aligned} \tag{1.46}$$

Let us add these waves and find the intensity of the resulting pattern, assuming as previously that $\delta_1 - \delta_2 = 0$,

$$I = (\underline{E}_1 + \underline{E}_2)(\underline{E}_1 + \underline{E}_2)^* = a_1^2 + a_2^2 + 2a_1a_2\cos(\vec{\kappa}\cdot\vec{r} - \Omega t) \quad . \tag{1.47}$$

Here, as previously, $\vec{\kappa} = \vec{k}_1 - \vec{k}_2$, and $\Omega = \omega_1 - \omega_2$. This equation describes a plane running interference pattern — a "wave of intensity" that propagates in the direction of the vector $\vec{\kappa}$ and having the frequency Ω.

At each fixed moment of time t_1, the position of the antinodes (and nodes) will be described by an expression similar to (1.36)

$$(\vec{\kappa}\cdot\vec{r}) = 2\pi m + \Omega t_1 \tag{1.48}$$

which, as in the preceding case, is an equation of a system of planes perpendicular to the vector $\vec{\kappa}$. In this case, however, the vectors $\vec{k}_1$ and $\vec{k}_2$ differ not only in direction, but also in magnitude. Accordingly, the planes of the antinodes are not parallel to the bisector of the angle α but make an angle χ with it, which, as follows from geometrical considerations (Fig.1.4), satisfies the equation

$$\tan\chi = \frac{k_2 - k_1}{k_2 + k_1}\cot\frac{\alpha}{2} = \frac{\lambda_1 - \lambda_2}{\lambda_1 + \lambda_2}\cot\frac{\alpha}{2} \quad . \tag{1.49}$$

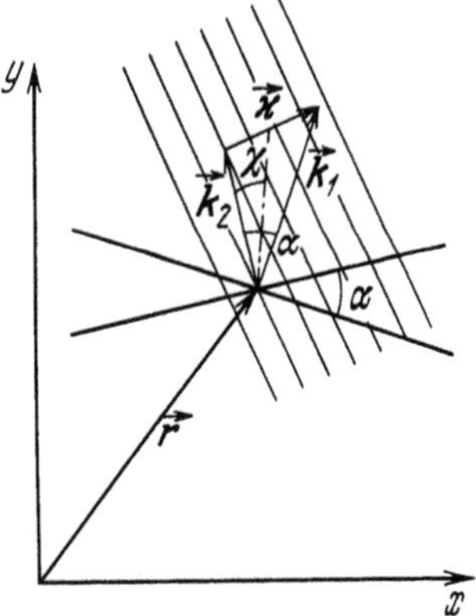

Fig.1.4. Addition of plane waves of different frequencies

The distances between the planes, as previously, will be

$$d = \frac{2\pi}{|\vec{\kappa}|} = \frac{2\pi}{\sqrt{k_1^2 + k_2^2 - 2k_1k_2\cos\alpha}} \quad . \tag{1.50}$$

The distance to the m-th plane of an antinode from the origin of coordinates in this case will be

$$a_m = \frac{2\pi m + \Omega t}{|\vec{\kappa}|} \quad . \tag{1.51}$$

We shall find the velocity of propagation of the "intensity wave" by differentiating this quantity with respect to time,

$$v_{int} = \frac{da_m}{dt} = \frac{\Omega}{|\vec{\kappa}|} = \frac{\Omega}{\sqrt{k_1^2 + k_2^2 - 2k_1k_2 \cos\alpha}} = \frac{v(\lambda_2 - \lambda_1)}{\sqrt{\lambda_1^2 + \lambda_2^2 - 2\lambda_1\lambda_2 \cos\alpha}} \quad . \tag{1.52}$$

It is not difficult to see that when $k_1 = k_2$ (waves of the same frequency) (1.50) transforms into (1.40), and the velocity v_{int} and the angle χ become equal to zero.

1.1.7 Interference of Spherical Waves

Let us now consider the superposition of two spherical waves that propagate from the point sources 0_1 and 0_2 (Fig.1.5a). We shall assume that these sources emit monochromatic waves of the same frequency. We shall also assume that the vectors of the electric field intensity at the points 0_1 and 0_2 are perpendicular to the plane of the drawing and are identical in amplitude.

We shall write the complex amplitude vector at an arbitrary point P of this plane, for the wave that emerges from 0_1, in the form

$$\vec{A}_1(\vec{r}_1) = \vec{a}_1(\vec{r}_1)e^{-i(\delta_1 - kr_1)} \quad . \tag{1.53}$$

Similarly, for the other wave which emerges from the point 0_2, we have

$$\vec{A}_2(\vec{r}_2) = \vec{a}_2(\vec{r}_2)e^{-i(\delta_2 - kr_2)} \quad . \tag{1.54}$$

Here $\vec{r}_1$ and $\vec{r}_2$ are the position vectors of the point P relative to 0_1 and 0_2.

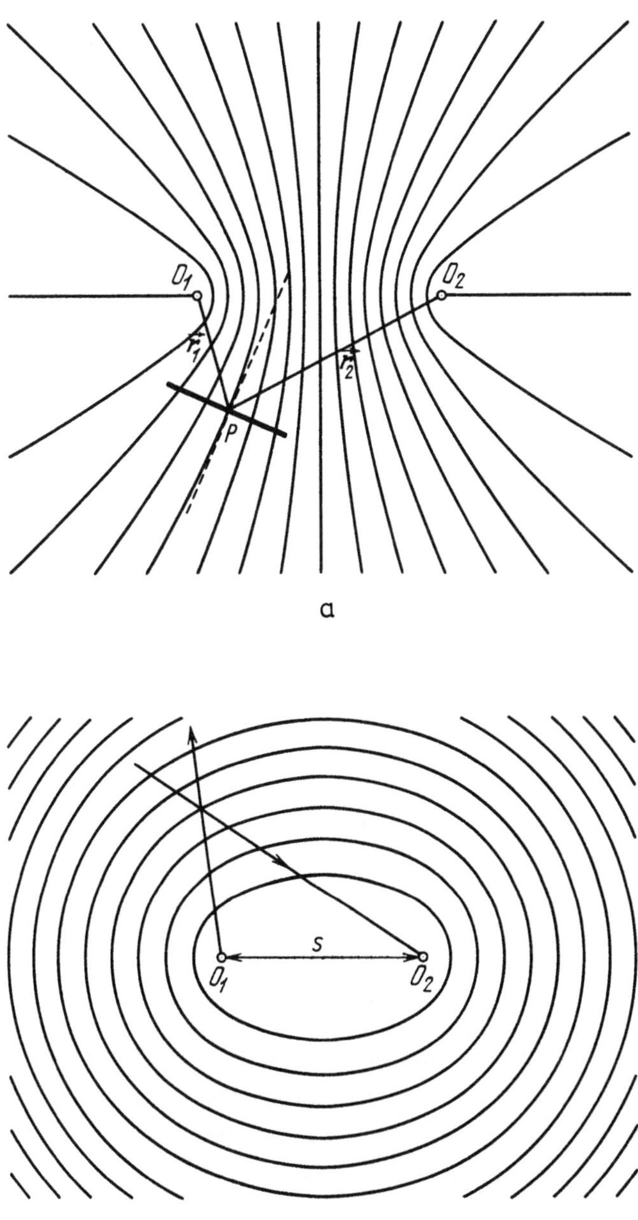

Fig.1.5. Interference of two spherical waves: (a) emerging from two point sources; (b) one emerges from the point O_1, and the second is collected at the point O_2

The intensity of the interference structure is determined as

$$I(P) = [\vec{A}_1(\vec{r}_1) + \vec{A}_2(\vec{r}_2)][\vec{A}_1(\vec{r}_1) + \vec{A}_2(\vec{r}_2)]^* \quad , \tag{1.55}$$

where $\vec{A}_1(\vec{r}_1)$ and $\vec{A}_2(\vec{r}_2)$ are given by (1.53, 54).

Hence

$$I(P) = a_1^2(\vec{r}_1) + a_2^2(\vec{r}_2) + + 2\vec{a}_1(\vec{r}_1)\vec{a}_2(\vec{r}_2)\cos[k(r_1 - r_2) - (\delta_1 - \delta_2)] \quad . \tag{1.56}$$

Assuming, as previously, that $\delta_1 - \delta_2 = 0$, we shall find the conditions for the formation of the antinodes of the interference structure,

$$k(r_1 - r_2) = 2\pi m \tag{1.57}$$

or

$$r_1 - r_2 = m\lambda \quad , \tag{1.58}$$

where m is an integer.

These equations describe a family of hyperboloids of revolution — surfaces that are the locus of points, the difference of whose distances from the two given points 0_1 and 0_2 is constant.

A family of hyperboloids also corresponds to the node surface, namely,

$$r_1 - r_2 = \frac{2m + 1}{2} \quad . \tag{1.59}$$

It is not difficult to find the total number of hyperboloids of the antinodes N_s, having in mind that $r_1 - r_2$ does not exceed the length of segment s — the distance between the sources 0_1 and 0_2. Therefore, $m_{max} = [s/\lambda]$, where the brackets signify that the integer part of the quantity in them must be taken, and, since $m = 0, \pm 1, \pm 2, \ldots, \pm m_{max}$, we have

$$N_s = 2m_{max} + 1 = \left[\frac{2s}{\lambda} + 1\right] \quad . \tag{1.60}$$

Figure 1.5a shows a section of the hyperboloids by the plane of the drawing. In this section, the lines of the nodes (and antinodes) are families of hyperbolas. It is general knowledge (see e.g. [1.5]) that a tangent to a hyperbola at any of its points is the bisector of the internal angle between the vectors $\vec{r}_1$ and $\vec{r}_2$, i.e., as for plane waves having the same frequency, the surfaces of the antinodes are arranged along the bisector of the angle between the wave vectors. Hence, the spatial frequency of an interference pattern formed by spherical waves can be calculated by (1.41), which therefore has general significance.

It follows from this equation that the maximum frequency of a pattern $\nu = 2/\lambda$ will be observed when $\alpha = \pi$. This is true for points on the segment $0_1 0_2$. The minimum frequency of the pattern $\nu = 0$ corresponds to $\alpha = 0$, i.e., to points that lie on extensions of the straight line $0_1 0_2$ outside of the segment $0_1 0_2$.

Surfaces of nodes and antinodes that have the shape of hyperboloids are formed by interference of two spherical waves that emerge from the points 0_1 and 0_2. We can also consider the case when one of these waves emerges from the point 0_1 as previously, while the other one is collected at the point 0_2, i.e., the wave vector of this waye $\vec{k}_2$ is now directed not from the point 0_2 to the point of observation A, but in the opposite direction. In this case, the complex amplitude $\vec{A}_2(\vec{r}_2)$ has the form

$$\vec{A}_2(\vec{r}_2) = \vec{a}_2(\vec{r}_2)e^{-i(\delta_2+kr_2)} \quad . \tag{1.61}$$

Upon the summation of $\vec{A}_1(\vec{r}_1)$ and $\vec{A}_2(\vec{r}_2)$, described by the relationships (1.53, 61), we get the following distribution of the intensities

$$\begin{aligned} I(P) = {} & a_1^2(r_1) + a_2^2(r_2) + \\ & + 2\vec{a}_1(\vec{r}_1)\cdot\vec{a}_2(\vec{r}_2)\cos[(\delta_1 - \delta_2) - k(r_1 + r_2)] \quad . \end{aligned} \tag{1.62}$$

The equations of the surfaces of the antinodes and nodes have the form

$$r_1 + r_2 = m\lambda \tag{1.63}$$

$$r_1 + r_2 = \frac{2m + 1}{2}\lambda \quad . \tag{1.64}$$

We have assumed in (1.63, 64), as previously, that $\delta_1 - \delta_2 = 0$. The surfaces described by (1.63, 64) are families of ellipsoids of revolution (their sections with a plane are shown in Fig.1.5b), and the points O_1 and O_2 are the focuses of these ellipsoids.

Equation (1.56), which describes the distribution of the intensity in an interference pattern, differs from the similar equation (1.32) in that the amplitudes a_1 and a_2 depend on the coordinates, whereas the last term that determines the visibility of the pattern contains the scalar product of the amplitude vectors $\vec{a}_1$ and $\vec{a}_2$. The expression for the visibility of a pattern therefore has the form

$$p = \frac{2\vec{a}_1(\vec{r}_1)\cdot\vec{a}_2(\vec{r}_2)}{a_1^2(\vec{r}_1) + a_2^2(\vec{r}_2)} = \frac{2\sqrt{\alpha}}{\alpha + 1}\,(\vec{e}_1\cdot\vec{e}_2) \quad , \tag{1.65}$$

where α, as previously, is the ratio of the intensities of the two waves at the given point, and $\vec{e}_1$ and $\vec{e}_2$ are unit vectors that coincide in direction with the vectors $\vec{a}_1$ and $\vec{a}_2$, i.e., $\vec{e}_1 = \vec{a}_1/a_1$ and $\vec{e}_2 = \vec{a}_2/a_2$.

An electric dipole is the simplest source of polarized electromagnetic waves. At distances much greater than its length, which corresponds to the model of a point source, the amplitude of a spherical wave emitted by it in a non-absorbing medium changes as follows [1.7]:

$$a(\vec{r}) \propto \sin\frac{\varphi}{r} \quad , \tag{1.66}$$

where φ is the angle between the axis of the dipole and the vector $\vec{r}$. According to this expression, each of the sources O_1 and O_2 has directions $\varphi_1 = 0$ and $\varphi_2 = 0$ in which they emit no radiation. For these directions, and also for the points of space where the vectors of the amplitudes of the waves being summated are mutually perpendicular, i.e., $(\vec{e}_1\cdot\vec{e}_2) = 0$, the visibility of the pattern is zero.

In the plane of the drawing (Fig.1.5), which is perpendicular to the intensity vectors of the electric field produced by the two sources, the vectors $\vec{a}_1$ and $\vec{a}_2$ are parallel and, consequently, $(\vec{e}_1\cdot\vec{e}_2) = 1$. In this plane, the angle φ equals $\pi/2$ everywhere. Consequently, in accordance with (1.66), we have

$$\alpha = \frac{a_1^2(\vec{r}_1)}{a_2^2(\vec{r}_2)} = \frac{r_2^2}{r_1^2} \quad . \tag{1.67}$$

Hence,

$$p = \frac{2r_1r_2}{r_1^2 + r_2^2} \quad . \tag{1.68}$$

The function (1.68) has a maximum when $r_1 = r_2$; i.e., the greatest visibility of the interference pattern, equal to unity is observed near a normal drawn through the middle of the segment O_1O_2 (Fig.1.5). At all of the remaining points of the plane of the drawing, the visibility of the pattern is less than unity, but tends to it as the point of observation moves away from the sources O_1 and O_2 when the distances r_1 and r_2 become much greater than the length of the segment O_1O_2.

1.1.8 Nonmonochromatic Point Sources

Let us now consider the case when each of the sources O_1 and O_2 emit light with the same power at two frequencies, ν' and ν'' (the corresponding wavelengths are λ' and λ''). We shall retain the other restrictions – let the sources be point ones as previously, and emit light at each of the frequencies with a constant phase difference, equal to zero.

It is easy to see that in the space that surrounds O_1 and O_2 two systems of stationary hyperboloids will now be observed[1], one of which corresponds to λ^* and the other to λ'' (Fig.1.6). The position of only one of these antinode surfaces, however, namely that which corresponds to $m = 0$, will coincide for both wavelengths. The other surfaces will be displaced relative to one another by an amount that increases with increasing m and $\Delta\lambda = \lambda' - \lambda''$ (Fig.1.6). Consequently, a growth of m will be attended by diminishing of the visibility of the observed fringes. The visibility vanishes completely when the surface of the antinodes for one wavelength coincides with the surface of the nodes for the other one. Equations (1.58, 59) show that this condition is observed when

$$m\lambda' = \frac{2m + 1}{2}\lambda'' \quad , \tag{1.69}$$

[1] It is quite evident that two running interference patterns formed by the waves of different frequency also appear.

whence

$$m = \frac{1}{2}\frac{\lambda''}{\lambda' - \lambda''} \approx \frac{1}{2}\frac{\lambda}{\Delta\lambda} \quad . \tag{1.70}$$

Upon further increase in the path difference, an interference pattern will again appear, will reach the maximum visibility when $m = \lambda/\Delta\lambda$, will again vanish when $m = 3/2(\lambda/\Delta\lambda)$, and so on.

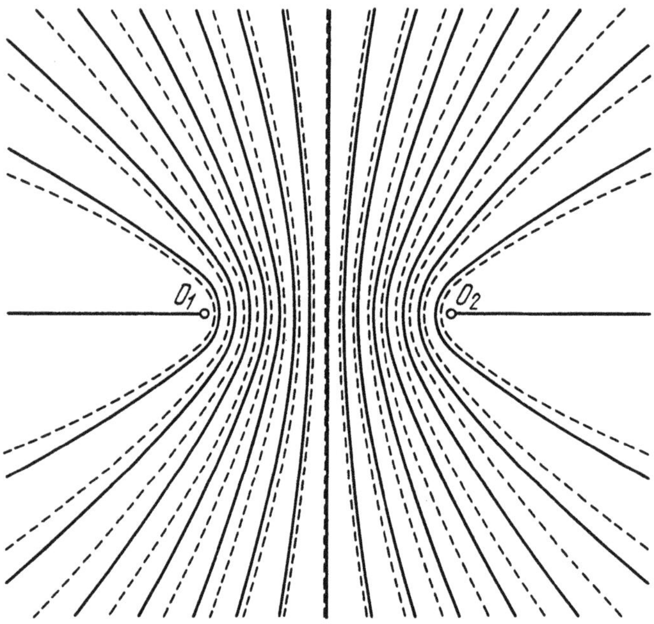

Fig.1.6. Interference of two nonmonochromatic spherical waves

Let us assume that the light sources O_1 and O_2 emit a continuous spectrum of constant brightness within the interval from λ' to λ''. In this case, the interference pattern will vanish when the surfaces of the antinodes formed by the monochromatic components into which the interval between λ' and λ'' can be divided completely fill all the gaps between the surfaces of the antinodes formed by λ'. For this purpose, the condition $m\lambda' = (m + 1)\lambda''$ must be observed, whence

$$m = \frac{\lambda}{\Delta\lambda} \quad . \tag{1.71}$$

The path difference $(r_1 - r_2)_0$ at which the interference pattern vanishes can easily be found by using (1.58, 71),

$$l_c = (r_1 - r_2)_0 = \frac{\lambda^2}{\Delta\lambda} \quad . \tag{1.72}$$

The quantity l_c is usually called the coherence length. To obtain a visible interference pattern, the optical path difference $r_1 - r_2$ must never exceed this quantity. The coherence time which characterizes the temporal coherence of a source is associated with the coherence length. It is determined by the expression

$$\tau = \frac{(r_1 - r_2)_0}{c} = \frac{\lambda^2}{c\Delta\lambda} \quad . \tag{1.73}$$

The coherence time establishes the maximum tolerated lag of one of the interfering waves behind the other.

1.1.9 Coherence

Expressions (1.72, 73) were obtained for idealized sources that are hardly real, physically. Actually the radiation of most natural sources is the result of superposition of the fields of separate emitters – excited atoms. The emission of each atom, in accordance with the classical theory, has the form of a sequence of wave packets or trains, i.e., oscillations described by the equation

$$\underline{\vec{E}}(t) = \vec{a}_0 e^{-i\varphi} e^{-i\omega t} e^{-t/\tau} \tag{1.74}$$

and differs from those described by (1.23) only by the factor $\exp(-t/\tau)$ which diminishes with time. This factor reflects the real fact of the finite duration of radiation of any atom. The second difference is that, after emitting one train (1.74), the same atom emits another train that differs from the preceding one both in the initial phase φ and, in general, the frequency of radiation ω. The independent emission of all of the excited atoms has the result that a real light field is a superposition of trains of the type (1.74) whose phases are independent and random, and whose distribution of frequencies may be most diverse, from a very broad band (a heated body) to close to monochromatic (the emission of gas-discharge sources on separate spectral lines).

The radiation of one atom can be written as the superposition of monochromatic oscillations (Fourier's expansion)

$$\underline{E}_a(t) = \frac{1}{2\pi}\int_{-\infty}^{\infty} f(\omega)e^{-i\omega t}d\omega \quad , \tag{1.75}$$

where $f(\omega)$ is the spectrum of radiation.

If radiation has a quasimonochromatic nature, i.e., is concentrated within a narrow interval, $(\omega_0 - \Delta\omega/2)) < \omega < (\omega_0 + \Delta\omega/2)$, where $\Delta\omega \ll \omega_0$, then separating in (1.75) the central radiation frequency, we arrive at the expression [1.7,8]

$$\underline{E}_a(t) = \frac{1}{2\pi} e^{-i\omega_0 t}\int_{-\infty}^{\infty} f(\omega)e^{i(\omega_0-\omega)t}d\omega = A(t)e^{-i\omega_0 t} \quad . \tag{1.76}$$

Equation (1.76) differs from (1.23) in that the amplitude a is not constant in time, but is the instantaneous value of a varying complex function.

What laws does the interference of light waves having the structure given by (1.76) obey?

Let us return to the experiment shown in Fig.1.5, bearing in mind that emitters that consist of atoms emitting light waves in accordance with (1.76) are located at the points 0_1 and 0_2, instead of the monochromatic sources. In this case, calculation of the intensity by (1.55) leads not to (1.56), as for monochromatic sources, but to [1.7,8]

$$I(P) = \langle A_1A_1^* \rangle + \langle A_2A_2^* \rangle + 2\mathrm{Re}\left\{\langle A_1\left(t + \frac{r_1 - r_2}{c}\right) A_2^*(t) \rangle\right\} \quad . \tag{1.77}$$

In (1.77), as in (1.11), the brackets $\langle\ \rangle$ signify averaging, which is inevitable when any photoreceiver is used, because the response time τ considerably exceeds the period of oscillations $T = 2\pi/\omega$, i.e., $\tau \gg T$.

$$< A(t)A^*(t) > = \lim_{\tau\to\infty} \frac{2}{\tau}\int_{-\tau}^{\tau} |A(t)|^2 dt$$

Equation (1.77) can be transformed to the form

$$I(P) = I_1(P) + I_2(P) + 2\sqrt{I_1}\sqrt{I_2}\mathrm{Re}\left\{\underline{\mu}_{1,2}\left(\frac{r_1 - r_2}{c}\right)\right\} \quad , \tag{1.78}$$

where I_1 and I_2 are the intensities produced at the point A by the sources O_1 and O_2 acting separately, and

$$\underline{\mu}_{1,2}\left(\frac{r_1 - r_2}{c}\right) = \left\langle \frac{A_1\left(t + \frac{(r_1 - r_2)}{c}\right) A_2^*(t)}{\sqrt{I_1}\sqrt{I_2}} \right\rangle \tag{1.79}$$

is the complex degree of coherence of the light waves that arrive at the point A from the sources O_1 and O_2.

The magnitude of a complex function such as (1.79) always satisfies the inequality [1.8]

$$\left|\underline{\mu}_{1,2}\left(\frac{r_1 - r_2}{c}\right)\right| \leq 1 \quad .$$

The limiting case $|\underline{\mu}_{1,2}| = 1$ corresponds to two coherent sources [see (1.56)]. If $|\underline{\mu}_{1,2}| = 0$, then the waves that emerge from the sources O_1 and O_2 are said to be incoherent, and no interference occurs, i.e., simple summation of the intensities occurs,

$$I(P) = I_1 + I_2 \quad .$$

When $0 < |\underline{\mu}_{1,2}| < 1$, the sources are called partly coherent.

Like any complex quantity, the degree of coherence can by expressed in the exponential form

$$\underline{\mu}_{1,2} = |\underline{\mu}_{1,2}|e^{i\Phi} = |\underline{\mu}_{1,2}|(\cos\Phi + i\sin\Phi) \quad , \tag{1.80}$$

where $\Phi[t, (r_1 - r_2)/c]$ is a formally introduced phase. Hence, using (1.42) for the visibility of the interference pattern p, we get, recalling (1.78),

$$p = \frac{I_{max} - I_{min}}{I_{max} + I_{min}} =$$

$$= \frac{I_1 + I_2 + 2\sqrt{I_1}\sqrt{I_2}|\underline{\mu}_{1,2}| - I_1 - I_2 + 2\sqrt{I_1}\sqrt{I_2}|\underline{\mu}_{1,2}|}{2(I_1 + I_2)} =$$

$$= \frac{2\sqrt{I_1}\sqrt{I_2}}{I_1 + I_2}|\underline{\mu}_{1,2}| = \frac{2\sqrt{\alpha}}{\alpha + 1}|\underline{\mu}_{1,2}| \quad . \tag{1.81}$$

Let us determine the conditions in which the visibility is maximum. It can be shown [1.7] that

$$\underline{\mu}_{1,2}\left(\frac{r_1 - r_2}{c}\right) = \frac{1}{\sqrt{I_1}\sqrt{I_2}}\frac{\pi}{T}\int_{-\infty}^{\infty} f_1(\omega)f_2^*(\omega)e^{\frac{i\omega(r_1-r_2)}{c}}\,d\omega \quad , \tag{1.82}$$

where $f_1(\omega)$ and $f_2(\omega)$ are the spectra of the sources 0_1 and 0_2 determined by relationships such as (1.76).

Equation (1.82) can be represented as the product of two factors,

$$\underline{\mu}_{1,2}\left(\frac{r_1 - r_2}{c}\right) =$$

$$= \frac{e^{-i\omega_0\left(\frac{r_1-r_2}{c}\right)}}{\sqrt{I_1}\sqrt{I_2}}\frac{\pi}{T}\int_{-\infty}^{\infty} f_1(\omega)f_2^*(\omega)e^{i(\omega-\omega_0)\frac{(r_1-r_2)}{c}}\,d\omega \quad . \tag{1.83}$$

The integral in (1.83) differs from zero only if

$$(\omega - \omega_0)\left(\frac{r_1 - r_2}{c}\right) = \frac{\Delta\omega(r_1 - r_2)}{c} \ll 1,$$

which is characteristic of integrals with infinite limits that contain an oscillating factor [1.9].

Because $\omega = 2\pi c/\lambda$ and $\Delta\omega = 2\pi c\Delta\lambda/\lambda^2$, this gives us the requirement for the path difference $r_1 - r_2$ for which interference is observed,

$$|r_1 - r_2| \ll |r_1 - r_2|_0 = \frac{\lambda^2}{\Delta\lambda} \quad . \tag{1.84}$$

When the condition (1.84) is satisfied, the degree of coherence $|\underline{\mu}_{1,2}| \to 1$, and the visibility of the interference pattern is maximum. Therefore, in the general case of quasimonochromatic radiation too, the coherence is determined by (1.72).

1.1.10 Extended Sources

The coherence length $|r_1 - r_2|_0$ and the coherence time

$$\tau = |r_1 - r_2|_0/c = \frac{\lambda^2}{c\Delta\lambda} \tag{1.85}$$

corresponding to it determine the visibility of the interference pattern if sources of phased radiation, i.e., obeying the same law, of the type given by (1.76), are at the points 0_1 and 0_2. This usually happens if the sources 0_1 and 0_2 are virtual or real images of the same initial point source. In this case, the visibility of the interference fringes diminishes if the difference between the distances from the points 0_1 and 0_2 to the region where interference is observed exceeds $|r_1 - r_2|_0$.

The visibility also becomes poorer in another case, if instead of the initial point source we use an extended quasimonochromatic source, for example the outlet slit of a spectroscopic instrument, a portion of a scattering surface illuminated with monochromatic light, etc.

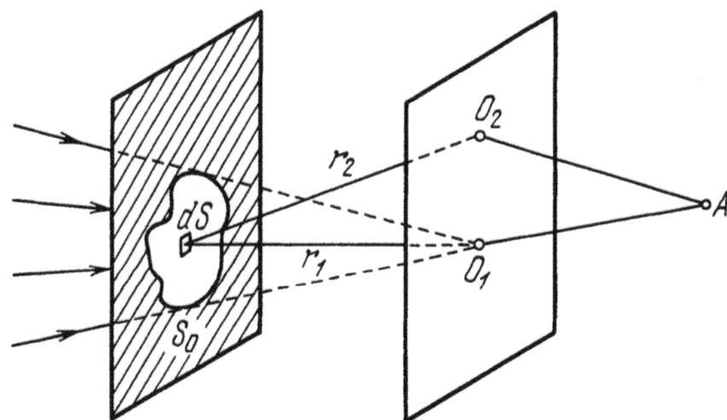

Fig.1.7. Young's experiment with an extended source

Let us consider Young's experiment for an extended source S_0 (Fig.1.7). The coherence of the radiation of the source S_0 in the region $0_1 0_2$ can be assessed in a paraxial approximation, i.e., considering that both the dimensions of this source and the distance $0_1 0_2$ are small in comparison with r_1 and r_2.

Taking into consideration that our extended source can be divided into N independently emitting regions, the size of each of them being small in comparison with the average wavelength $\lambda_{av} = 2\pi c/\omega_0$, and calculating the function of the mutual correlation $\mu(O_1, O_2)$ of the fields that arrive from the source at the points O_1 and O_2, we get the degree of spatial coherence [1.7]

$$\underline{\mu}(O_1,O_2) = \frac{1}{I(O_1)I(O_2)} \int_{S_0} \frac{I(S)\exp[i\bar{k}(r_1 - r_2)]}{r_1 r_2} dS \quad , \tag{1.86}$$

where $I(O_1)$ and $I(O_2)$ are the values of intensity measured independently at the points O_1 and O_2, $I(S)$ is the distribution of radiance over the source, and r_1 and r_2 are the distances from an element of the source dS to the points O_1 and O_2, respectively.

Attention is drawn to the similarity of (1.86) to the Kirchhoff-Fresnel diffraction integral. Consequently, the degree of spatial coherence, i.e., the correlation of the oscillations at the points O_1 and O_2 in a plane illuminated by the quasimonochromatic source, equals the normalized complex amplitude at the corresponding point O_2 of a certain diffraction pattern with its center at O_1. This pattern is obtained if we replace the source with an aperture of the same size and fill it with a spherical wave that converges at the point O_1 and that has a distribution of amplitude over the wavefront proportional to the distribution of radiance over the source (the van Cittert-Zernike theorem).

If the source of radiation is homogeneous, $I(S) = \text{const}$, this theorem makes it possible to assess the spatial coherence of its radiation by considering the diffraction of a spherical wave of constant amplitude on an aperture whose size and shape are the same as those of the source. In the paraxial approximation, this signifies that the quantity $\underline{\mu}(O_1,O_2)$ is proportional to the Fourier transform of the function that describes the shape of the source.

Using what has been stated above, we can show [1.7] that a homogeneous quasimonochromatic source of angular radius $\alpha = \rho/r_1$ (here ρ is the radius of the source) practically coherently ($|\underline{\mu}| \geqslant 0.8$) illuminates a circular area of diameter $0.2\lambda/\alpha$. The quantity $|\underline{\mu}|$ decreases to zero on a circle whose center is at O_1 and whose diameter is $1.22\lambda/\alpha$.

1.1.11 General Requirements on the Elements of an Interference Setup

The characteristics of a spatial interference pattern formed by two plane or spherical waves considered in this section allow us to determine the general requirements on interference setups, sources of radiation, and also on the media and on devices that record the interference pattern.

The spatial frequency of an interference pattern as a function of the angle at which the interfering waves converge [see (1.41)], determines the requirements on the spatial resolving power of the recording device. Particularly, when photographic recording is used, the requirements on the resolving power of the photosensitive layer follow from (1.41).

The temporal coherence associated with the monochromatic nature of the source [see (1.73)] determines the maximum permissible path differences of the interfering waves. For observation of an interference pattern, it does not matter whether the source emits light within a narrow spectral band $\Delta\lambda$ or whether it emits within a broad range and the band $\Delta\lambda$ is determined by a selective detector. In the latter case, (1.73) can be considered as a requirement that must be satisfied by the spectral resolving power of the detector.

The difference between the frequencies of the sources determines, in accordance with (1.52), the velocity of displacement of the interference pattern and, consequently, the requirements on the temporal resolving power of the detector. It is necessary, in order to ensure recording, that the interference pattern be recorded in a time during which it is displaced by not more than a fraction of the spatial period. In practice, the interference of waves of different frequencies is usually encountered when a reflecting mirror moves in an interferometer, as a result of the vibration or deformation of separate parts of a holographic or interferometric setup, or when holograms of moving objects are recorded. The wave frequency in all of these cases is displaced by the Doppler effect, and we observe movement of the interference fringes.

The spatial coherence of interfering waves (associated with the size of the sources) determines the regions of space in which a visible interference pattern is formed.

The visibility of an interference pattern is determined by the ratio of the intensities of the interfering waves, the mutual orientation of their polarization planes [see (1.65)], and also by the spatial and temporal coherence of the interfering waves. The visibility of a pattern determines the minimum relative change of the signal that the receiver must register for the interference pattern to be detected. It is quite to be expected that this quantity depends not only on the properties of the detector, but also on the noise levels of the sources O_1 and O_2 determined by their power and stability.

The spectral radiance of a source, i.e., the power emitted by a unit area of it in a unit solid angle and a unit spectral interval, is an exceedingly important characteristic of a source that in the long run determines whether or not it is possible to observe an interference pattern. Indeed, having at our disposal a source with an adequate spectral radiance, we can separate a narrow spectral band with the aid of a filter or monochromator, thus meeting the requirement of temporal coherence. By focusing the radiation of such a source on an aperture of a sufficiently small size, we can also meet the requirements of spatial coherence. For conventional (not laser) sources of light, however, this leads to great losses of energy. The spectral radiance of the source also determines the minimum time needed to record the interference pattern and, consequently, the requirements to the stability of the elements of the interference setup, and also the possibility of obtaining interferograms of moving objects. The spectral radiance of the source determines the requirements on the sensitivity of the detectors of the radiation and of the recording media. Finally, the spectral radiance of the source determines the relative level of noise and, consequently, the minimum visibility of the interference pattern needed for its detection.

Lasers are superior to conventional sources with respect to their spectral radiances, by factors of millions and hundreds of millions. Unlike conventional sources, a laser emits light within a very small solid angle. Hence, when we focus laser radiation on a small aperture, we lose virtually no energy. Laser radiation, as a rule, is highly monochromatic, which makes it possible to meet also the requirement of temporal coherence without any appreciable losses. This is why the invention of lasers opened up new broad possibilities in conventional and, especially, in holographic interferometry.

1.2 Optical Interferometry

In the preceding section, we saw how the shape of interfering fronts, their mutual orientation, and also the spectral composition of the radiation and the angular dimensions of the light sources affect the form of the interference pattern – the arrangement of the node and antinode surfaces and the distribution of the intensity. It is also possible in principle to solve the reverse problem – knowing the shape of one of the interfering wavefronts, we can determine the shape of the second one, according to the form of the interference pattern. By investigating the distribution of the intensity in an interference pattern, we can determine the spectral composition of the radiation or the angular dimensions of the source. By investigating the displacements in space of the interference pattern, we can determine the difference between the frequencies of the interfering waves, which is due, as a rule, to the displacement of the optical elements of the interference setup.

All of these problems are solved in optical interferometry. The instruments intended for such investigations are called interferometers.

Holographic interferometry, which our book is devoted to, developed on the basis of optical interferometry. Therefore, before passing over to a treatment of the fundamentals of holographic interferometry, we shall briefly deal with the fundamentals and methods of classical optical interferometry. We consider it expedient to treat here not only the methods whose analogues are already being used successfully in holographic interferometry, but also the directions which to date have not yet any holographic analogues. It is quite possible that this will prompt our readers to undertake research in these directions. A detailed treatment of the principles and methods of optical interferometry is given in [1.7,8-12].

1.2.1 Interferometers and Their Classification

All interferometers from separate, usually spatially separated beams from the initial light beam. These beams are then made to coincide and produce an interference pattern in the region of their intersection.

Interferometers are usually classified according to the number of interfering beams and to the way of dividing the initial beam. Twin-wave, triple-wave, and multiple-wave interferometers are distinguished according to the number of interfering beams. According to the way the initial beam is divided, interferometers are classified as instruments in which the interfering beams are formed from different sections of the initial wavefront (wavefront division), and instruments in which the entire surface of the initial wavefront participates in the formation of each of the interfering waves (amplitude division) (Fig.1.8).

In holography, as will be shown in Sect. 1.3, an interference pattern is recorded that is formed by an object and a reference beam. Thus, a holographic setup can be considered as an interferometer. The same methods of wavefront division and amplitude division are used in holography to form the object beam and the reference beam.

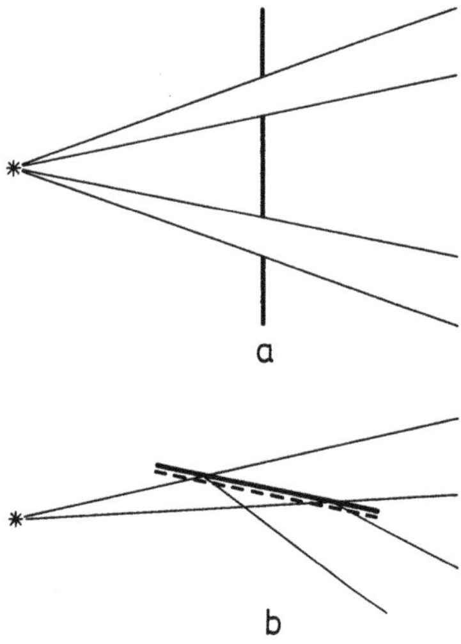

Fig.1.8. Division of an initial wave: (a) wavefront division; (b) amplitude division

1.2.2 Twin-Wave Interferometers with Wavefront Division

Let us first consider twin-wave interference setups, in which wavefront division of the initial wave is used. They include Young's setup, already described in Sect. 1.1 (see Fig.1.1), the Rayleigh interferometer (Fig. 1.9a), Fresnel's double mirror (Fig.1.9b), the setup with Lloyd's mirror (Fig.1.9c), Fresnel's biprism (Fig.1.9d), various kinds of bilenses or split lenses (Fig.1.9e,f), and Michelson's stellar interferometer (Fig. 1.9g).

Interference fringes are observed in these setups where the interfering beams overlap (these zones are hatched in Fig.1.9). In the majority of these setups, the same device is used to divide a beam and bring its parts together, but in some of them (for instance, in the Rayleigh inter-

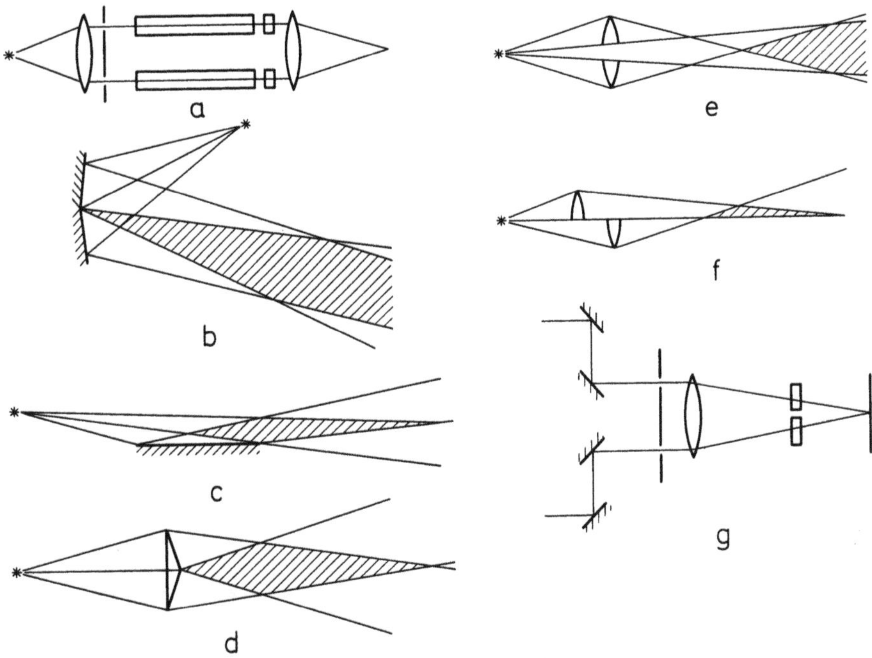

Fig.1.9. Arrangements of twin-wave interferometers with wavefront division: (a) Rayleigh interferometer; (b) Fresnel's double mirror; (c) arrangement with Lloyd's mirror; (d) Fresnel's biprism; (e) Billet split lens; (f) Meslin split lens; (g) Michelson's stellar interferometer

ferometer and Michelson's stellar interferometer) additional lenses are used to bring the beams together.

The common feature of all of these kinds of interferometers is that the waves that interfere in them are emitted from different parts of the initial wavefront (waves emitted by the light source in different directions). Consequently, the visibility of the interference pattern depends appreciably on the spatial coherence of the light that enters the interferometer. To obtain a visible (high-contrast) interference pattern, it is necessary to limit the angular dimensions of the source of radiation. With conventional sources, this reduces the intensity of the interference fringes. On the other hand, by studying the visibility of the fringes in the interference pattern, we can assess the spatial coherence of the radiation. Young's setup is widely used for this purpose, as well as Michelson's stellar interferometer, which is specially designed for determining the angular dimensions of stars from the spatial coherence of light from them.

Interferometers with wavefront division do not produce a localized interference pattern because waves emitted by the source in the same direction do not meet anywhere in the region of intersection.

An advantage of setups with wavefront division is their simplicity and relatively low sensitivity to vibrations.

1.2.3 Twin-Wave Interferometers with Amplitude Division

These include numerous setups in which the initial wave is divided by semitransparent mirrors, as a result of the reflection and refraction of light on the interface between two media, and also of double refraction. Amplitude division can also be accomplished by use of a diffraction grating (although the grating itself is a multiple-wave interference instrument with wavefront division).

Figure 1.10 shows interference setups with amplitude division that are in the greatest favor. In most of these setups, the same device (a semi-

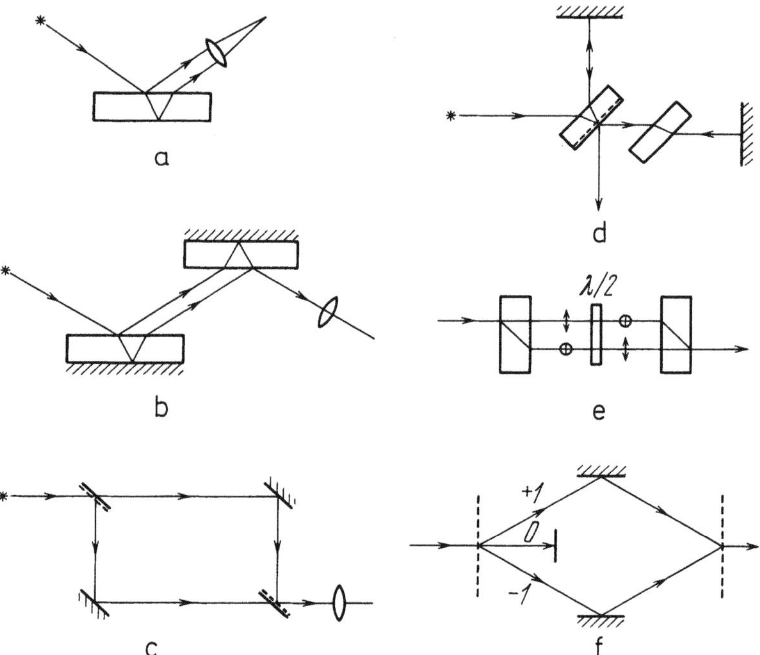

Fig.1.10. Twin-wave interference setups with amplitude division: (a) interference of light reflected from a plane-parallel plate; (b) Jamin interferometer; (c) Mach-Zehnder interferometer; (d) Michelson interferometer; (e) Lebedev polarization interferometer (the arrows show the directions of polarization of the beams); (f) interferometer with diffration grating

transparent mirror, a polarization prism, a diffraction grating) is used to join the interfering beams as that which divides them.

A feature of interferometers with amplitude division is that they permit the use of extended light sources for obtaining a high-contrast interference pattern. When an extended source is used, however, a high-contrast interference pattern is not observed in the entire region where the light beams overlap, but only in the region where corresponding rays meet, i.e., rays that emitted by the same point of the source in the same direction. In other words, the fringes formed in such interferometers are localized.

Owing to the use of extended sources, the intensity of the interference maxima is much higher in such interferometers than in ones with wavefront division.

1.2.4 Equivalent Scheme of a Twin-Wave Interferometer with Amplitude Division

We can compare each of the interferometers with amplitude division treated above with an equivalent scheme in which, paying no attention to the specific arrangement of the light-splitting elements and mirrors of the interferometer, we can consider the interference of light from two images of the source whose arrangement is given by the adjustment of the instrument (Fig.1.11). Such a scheme allows us to deal with the requirements on

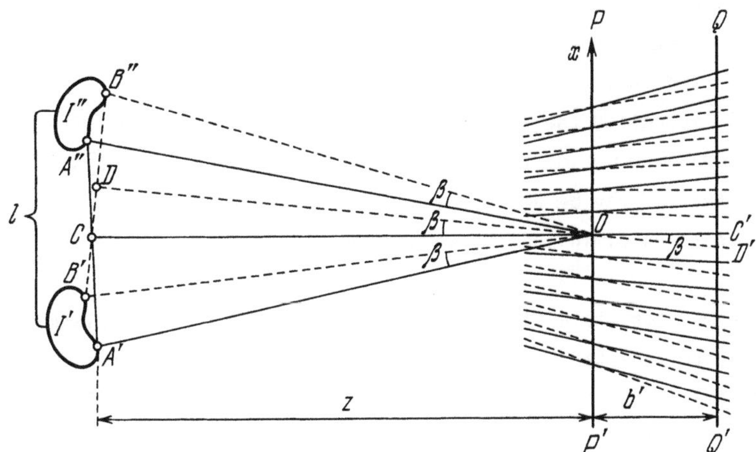

Fig.1.11. Equivalent setup of twin-wave interferometer with amplitude division

the spatial coherence of the light and to determine the position of the zone of localization of the fringes.

Let us first consider the interference pattern formed by the two corresponding points A' and A" of the virtual sources I' and I", meanwhile disregarding the radiation of other points of the sources. Let l be the interval between the virtual sources, and z the distance from them to the plane of observation.

We can find the coordinates of the maxima and minima by (1.58,59), taking into account that $r_1 = \sqrt{z^2 + (l/2 + x)^2}$ and $\sqrt{z^2 + (l/2 - x)^2}$. When $x, l \ll z$, we have

$$x_{max} = m \frac{\lambda z}{l} , \qquad x_{min} = \frac{2m + 1}{2} \frac{\lambda z}{l} . \tag{1.87}$$

The position of the maxima in the drawing is shown by solid lines.

In a similar way, for a pair of other corresponding points B' and B" of the sources, we can find the positions of the maxima of the interference pattern they form. They are shown by dashed lines in the drawing. It is easy to see that this pattern is identical to the previous one, but is turned, relative to it, through the angle β that corresponds to the angular distance between the points of the source. The axis of rotation passes through the point O that is the intersection of the perpendiculars CC' and DD' drawn through the middles of the segments A'A" and B'B". Inspection of the drawing shows that the interference maxima formed by the two pairs of corresponding points coincide only on the surface PP' that passes through the point O and is perpendicular to the bisector of the angle β betwenn CC' and DD'.

The surface PP' is called the surface of localization. It is the locus of the points of intersection of the corresponding rays (i.e., the rays obtained from one initial ray by amplitude division). The visibility of the interference pattern diminishes as we move away from this furface in both directions.

The fringes become completely blurred at the distance b' from the plane of localization. At this distance, the maxima of one pattern are superposed onto the minima of another. This occurs (as follows from Fig.1.11) when $b'\beta = d/2$, where d is the distance between adjacent maxima, i.e., the period of the interference pattern, which, according to (1.40), equals

$d \approx \lambda z/l$. Hence,

$$b' = \frac{\lambda z}{2l\beta} \quad .$$

If we consider the interference of light from all of the points of the source confined within the angle β, we can show that the fringes will vanish at a double distance from the plane of localization, i.e.,

$$b = \frac{\lambda z}{l\beta} \quad .$$

The depth of the localization zone can be taken as the quantity

$$L = 2b = \frac{2\lambda z}{l\beta} = \frac{2}{\beta\nu} \quad ,$$

which corresponds to the displacements in both directions from the surface of localization.

Thus, the depth of the localization zone is directly associated with the angular dimensions of the light source and can be a measure of the spatial coherence of the light that it emits [1.13].

1.2.5 Distribution of Intensities in a Twin-Wave Interference Pattern

The form of intensity distribution in an interference pattern depends appreciably on the number of interfering beams. Here we shall consider a twin-wave interferometer.

We showed in the preceding section that for the interference of two coherent waves of identical frequency the intensity distribution is given by (1.32,56). To define the visibility of an interference pattern, introduced earlier [see (1.42,45)], expressions (1.32,56) can be written in the form

$$I = I_{av}(1 + p\cos\delta) \quad , \tag{1.89}$$

where $I_{av} = I_1 + I_2$, and δ is the phase difference of the interfering waves.

Thus, the distribution of the intensity in the interference pattern obtained in a twin-wave interferometer has a sinusoidal character, and the visibility determines the depth of modulation (Fig.1.12a). In the particular case of the equality of the intensities of the interfering beams ($I_1 = I_2$), the visibility p is unity, and (1.89) can be written in the form

$$I = 2I_{av} \cos^2\left(\frac{\delta}{2}\right) \quad . \tag{1.90}$$

The corresponding distribution of intensity is shown in Fig.1.12b.

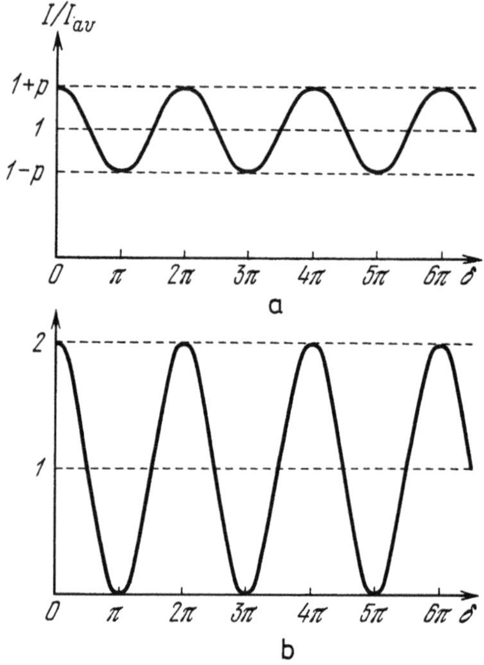

Fig.1.12. Distribution of intensity in an twin-wave interference pattern with an arbitrary ratio of beam intensities (a) and when they are equal (b)

For spatially incoherent radiation, (1.89,90) give the distribution of the intensity on the localization surface of the interference fringes. Equation (1.89) may also be used outside this surface, but the visibility of the fringes will be determined not only by the ratio of the intensity of the beams, but also by the degree of their mutual coherence $\underline{\mu}_{1,2}$

$$P = p|\underline{\mu}_{1,2}| \quad . \tag{1.91}$$

Our treatment up to now was for monochromatic radiation. If the radiation of a source consists of several monochromatic compoments that have different wavelengths (or in general has a complicated spectral composition), then the superposition of interference patterns that correspond to each of the components will be observed. As we showed in the preceding paragraph, the visibility of the interference pattern decreases with increase of distance from the center line, which corresponds to zero path difference. Figure 1.13 shows the nature of the change of the visibility of fringes with increase of the path difference, for radiation consisting of two monochromatic components that have equal intensities and for a band of a continuous spectrum of constant spectral intensity distribution.

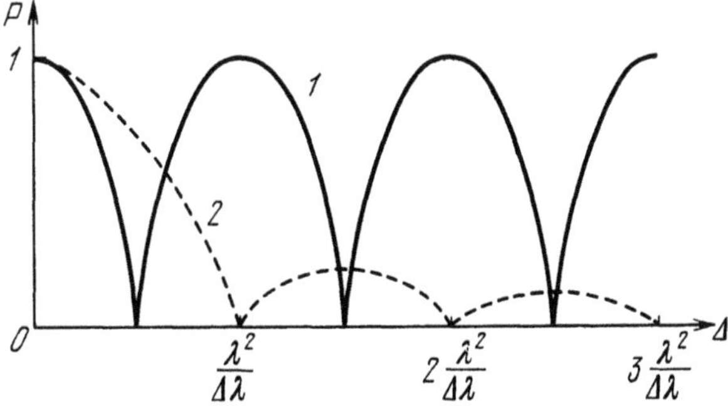

Fig.1.13. Dependence of visibility of interference fringes on the path difference: 1 - two monochromatic lines with a wavelength difference of $\Delta\lambda$ $\left(P = \left|\cos\pi \frac{\Delta}{\lambda^2/\Delta\lambda}\right|\right)$; 2 - band of a continuous spectrum of width $\Delta\lambda$ with a constant intensity within this interval $\left(P = \left|\frac{\sin\pi \frac{\Delta}{\lambda^2/\Delta\lambda}}{\pi \frac{\Delta}{\lambda^2/\Delta\lambda}}\right|\right)$

1.2.6 Application of Twin-Wave Interferometers

We shall consider the following three groups of applications of twin-wave interferometers:

1) Study of the shape of a reflecting surface, as well as the distribution of refractive index or of thickness of transparent objects, according to the shape of the interference fringes;

2) Study of the movements of objects, according to the motion of the interference fringes; and a

3) Study of the spectral composition of radiation, according to the distribution of intensity in the interference pattern.

1.2.7 Study of the Shape of Wavefronts

Let us consider as an example of the first group of applications the study of the shape of reflecting and transparent optical components by use of a Twyman-Green interferometer, which is a modification of the Michelson instrument (see Fig.1.10d).

The reflecting object being studied, for example a mirror for which it is necessary to find the deviation of its shape from a plane, is placed in one of the arms of the interferometer (Fig.1.14a). As a result of reflection from the mirror being studied, the front of one of the interfering waves becomes distorted. At the outlet from the interferometer, a pattern is observed of the interference of this wave with the plane wave that has passed through the other arm – the reference arm.

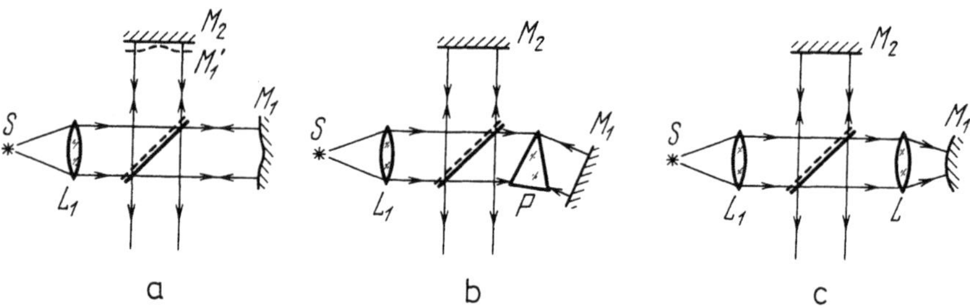

Fig.1.14. Study of optical components with the aid of a Twyman-Green interferometer: (a) mirror M_1; (b) prism P; (c) lens L

When the light source is sufficiently small and the lens L_1 is of a high quality, then the change of the phase difference over the field is due only to the difference of thickness of the air gap between the mirror M_2 and the virtual image M_1' of the mirror M_1 constructed by the light-splitting surface. Such fringes are called fringes of equal thickness. If the mirror M_1 being tested has an ideal flat surface (like that of the interferometer mirror M_2) and is installed so that M_2 and M_1' are parallel, then we shall see no interference fringes — the entire field has a constant intensity corresponding to the constant thickness of the gap between M_1' and M_2'.

If the mirror M_1 has local deviations from a plane, interference fringes will appear. Each of them unites the points of the mirror being studied for which the thickness of the air gap is the same. A change of the path difference of λ (a change in the thickness of the gap by $\lambda/2$) corresponds to adjacent fringes. Such fringes, however, do not make it possible to determine the sign of the deviation of the surface from a plane, i.e., distinguish a convexity from a concavity, because in both cases the shape of the interference fringes is the same (Fig.1.15b). To determine not only the shape, but also the sign of the deviation, one of the interfering wavefronts is inclined relative to the other when the interferometer is first adjusted. If the mirror M_1 being studied has an ideal flat surface, we shall see equidistant straight fringes whose frequency is determined by the angle between the interfering beams [see (1.41)].

Deviations of the mirror M_1 being studied from a plane result in the distortion and displacement of such fringes, the direction of the displacement being different for a convexity and a concavity (see Fig.1.15c). The

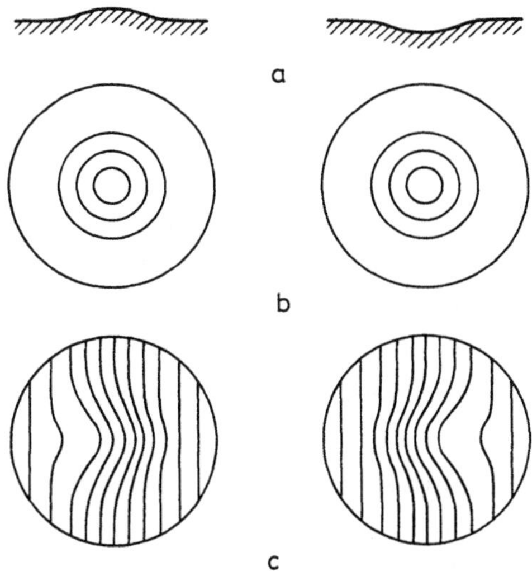

Fig.1.15. Interference patterns when studying the profiles of mirrors in fringes of infinite (b) and finite (c) width; (a) profiles of the mirrors

interference pattern obtained in this case is called a pattern of fringes of a finite width. Indeed, the period of the fringes set by the adjustment of the interferometer is finite in this case and equals

$$d = \frac{\lambda}{2 \sin(\alpha/2)} \quad . \tag{1.92}$$

When α tends to zero, the distance between the fringes tends to infinity, therefore the interference pattern obtained with such adjustment (Fig. 1.15b) is called a pattern of fringes of an infinite width.

Not only flat mirrors, but also other optical components, such as transparent plates, prisms, and lenses can be studied in such an interferometer.

If the component being studied is a prism, then the inclination of the wavefront due to twofold passage of the object beam through the prism is compensated by correspondingly inclining the mirror (see Fig.1.14b). The shape of the interference pattern in this case gives the deviations of the optical thickness of the prism being studied from that of the ideal compensating air prism set by the inclination of the mirror M_1 (see Fig. 1.14b), regardless of whether these deviations are due to faults in processing the prism surfaces or to variations of the refractive index.

When a lens is being investigated in such an interferometer, the plane mirror M_1 is replaced with a spherical mirror as shown in Fig.1.14c, or an ideal standard lens is inserted in the reference arm of the interferometer. In any case, when distortions are absent in the component being studied, both arms of the interferometer should form wavefronts of a strictly identical shape.

This, on one hand, limits the range of objects that can be studied to optical components of a regular shape (spherical and cylindrical lenses and mirrors, plane-parallel plates, etc.), and on the other, sets severe requirements to the quality of the optical elements of the interferometer, which must never introduce distortions greater than tenths of a wavelength into the wavefronts.

Other transparent phase inhomogeneities – air streams, eddies, shock waves, plasma and the like – can be studied in a similar way (Fig.1.16). Such investigations, as a rule, are conducted in a Mach-Zehnder interferometer. Unlike the Michelson interferometer, the object beam in this case passes through the object being studied only once. The Mach-Zehnder interferometer makes it possible to ensure coincidence of the zone of localization of the fringes with the object being studied by inclining the working mirrors.

A variation of this interferometer – the Rozhdestvensky interferometer – produces only fringes localized at infinity. It is customarily used for studying the spectral distribution of the optical thickness of the object introduced into one of the interferometer arms, and not its spatial distribution. For this purpose, the interference pattern is projected onto the slit of a spectrograph in whose focal plane fringes are obtained that extend over the spectrum (Fig.1.17b). This method is used for studying

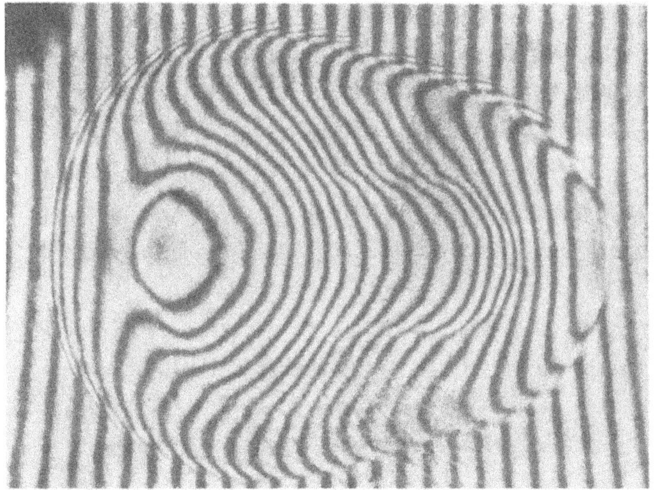

Fig.1.16. Interferogram of a laser spark obtained in a Mach-Zehnder interferometer

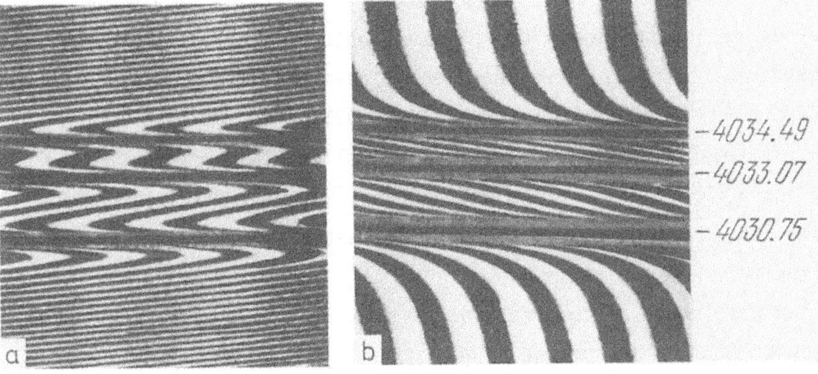

Fig.1.17. A pattern of hooks (a) and an anomalous dispersion (b) in manganese vapor

anomalous dispersion (Puccianti's method). The shape of the fringes corresponds to the dependence of the refractive index on wavelength.

When a transparent plate is introduced into the other arm of the interferometer, characteristic hooks are formed near the absorption lines (Fig.1.17a), and the distances from them to a line permit determination of the concentration of atoms, the transition probabilities, etc. (Rozhdestvensky's hook method) [1.14, 15].

1.2.8 Metrological Applications

Let us again revert to the Michelson interferometer (see Fig.1.10d) illuminated by a monochromatic source of light and adjusted to a fringe of an infinite width. When the mirror M_2 is moved along the axis of a beam, the intensity in the field of vision changes periodically in accordance with the change of the path difference between the beams. A change of the path difference by $\lambda/2$, which occurs when the mirror is moved through $\lambda/4$, corresponds to a transition from a light field to a dark field or vice versa.

Thus, by recording the periodic changes of the intensity, we can determine the number of wavelengths to which the displacement of the interferometer mirror corresponds. This makes it possible to determine the length of a light wave, according to a known displacement, or the displacement of the mirror if we know the wavelength.

By this method, Michelson performed the first accurate measurement of the wavelength of the red line of cadmium. At present, more accurate multiple-wave interference methods [1.16] are used for measuring wavelengths.

Interferometric measurements of the wavelength of the red line of cadmium performed by Fabry, Perot and Benoit gave the value of $\lambda = 6438.4696 \times 10^{-10}$m in dry air containing 0.03% of CO_2 at 15°C and 760 torr pressure. Air in these conditions is called standard. These measurements, for a long time, served as the basis for most wavelength determinations, which were compared not with the meter, but with the wavelength of the red line of cadmium.

Later, the wavelength of the red line of cadmium was compared very accurately with that of the orange line of one of the krypton isotopes ^{86}Kr($\lambda = 6057.80211 \times 10^{-10}$m for a vacuum), which is now the basic unit of length instead of the old standard – the meter [1.16]. Thus, by definition, 1m = 1 650 763.73 λ_{vac} ^{86}Kr. For standard air, λ_{air} ^{86}Kr = 6056.12525 Å.

The meter was replaced by the new standard of length by decision of the eleventh Conférence Générale des Poids et Mesures in 1960. This decision arose from the desire to have a standard of length that is not associated with a possible unstable specimen, but with an unchanging atomic constant.

For determination of translations by counting interference fringes, in recent years, by use of lasers that make it pissible to observe interference fringes with great path differences, as great as tens and hundreds of meters, these methods have been developed with great success. The counting of the interference fringes in such laser interferometers is completely automated.

Laser interferometers are used in high-precision machine tools for marking workpieces, in controlling the cutting of diffraction gratings, etc.

1.2.9 Spectroscopic Applications

We showed above that the dependence of the visibility of interference fringes on the path difference between interfering beams is determined by the spectral composition of the radiation. Thus, by studying the change of visibility of fringes while moving the mirror of a Michelson interferometer, we can determine the spectrum of the source. Investigations of this kind were first performed by FIZEAU [1.17] (who determined the structure of the sodium doublet according to the periodic change of the visibility of Newton's rings) and MICHELSON [1.10] who studied the structure of many spectral lines in great detail. At present, this principle underlies the so-called Fourier spectroscopy [1.15,18,19].

Figure 1.18 shows schematically a very simple Fourier spectrometer constructed on the basis of the Michelson interferometer in whose exit beam a circular diaphragm D and photocell P are installed. When one of the interferometer mirrors is moved with the velocity v, the intensity of each of the monochromatic components of the radiation of which the source spectrum consists changes periodically according to

$$I_\nu(t) = I_\nu\left(1 + \cos\frac{4\pi v\nu}{c}\,t\right) \quad . \tag{1.93}$$

It is evident that the depth of modulation of the intensity for all the wavelengths is identical. The frequency of the modulation $\nu_{mod} = 2\nu v/c = 2v/\lambda$ differs for different wavelengths.

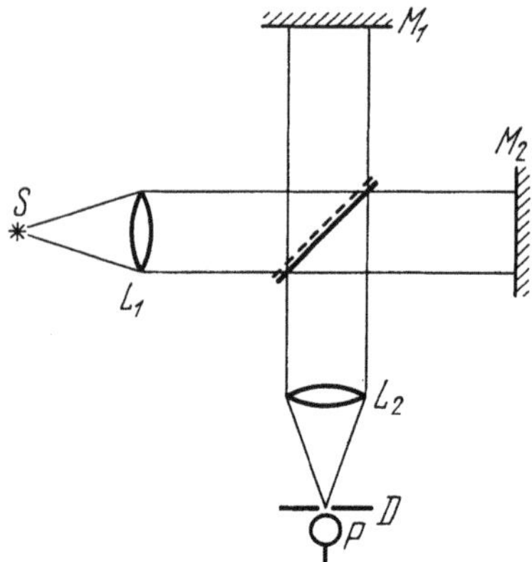

Fig.1.18. A very simple Fourier spectrometer

Let us put the signal from the detector into a number of narrow-band amplifiers. Each of them yields a signal that corresponds only to light modulated with the frequency ν_i. Signals will be reproduced by the amplifiers that in combination form a spectrum measured at a number of points $\nu_1, \nu_2, \ldots, \nu_N$ (N is the total number of amplifiers). It is quite obvious that such a registration procedure can be effective only for studying the distribution of energy in a spectrum consisting of a small number of lines. Actually, the complete signal is registered. After being amplified by an ac broad-band amplifier, it is inscribed on the recorder tape.

The energy flux $d\psi$ incident on the detector within the interval from ν to $\nu + d\nu$, where $d\nu$ is the resolution limit of the instrument, is

$$d\psi = I_\nu \cos\left(4\pi \frac{\nu v t}{c}\right) d\nu + I_\nu d\nu \quad . \tag{1.94}$$

Integrating over the entire spectrum, we get the total flux $\psi(t)$ incident on the detector,

$$\psi(t) = \psi' + \psi''(t) = \int_0^\infty I_\nu d\nu + \int_0^\infty I_\nu \cos\left(4\pi \frac{\nu v t}{c}\right) d\nu \quad . \tag{1.95}$$

The first constant term is not passed by the ac amplifier. The signal at the outlet will be proportional to the second term — the Fourier transform of the function I_ν. The observed function $\psi''(t)$ according to the main property of Fourier transformation is related to the desired energy distribution function of frequencies by

$$I_\nu \propto \int_0^\infty \psi''(t) \cos\left(\frac{4\pi\nu v}{c}\, t\right) dt \quad . \tag{1.96}$$

Thus, to extract the required spectral distribution from the observed signal, the corresponding Fourier transformation must be performed. This operation can be performed by an electronic computer.

Another spectroscopic application of twin-wave interferometers are spectrometers with interference selective amplitude modulation - SISAM's.

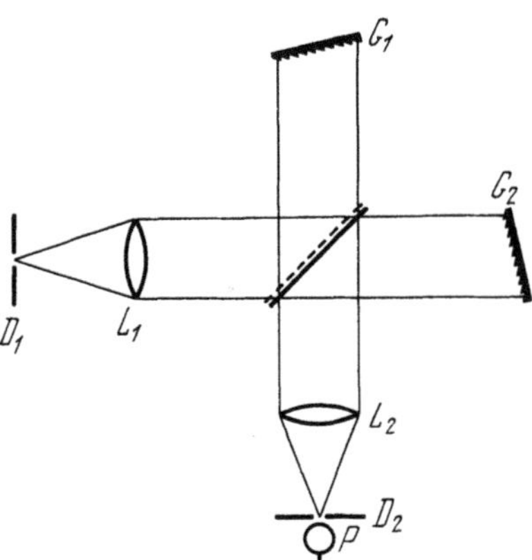

Fig.1.19. A very simple spectrometer with interference selective amplitude modulation (SISAM)

The setup of a very simple SISAM is similar to that of a Michelson interferometer in which the mirrors M_1 and M_2 have been replaced by reflecting diffraction gratings G_1 and G_2 (Fig.1.19).

For the wavelength λ_a that corresponds to the condition of autocollimation (the angle of incidence equals the angle of diffraction), the interferometer will be adjusted to form fringes of infinite width. For other wavelengths, fringes will be observed in the interference field, whose frequency is greater, the more distant the wavelength is from λ_a. Consequently, by continuous change of the path difference between beams, caused by parallel movement of one of the gratings, the depth of modulation of the signal is greatest for λ_a and decreases sharply with increasing distance from that wavelength.

The detector signal registered by an ac amplifier is proportional to the intensity at the given wavelength and the depth of modulation. By synchronously turning both gratings, the wavelength to which the instrument is tuned is changed.

1.2.10 Multiple-Wave Interferometers

Like their twin-wave counterparts, multiple-wave interferometers can be divided into two kinds: interferometers with wavefront division and interferometers with amplitude division of the initial wave.

The first kind includes a diffraction grating and a Michelson echelon (Fig.1.20). The second kind includes the Fabry-Perot interferometer and the Lummer-Gehrcke plate (Fig.1.21).

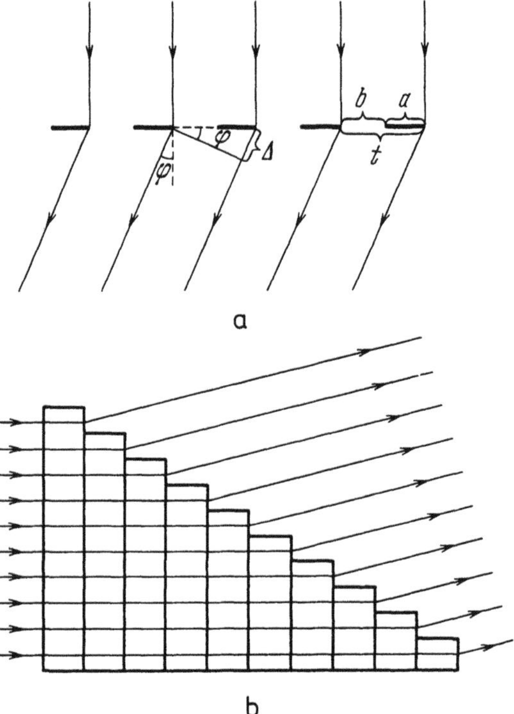

Fig.1.20. Multiple-wave interferometers with wavefront division; (a) diffraction grating; (b) transparent Michelson echelon

In all kinds of these interferometers, waves interfere whose phases consecutively increase by the same rate. In the Fabry-Perot interferometer, for example, each consecutive beam acquires an additional path difference as a result of the twofold passage of the distance between the mirrors.

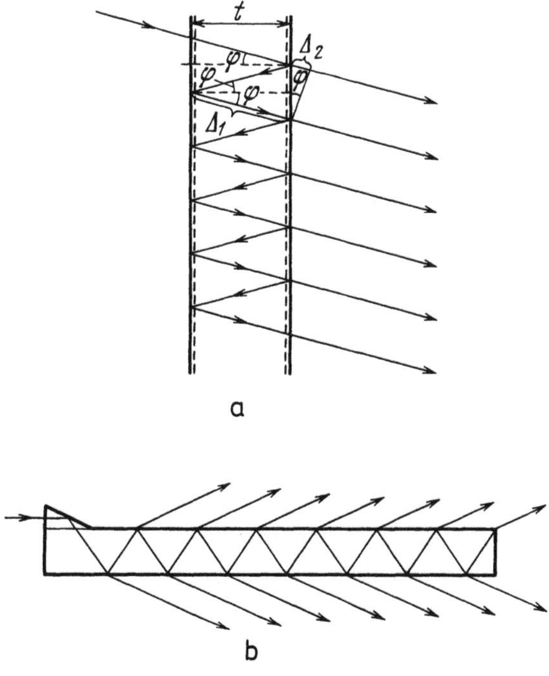

Fig.1.21. Multiple-wave interferometers with amplitude division: (a) Fabry-Perot êtalon; (b) Lummer-Gehrcke plate

1.2.11 Distribution of Intensities in a Multiple-Wave Interference Pattern

Let us consider the interference of N coherent beams. We shall assume that all of the interfering waves have the same direction of the electric vector, i.e., we shall limit ourselves to the scalar summation of complex amplitudes.

According to (1.29), the resultant complex amplitude in this case is found by summation of the complex amplitudes of the interfering waves

$$A = \sum_{m=1}^{N} A_m \quad . \tag{1.97}$$

We shall write the amplitudes of the beams that emerge from a multiple-wave interferometer in the form

$$\begin{aligned} A_1 &= a_1 e^{i\varphi_1} \ , \\ A_2 &= a_1 r e^{i(\varphi_1+\delta)} \ , \\ A_3 &= a_1 r^2 e^{i(\varphi_1+2\delta)} \ , \\ &\cdots\cdots\cdots\cdots \\ A_N &= a_1 r^{N-i} e^{i[\varphi_i+(N-i)\delta]} \ , \end{aligned} \tag{1.98}$$

where a_1 and φ_1 are the amplitude and phase of the first beam, r is a coefficient equal to the ratio of the amplitudes of the following two beams, and δ is the additional phase difference acquired by each successive beam with respect to the preceding one.

Introducing (1.98) into (1.97), we have

$$A = a_1 e^{i\varphi_1} \sum_{m=1}^{N} r^{m-1} e^{i(m-1)\delta} = \frac{a_1 e^{i\varphi_1}[r^N e^{iN\delta} - 1]}{re^{i\delta} - 1} \quad : \tag{1.99}$$

To find the distribution of intensities in a multiple-wave interference pattern, we multiply A by A*, and after simple transformations we get

$$I = AA^* = a_1^2 \frac{(1 - r^N)^2 + 4r^N \sin^2 \frac{N\delta}{2}}{(1 - r)^2 + 4r \sin^2 \frac{\delta}{2}} \quad . \tag{1.100}$$

Let us consider the particular case when all of the interfering beams have the same amplitude, i.e., r = 1. Equation (1.100), in this case, becomes

$$I = a_1^2 \frac{\sin^2(N\delta/2)}{\sin^2(\delta/2)} \quad . \tag{1.101}$$

This function has its main maxima when $\delta = 2k_1\pi$, where k_1 is an integer. Between adjacent main maxima there are N - 1 equidistant minima corresponding to $N = 2k_2\pi$, where k_2 is an integer that is not a multiple of N.

The interference of a number of beams of an identical amplitude is carried out with a diffraction grating and a Michelson echelon.

When a grating is used, the phase difference between adjacent beams for the normal incidence of light, as follows from Fig.1.20a, is

$$\delta = \frac{2\pi}{\lambda}\Delta = \frac{2\pi t}{\lambda}\sin\varphi \quad , \tag{1.102}$$

where φ is the angle of diffraction of the light, and t = a + b is the grating constant.

Thus, (1.101) by use of (1.102) gives the angular distribution of the diffracted light. Although all of the interfering beams in a given direction have the same amplitude, in different directions these amplitudes will be different, in accordance with the angular distribution of the intensity of the light diffracted at each of the slits that form the grating.

It is known that

$$a_1^2 = a_0^2 \frac{\sin^2(u/2)}{(u/2)^2} \quad , \tag{1.103}$$

where

$$u = \frac{2\pi}{\lambda} b \sin\varphi \quad . \tag{1.104}$$

Thus, the angular distribution of the intensity given by the grating has the form

$$I = a_0^2 \frac{\sin^2(u/2)}{(u/2)^2} \frac{\sin^2(N\delta/2)}{\sin^2(\delta/2)} \quad , \tag{1.105}$$

where δ and u are functions of the angle of diffraction determined by (1.104,102).

Figure 1.22 shows the angular distributions of the intensity of the light diffracted in form 1 to 6 slits.

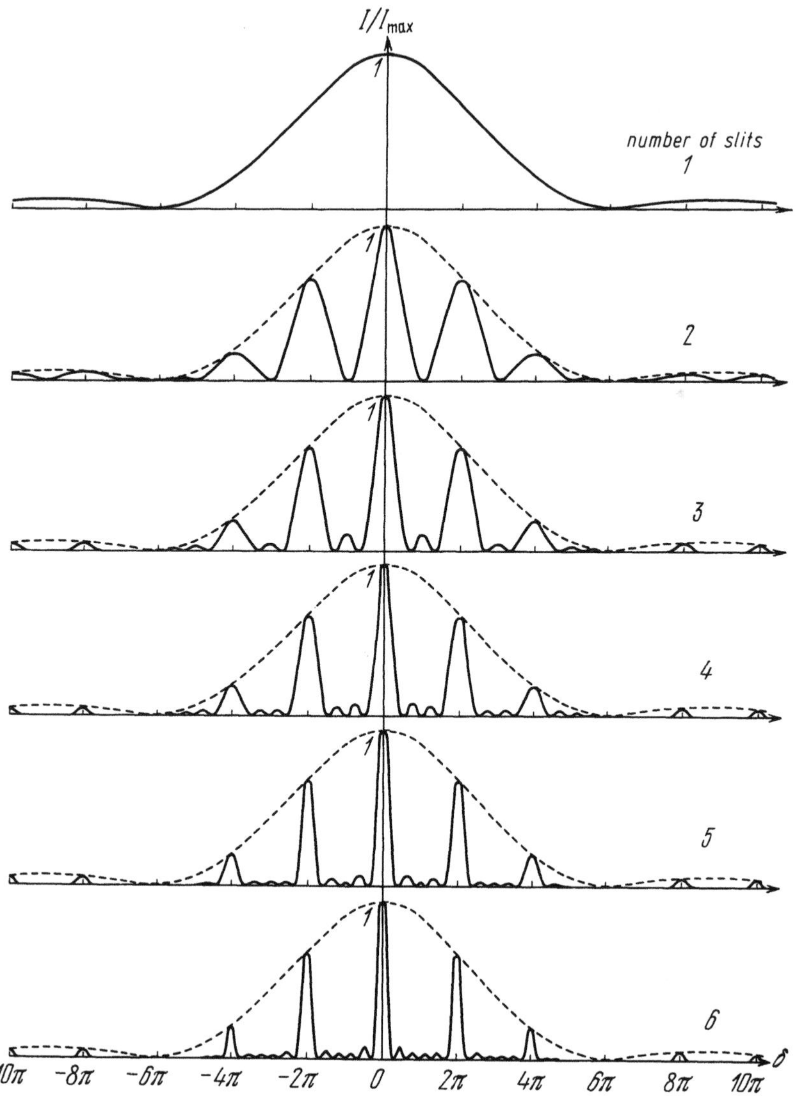

Fig.1.22. Angular distribution of intensity of light diffracted on 1 to 6 slits for t = 3b

As can be seen from (1.105) and from the figure, with an increase in the number of slits N the main peaks become narrow, and the intensities of the secondary maxima decrease. It can be shown (see [1.15]) that the angular width of the main peaks is

$$\delta\varphi = \frac{\lambda}{tN\cos\varphi} , \tag{1.106}$$

i.e., it is inversely proportional to the number of interfering beams.

Now let us consider the case that corresponds to $r < 1$ and $N \to \infty$, which is realized in the Fabry-Perot etalon. As follows from (1.100), the corresponding distribution of the intensity will be

$$I = \frac{a_1^2}{(1-r)^2 + 4r\sin^2(\delta/2)} . \tag{1.107}$$

This expression is known as the Airy formula. The quantities in it, when dealing with a Fabry-Perot etalon, have the following meanings,

$$\delta = \frac{2\pi}{\lambda}\Delta = \frac{2\pi}{\lambda}(2\Delta_1 - \Delta_2) = \frac{4\pi t\cos\varphi}{\lambda} , \tag{1.108}$$

where Δ is the path difference between consecutive beams, t is the thickness of the air gap, and φ is the angle of incidence (see Fig.1.21a).

It is not difficult to see that the quantity r is the amplitude reflectance of the interferometer mirrors. The intensity of consecutive beams diminishes as a result of twofold reflection from the mirrors, i.e., $I_{k+1}/I_k = r^2$, while the corresponding ratio of the amplitudes of consecutive beams is $a_{k+1}/a_k = r$.

It follows from (1.108) that the path difference between adjacent beams is identical for all the rays that fall on the interferometer at the same angle φ. The interference pattern obtained in this case is called isoclinic fringes. The beams that leave the interferometer, formed from one initial beam by amplitude division, are parallel. In other words, the pattern is localized at infinity. To observe it, an objective is placed in the exit beam of the interferometer, and the interference pattern is observed in its focal plane. If the radiation that falls on the interferometer contains rays of diverse directions, then the pattern has the form of concentric fringes whose angular radii, as follows (1.107,108), are $\varphi = \arccos(k\lambda/2t)$ (Fig.1.23).

The distribution of the intensity in the interference pattern is determined by the reflectance of the mirrors. It is given in Fig.1.24 for several values of r. A glance at the figure shows that the peaks become narrower with increase of reflectance. This is due to the increase of the effective number of interfering beams, which equals $\pi\sqrt{r}/(1 - r)$ [1.15].

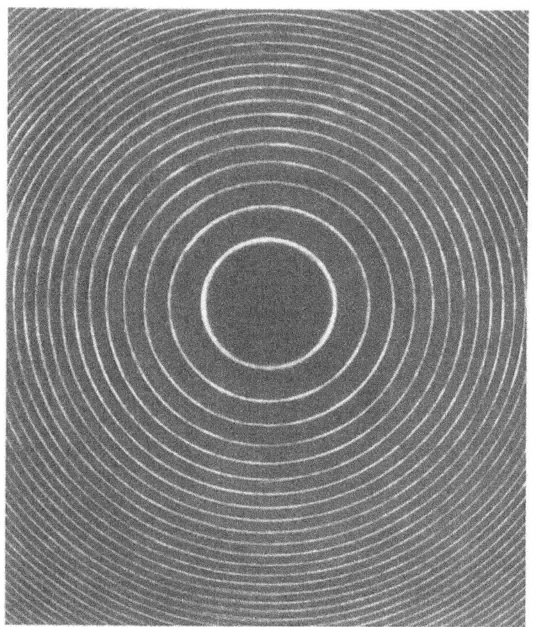

Fig.1.23. Photograph of the interference fringes obtained with the aid of a Fabry-Perot étalon (the light source was a helium-neon laser)

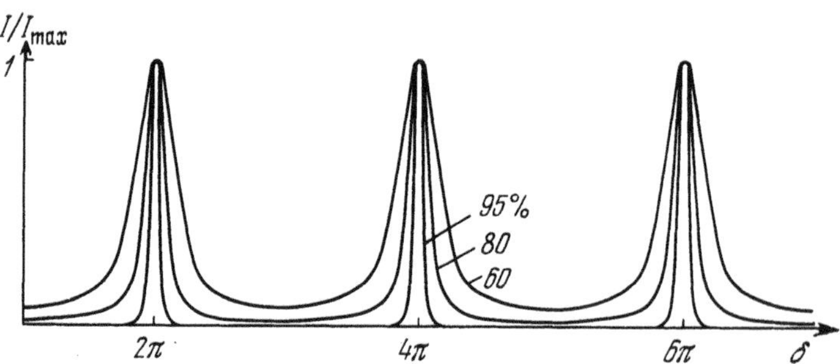

Fig.1.24. Distribution of intensity in the interference pattern of a Fabry-Perot étalon with different reflectances of the mirror: r = 95, 80, and 60%

A feature of an interference pattern obtaind with beams whose intensity attenuates is the absence of secondary maxima. The intensities of the maxima and minima of the interference pattern, as follows from (1.107), are

$$I_{max} = \frac{a_1^2}{(1 - r)^2} \quad \text{and} \quad I_{min} = \frac{a_1^2}{(1 + r)^2} \quad . \tag{1.109}$$

Hence, the visibility of the interference pattern is

$$p = \frac{I_{max} - I_{min}}{I_{max} + I_{min}} = \frac{2r}{1 + r^2} \quad . \tag{1.110}$$

It follows from this formula that the visibility of a pattern also increases with increase of reflectance of the mirrors.

1.2.12 Applications of Multiple-Wave Interferometers

One of the main applications of multiple-wave interferometers is high-resolution spectroscopy.

For all multiple-wave interference spectral instruments, the width of the instrument contour $\delta\lambda$ is inversely proportional, and the resolving power is directly proportional, to the product of the number of interfering beams and the path difference between adjacent beams

$$R = \frac{\lambda}{\delta\lambda} = \frac{N\Delta}{\lambda} \quad . \tag{1.111}$$

For a diffraction grating, great values of R are achieved by interference of a very great number of beams. N is the total number of grating rulings; in modern gratings it is as great as about 10^5. The path difference between adjacent beams is not great and does not exceed several wavelengths. Conversely, in a Fabry-Perot interferometer, a Michelson echelon, and a Lummer-Gehrcke plate, the number of interfering beams is comparativley small ($N = 10$ to 100), and high resolving power is achieved at the expense of great path differences between them (Δ reaches 50 cm).

An important characteristic of multiple-wave interference spectral instruments is the interval free of the superposition of spectra of adjacent orders.

Interference peaks for rays with a wavelength of λ are formed if the path difference is $\Delta = k\lambda$. A peak of the $(k - 1)$-th order for a wavelength of $\lambda' = \lambda + \Delta\lambda$ will correspond to the same path difference, i.e., $k\lambda = (k - 1)(\lambda + \Delta\lambda)$. Hence

$$\Delta\lambda \approx \frac{\lambda}{k} = \frac{\lambda^2}{\Delta} \quad . \tag{1.112}$$

For a grating operating in the second or third diffraction order, $\Delta\lambda$ equals thousands of angstroms, and for a Fabry-Perot interferometer several centimeters thick it is hundredths of an angstrom. It is easy to see that condition (1.112) corresponds to the fact that the path difference between adjacent beams must not exceed the coherence length [see (1.72)] of the radiation being studied.

Another application of multiple-wave interference closely related to the spectroscopic application includes construction of a variety of antireflection and reflecting interference coatings, and also interference filters [1.20]. Multiple-wave interferometers are also widely employed as laser resonators (the Fabry-Perot interferometer), and for selecting the mode composition of laser radiation as constituent elements of dispersion resonators (diffraction gratings, the Fabry-Perot interferometer). Multiple-wave interferometers of the Fabry-Perot type are used to study phase nonhomogeneities, gas flows, and plasma [1.21,22]. Multiple-wave interference setups are also used for studying the profiles of surfaces, and owing to the narrow interference peaks obtained, the sensitivity of such measurements can be as fine as a few angstroms [1.23]. More detailed information on multiple-wave interferometers and their numerous applications will be found in [1.7,12,15,20,21,24-26].

1.3 Holography

Holography is a way of recording and reconstructing waves that is based on recording the distribution of the intensity in an interference pattern formed by an object wave and a reference wave coherent with it. A recorded interference pattern is called a hologram.

A hologram illuminated by a reference wave creates in its plane the same amplitude and phase spatial distribution as was created by the object wave when it was recorded. Consequently, in accordance with the Huygens-Fresnel principle, a hologram transforms a reference wave into a copy of the object wave. Such a transformation is achieved practically independently of how the intensity distribution is recorded in the interference pattern – in the form of variations of the absorption or reflectance (an amplitude hologram) or in the form of variations of the refractive index or relief (a phase hologram).

A hologram can be recorded on a surface (two-dimensional recording) or in a volume (three-dimensional holograms). Of course, many features of the recording and reconstruction processes are different in these various cases, but the main property of a hologram – the transformation of a reference wave into a copy of the object beam – remains.

It will be simplest to explain the functioning of a hologram on the basis of the following elementary treatment. Let us imagine an interference pattern formed by an object wave and a reference wave recorded as a photographic positive, in the form of an amplitude two-dimensional hologram. Hence, the sections of the hologram with the maximum transmittance correspond to the sections of the object wavefront in which its phase was identical with that of the reference wave. The transparency of these sections increases with increased intensity of the object wave. Therefore, when such a hologram is subsequently illuminated with the reference wave, the same distribution of the amplitude and phase will be formed in its plane that the object wave had, and this ensures reconstruction of the latter.

1.3.1 Brief History of Development of Holography

The foundation of holography was laid in 1948 by the British physicist GABOR [1.27-29]. The word holography itself originates from the Greek "ὅλοσ" meaning "the whole". The inventor of holography, in using this word, wanted to stress that it records complete information about a wave – both about its amplitude and its phase.

GABOR set himself the task of improving the principles of electronic microscopy. He proposed to record, apart from amplitude information, phase information of electron waves by the superposition of a coherent reference wave. But owing to the absence of coherent electron beams he performed optical-model experiments that were the beginning of holography. The difficulties associated with the absence of powerful sources of coherent light were so significant, however, that holography remained only an "optical paradox".

Holography was born the second time in 1962-1963, when the U.S. scientists LEITH and UPATNIEKS [1.30] used lasers, which had been invented by that time, for holography, and when they proposed their split-beam (also called two-beam or off-axis) method of preparing holograms [1.31], and when the Soviet physicist DENISYUK [1.32,33] obtained the first holograms by recording in a three-dimensional medium, thus combining Gabor's idea of holography with Lippmann's method of color photography.

The development of holography after that was very tempestuous. By 1965-1966, its theoretical and experimental foundations were laid, and in the following years holography developed mainly along the path of improving its applications.

The fundamentals of holography and many of its applications are treated in a number of books [1.34-47], which we recommend to our readers for a detailed acquaintance with this rapidly developing branch of optics.

1.3.2 Fundamental Equations

Because a hologram is an interference pattern formed by an object wave and a reference wave, we can take advantage of all of its properties, considered in Sect. 1.1.

Let us now turn to Fig.1.25 showing schematically a setup for preparing a hologram. Light waves from an object and from a reference source fall on a photographic plate. We shall consider that the complex amplitude vectors of these waves are perpendicular to the plane of the drawing. Therefore, as previously, we shall limit ourselves to a scalar treatment. The complex amplitudes of the object and reference waves in the plane of the hologram (x, y) can be written as

$$A_0 = a_0 e^{i\varphi_0} \quad \text{and} \quad A_r = a_r e^{i\varphi_r} , \tag{1.113}$$

where a_0 and a_r, φ_0 and φ_r are the amplitudes and phases of the object and reference waves, which are functions of the coordinates x, y.

If these waves are coherent, they form an interference pattern on the photographic plate (hologram) in which the distribution of the illumination is

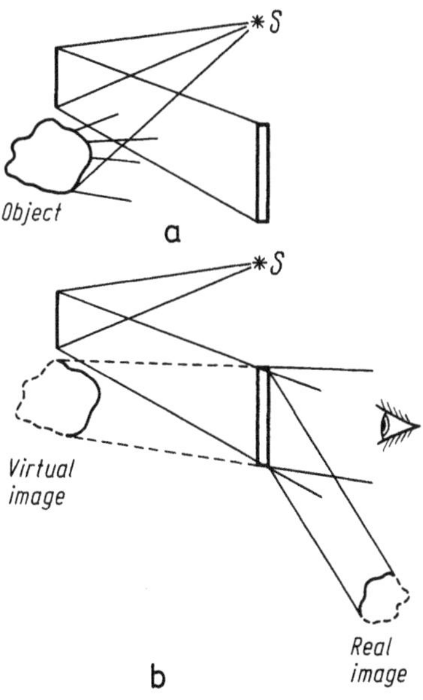

Fig.1.25. Schematic diagram for preparing a hologram (a) and reconstructing the wavefront (b)

$$I(x,y) = |A_0 + A_r|^2 = (A_0 + A_r)(A_0^* + A_r^*) =$$

$$= a_0^2 + a_r^2 + A_0A_r^* + A_0^*A_r \quad . \tag{1.114}$$

We have limited ourselves here to the case of purely amplitude recording of this distribution when the photolayer reacts to illumination only by its transmittance.

The photographic properties of a photoplate are generally described by the so-called characteristic (H & D) curve. It shows the dependence of the optical density of the plate, i.e., the quantity $d = \log(1/\tau)$, where τ is the transmittance of the developed photolayer with respect to intensity, on the logarithm of the exposure ($H = It$, where t is the exposure time), Fig.1.26a. The slope of the characteristic curve determines the contrast factor γ of the photographic emulsion.

For a holographic process, it is more convenient to represent the properties of a photoplate in the form of a curve showing how the amplitude transmittance of the emulsion $T = \sqrt{\tau}$ depends on the exposure (Fig.1.26b).

Within the limits of the section from H' to H", the curve can be approximated by a straight line of the form

$$T = T_0 = kIt \quad , \tag{1.115}$$

where the coefficient k determines the slope of the straight section (for a negative record, which Fig.1.26b corresponds to, the coefficient k is less than zero).

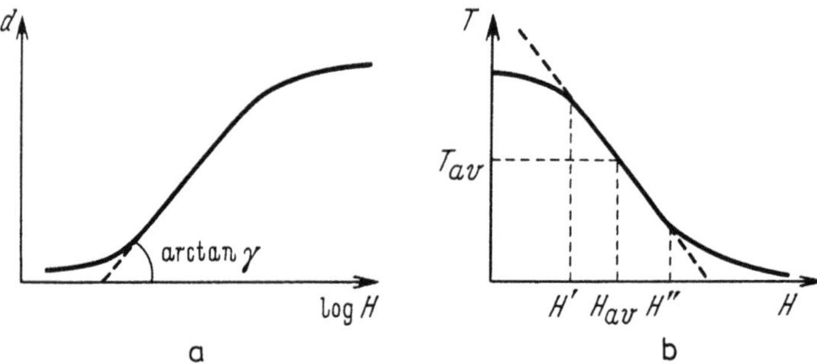

Fig.1.26. Characteristic curve of a photolayer (a) and the dependence of its amplitude transmittance on exposure (b)

Introducing into (1.115) the value of I determined by (1.114), we get the following equation for the distribution of the amplitude transmittance of a hologram:

$$T = T_0 + kt(a_0^2 + a_r^2) + ktA_0A_r^* + ktA_0^*A_r \quad . \tag{1.116}$$

Upon illuminating the hologram with the reference wave A_r, we have the following distribution of the complex amplitudes directly after its plane:

$$TA_r = [T_0 + kt(a_0^2 + a_r^2)]A_r + kta_r^2A_0 + ktA_r^2A_0^* \quad . \tag{1.117}$$

The first term in the right-hand side of this equation equals the complex amplitude of the reference wave with an accuracy up to the multiplier $T_0 + kt(a_0^2 + a_r^2)$ and corresponds to a zero-order wave. The second term differs from the complex amplitude of the object wave only in the real multiplier kta_r^2. This term describes the object wave reconstructed by the hologram (a wave of the +1st order). It forms a virtual three-dimensional image of the object at the same place where the object was during recording of the hologram (see Fig.1.25b). The third term of (1.117) differs from the wave that is the conjugate to the object wave in the complex multiplier ktA_r^2. It describes a wave of the -1st order, which forms a distorted real image of the object. To obtain an undistorted real image, the hologram has to be illuminated with a wave that is the conjugate to the reference wave.

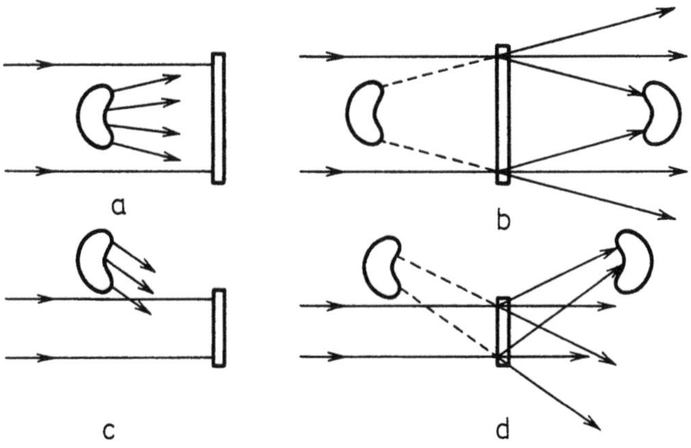

Fig.1.27. Formation of a hologram and reconstruction of the wavefront according to GABOR (a,b) and according to LEITH and UPATNIEKS (c,d)

The angles at which the waves of the zero and ±1st orders propagate are determined by the angles of incidence of the object and reference waves on the hologram. In Gabor's setup, the reference source and the object were arranged along the axis of the hologram (an in-line setup) (Fig. 1.27a). All three waves propagated behind the hologram in the same direction, creating mutual disturbance (Fig.1.27b). In the setup proposed by LEITH and UPATNIEKS, such disturbance was eliminated by using an inclined reference wave (the off-axis setup) (Fig.1.27c,d).

1.3.3 Classification of Holograms

Holograms can be classified according to the way of forming the object and reference waves and according to the way of recording the interference pattern. Let us consider these kinds of classification.

The object wave is formed as follows. The object is illuminated by a beam of coherent light. The light wave dispersed by the object and carrying information about it impinges on the hologram. Depending on the mutual arrangement of the object and the hologram, and also on the presence of optical elements between them, the relationship between the amplitude-phase distribution in the plane of the hologram and the corresponding distribution directly after the object can be described by the Fresnel (1.118) or Fourier (1.119) transformations

$$a(x,y) = \int\int f(x_1,y_1)\,\frac{1}{\lambda d}\,e^{\frac{i\pi}{\lambda}[(x_1-x)^2+(y_1-y)^2]}dx_1dy_1 \quad , \tag{1.118}$$

$$a(x,y) = \int\int f(x_1,y_1)e^{-\frac{2\pi i}{\lambda f}(x_1x+y_1y)}dx_1dy_1 \quad , \tag{1.119}$$

where x, y and x_1, y_1 are the coordinates in the plane of the hologram and the object, respectively, d is the distance from the object to the hologram, f is the focal length of the lens used to carry out the Fourier transformation.

If the object is in the plane of the hologram or is focussed onto it, the amplitude-phase distribution on the hologram will be the same as in the plane of the object (Fig.1.28a). The corresponding holograms are called image holograms.

When a hologram is at an infinite distance from the object, i.e., in the Fraunhofer diffraction region, it is called a Fraunhofer hologram. In this case, each point of the object sends a parallel light beam to the hologram, and the relationship between the amplitude-phase distribution of the object wave in the plane of the hologram and in the plane of the object is given by a Fourier transformation. To obtain such a hologram, the object must be sufficiently far from the photographic plate, or in the focus of the lens (see Fig.1.28b).

The most general kind of hologram is the Fresnel hologram (Fig.1.28c). It is formed when the recording medium is in the near-field (Fresnel) diffraction region. As the distance between the object and the hologram increases, the latter changes into a Fraunhofer hologram, whereas if that distance tends to zero, we get an image hologram.

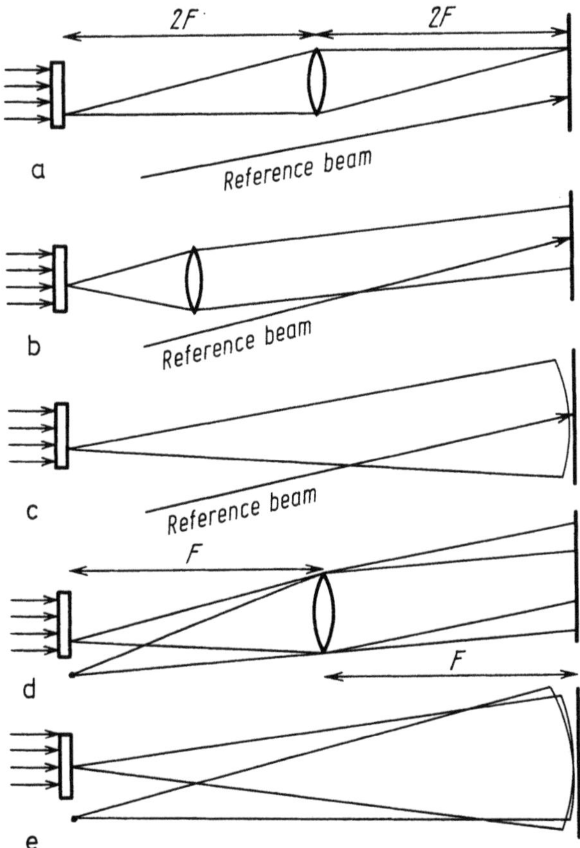

Fig.1.28. Arrangements for obtaining holograms of various kinds: (a) image hologram; (b) Fraunhofer hologram; (c) Fresnel hologram; (d) Fourier-transform hologram; (e) lensless Fourier-transform hologram

Because a hologram records the interference pattern formed by the object and reference waves, the shape of the wavefront of the latter is also important for classification of holograms.

If both the object and the point source of the reference wave are at infinity, then the distribution of the amplitudes of each of the waves in the plane of the hologram coincides with the Fourier transform of distribution of the amplitudes of the object and the reference source, respectively. Such a hologram is called a Fourier transform hologram. To obtain a Fourier transform hologram, the object and the reference source are usually arranged in the focal plane of a lens (see Fig.1.28d).

Another setup for obtaining a Fourier transform hologram is shown in Fig.1.28e. In this case, both the object and the reference source are placed at a finite distance from the photographic plate, but because they are at the same distance from it, the front of the reference wave and those of the elementary waves dispersed by separate points of the object have the same curvature. Consequently, the structure and the properties of such a lensless Fourier transform hologram are practically the same as for the hologram obtained using the setup of Fig.1.28d.

We considered the structure of a three-dimensional interference pattern formed upon the superposition of two spherical waves in Sect. 1.1. Accordingly, for a point reference source (0_1) and a point object (0_2), the surfaces of the maxima and minima are a system of hyperboloids of revolution whose section with the plane of the drawing is shown in Fig.1.29.

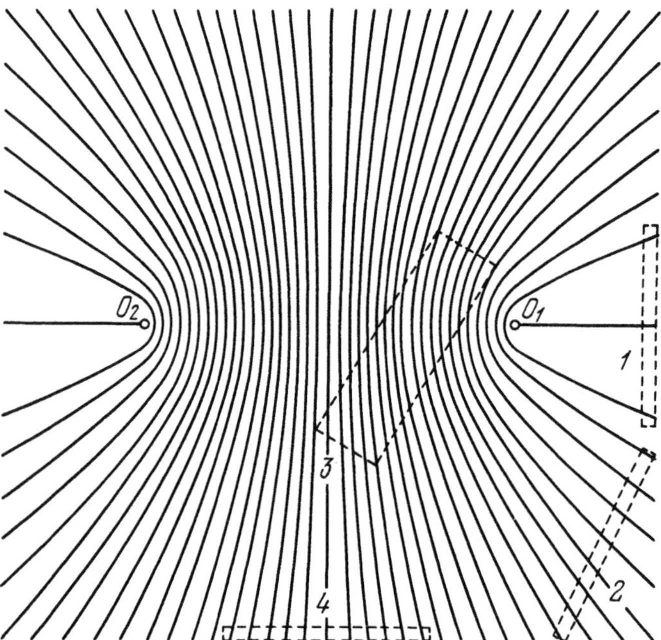

Fig.1.29. General arrangement for obtaining holograms of a point object 0_1: 1 - Gabor's in-line arrangement; 2 - arrangement with off-axis reference beam; 3 - three-dimensional hologram with recording in opposed beams; 4 - lensless Fourier-transform hologram

The spatial frequency ν of the pattern is determined by the angle α at which the light rays emerging from the reference source and the object meet at the given point [see (1.41)].

In Gabor's in-line setup, the reference source and the object are on the axis of the hologram. In this case, the angle α is close to zero, and the spatial frequency of the interference pattern is minimum. In-line holograms are sometimes also called single-ray holograms because Gabor's setup uses one beam of light, a part of which dispersed by the object forms the object wave, whereas the part of the initial beam that has passed through the object without distortion is the reference wave. The arrangement of the photographic plate in preparing an in-line hologram is shown in Fig. 1.29 by the reference number 1.

The setup of LEITH and UPATNIEKS for producing an off-axis hologram corresponds to reference number 2 in Fig.1.29. It is usual practice in this setup to form the coherent reference beam separately. This is why an off-axis hologram is sometimes called a split-beam or two-beam hologram. The spatial frequency of the interference pattern for an off-axis hologram is higher than for in-line holograms. Therefore, light-sensitive materials that have high spatial resolution are needed for recording such holograms.

Reference number 3 in Fig.1.29 corresponds to a hologram recorded in opposed beams. Here, the reference and object beams strike the sensitive layer from different sides, and the angle α between them is close to 180 degrees. The spatial frequency of the pattern is maximum and is close to $2/\lambda$. When holograms are recorded in opposed beams, the interference maxima are arranged along the surface of the emulsion inside of it. This setup was first proposed by DENISYUK [1.32,33]. Because upon the illumination of such a hologram with the reference beam, the reconstructed object wave propagates toward the illuminating beam, such holograms are also called reflection holograms.

Figure 1.29 also shows the position of a photographic plate (4) in recording a lensless Fourier transform hologram.

The classification given above relates only to the shape of the object and reference waves and their mutual orientation, which determines the interference structure of the hologram. The latter are also distinguished according to the way of recording this structure.

If the thickness of the sensitive layer is much greater than the distance between adjacent surfaces of the interference maxima, then the hologram should be considered as a volume or three-dimensional hologram. The three-dimensional properties of holograms manifest themselves especially clearly in the setup with opposed beams.

If the holographic pattern is recorded, not inside the recording layer but on its surface, or if the thickness P of the layer is sufficiently small in comparison with the distance d between adjacent elements of the pattern, then such holograms are called plane or two-dimensional holograms. The criterion of the transition from two-dimensional to three-dimensional holograms is determined by the expression [1.36]

$$P \approx \frac{1.6d^2}{\lambda} , \tag{1.120}$$

where λ is the wavelength of light in the recording medium.

An interference pattern can be recorded by a light-sensitive medium in one of the following ways:

- in the form of variations of the transmittance or reflectance. Such holograms modulate the amplitude of the illuminating wave upon reconstruction of the wavefront and are therefore called amplitude holograms;
- in the form of variations of the thickness or the refractive index. Such holograms modulate the phase of the illuminating wave and are therefore called phase holograms.

In many cases, both phase and amplitude modulation occur simultaneously. For example, a conventional photographic plate records the interference pattern in the form of variations of the blackening, refractive index, and relief. After bleaching such a hologram, only the phase modulation remains.

Generally, the pattern recorded on a hologram is retained for a long time, and the recording process is separated in time from that of reconstructing the wavefront. In this case, the hologram is called stationary. There are media, however (some dyes, crystals, metal vapors), whose phase and amplitude characteristics react almost without inertia to the action of illumination. In this case, a hologram exists only during the action on the medium of the object and reference waves, and reconstruction of the wavefront is performed at the same time as the recording, as a result of the interaction of the reference and object waves with the hologram pattern which they themselves form. Such holograms are called dynamic.

1.3.4 Basic Properties of Holograms

1) A basic property of a hologram which distinguishes it from a conventional photograph is that the latter registers only the distribution of the irradiance in the light wave falling on it, whereas a hologram, in addition, records the distribution of the phase of the object wave relative to that of the reference wave. Information on the amplitude of the object wave is recorded on a hologram in the form of the visibility of the interference pattern, and information on the phase – in the form of the shape and frequency of the interference fringes. As a result, a hologram when illuminated by the reference wave reconstructs a copy of the object wave with all its amplitude and phase details.

2) Amplitude holograms are usually recorded on negative photographic materials. The properties of a hologram remain the same as in a positive hologram – bright spots of the reconstructed image correspond to bright spots of the object, and dark spots to dark ones. It is easy to understand this property of a hologram by taking into account that the information on the amplitude of the object wave is contained in the visibility of the interference pattern whose distribution on the hologram does not change when a positive process is replaced with a negative one. With such a substitution, only the phase of the reconstructed object wave is shifted by π. This is not noticeable in visual observation, but manifests itself in some experiments in holographic interferometry.

3) When in recording a hologram, light from every point of the object falls on the entire surface of the hologram, each small portion of the latter is capable of reconstructing the entire image of the object. Of course, a smaller portion of the hologram will reconstruct a correspondingly smaller portion of the wavefront carrying information on the object. If this portion is very small, the quality of the reconstructed image is poor.

With image holograms, every point of the object sends light to the small portion of the hologram corresponding to it. Therefore a fragment of such a hologram reconstructs only the portion of the object to which it corresponds. Holograms of transparencies obtained without a diffuser have the same property.

4) The total range of brightnesses reproduced by a photographic plate, as a rule, does not exceed one or two orders of magnitude. Real objects, however, frequently have much greater brightness differences. A hologram,

which has focusing properties, uses the light falling on its entire surface for constructing the brightest portions of the image, and is capable of reproducing brightness ratios up to five or six orders of magnitude.

5) If the reference source (without displacement relative to the hologram) is used as the source of light illuminating the hologram in the reconstruction process, then the reconstructed virtual image coincides in shape and position with the object itself. This coincidence is violated when the position of the reconstructing source relative to the reference one is changed, when its wavelength is changed, and also when the orientation of the hologram and its scales are changed. Such changes, as a rule, are attended by aberrations of the reconstructed image.

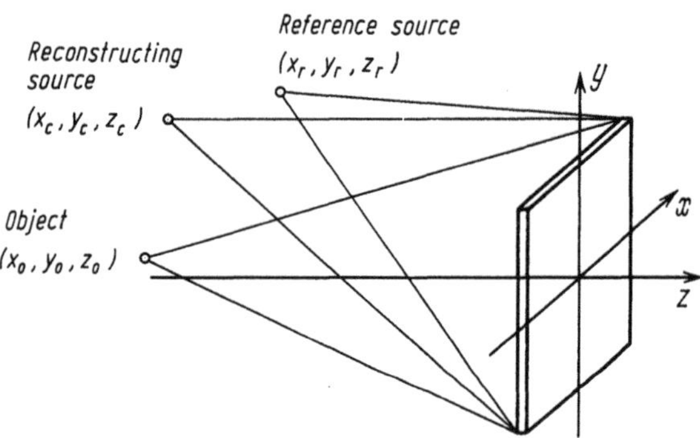

Fig.1.30. Notation used in (1.121)

Below are given formulas that can be used to calculate the position of the reconstructed image and its enlargement for any case [1.48]. We shall introduce the following notation (Fig.1.30): the hologram is in the plane x, y with z = 0 (it retains this position both during the exposure and when the wavefront is being reconstructed); the coordinates of the object are x_0, y_0, and z_0; those of the reconstructed image are x_i, y_i, and z_i; those of the point source used for reconstruction are x_c, y_c, and z_c. Assume that, prior to reconstruction, the hologram was enlarged m times, and the wavelength of the reconstructing source is μ times greater than that of the source of light used in the formation of the hologram. In this case,

$$\left.\begin{aligned} z_i &= \frac{m^2 z_r z_c z_0}{(m^2 z_r - \mu z_c) z_0 + \mu z_r z_c} \,, \\ x_i &= \frac{\mu m z_r z_c x_0 + (m^2 x_c z_r - \mu m x_r z_c) z_0}{(m^2 z_r - \mu z_c) z_0 + \mu z_r z_0} \,, \\ y_i &= \frac{\mu m z_r z_c z_0 + (m^2 y_c z_r - \mu m y_r z_c) z_0}{(m^2 z_r - \mu z_c) z_0 + \mu z_r z_c} \,. \end{aligned}\right\} \quad (1.121)$$

The angular enlargement of a hologram in any case, regradless of all of the quantities in (1.121), equals μ/m. Hence, the linear lateral enlargement is

$$M_{lat} = \frac{\mu}{m} \frac{z_i}{z_0} \,, \quad (1.122)$$

and according to (1.121)

$$M_{lat} = \frac{m}{1 + \frac{m^2}{\mu} \frac{z_0}{z_c} - \frac{z_0}{z_r}} \,. \quad (1.123)$$

If a hologram simultaneously forms two images of the object, the coordinates of the second image can be obtained by substituting $-\mu$ for μ in (1.121,123).

The longitudinal enlargement of a hologram, generally speaking, differs from the lateral enlargement. It is

$$M_{long} = \frac{\partial z_i}{\partial z_0} = \frac{\frac{m^2}{\mu}}{\left(1 + \frac{m^2}{\mu} \frac{z_0}{z_c} - \frac{z_0}{z_r}\right)^2} = \frac{M_{lat}^2}{\mu} \,. \quad (1.124)$$

Equations (1.121-1.124) are not absolutely accurate. They have been derived on the assumption that the distance from a hologram to the object is much greater than the lateral dimensions of the hologram.

6) The limiting resolving power of a hologram is determined by the diffraction of light by its aperture and can be calculated in the same way as for conventional optical systems. In accordance with the Rayleigh criterion, for a round hologram of diameter d, the angular resolution

$$\delta\varphi = 1.22\,\frac{\lambda}{d}\ , \tag{1.125a}$$

and for a square hologram with the side L

$$\delta\varphi = \frac{\lambda}{L}\ . \tag{1.125b}$$

The limiting resolution determined by (1.125a,b) usually cannot be achieved in practice, for a number of reasons.

Any image of a diffuse object obtained in coherent light, including an image reconstructed by means of a hologram, has a chaotic speckle pattern that appears as a result of interference of the light waves dispersed by separate elements of the microstructure of the object within the diffraction resilution of the optical system or hologram. The image of the object thus consists of separate spots or speckles whose average angular dimensions are given by (1.125). The resolving power of a hologram, as a result, is about half the diffraction resolving power.

In addition, for most setups used to produce holograms, the limiting size of the latter is determined by the resolving power of the recording medium, because an increase of the size of a hologram is attended by an increase of the angle between the object and reference beams and, consequently, by a higher spatial frequency of the interference pattern. An exception in this sense is the setup for lensless Fourier transform holography (see Fig.1.28e) in which the spatial frequency of the pattern does not increase with increase of the size of the hologram.

It must also be remembered that (1.125) holds only if the light from each point of the object is distributed over the entire surface of the hologram. Otherwise, the size of the portion of the hologram that is illuminated by the wave from each point of the object should be used in these equations, instead of the size of the entire hologram.

Figure 1.21 shows reconstructed images obtained with different hologram sizes. Examination of the figure shows how the dimensions of the elements of the speckle structure increase and how the resolving power becomes poorer with diminishing hologram size. Figure 1.32 shows how the resolution limit obtained depends on the hologram size.

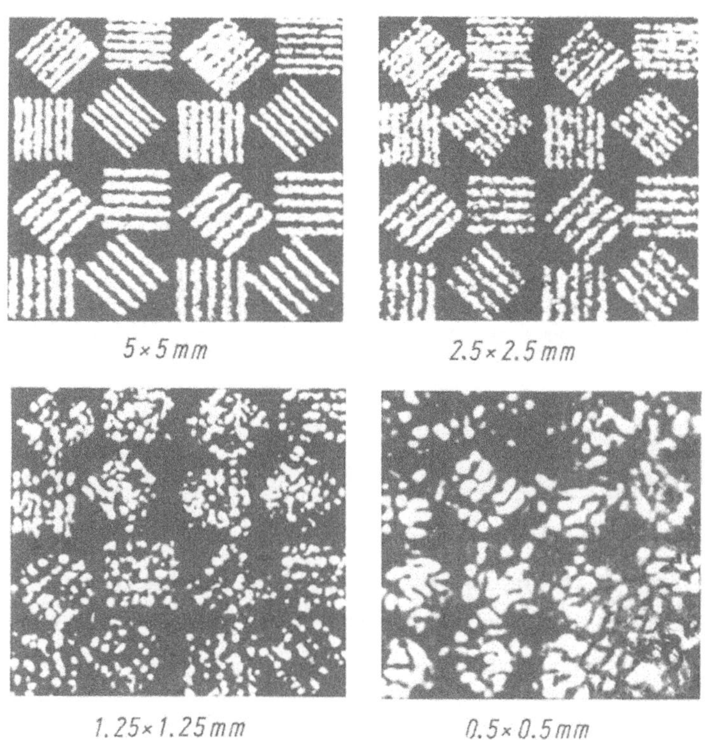

Fig.1.31. Images reconstructed by the same hologram when diaphragmed to different sizes [1.49]

7) A very important characteristic of holograms that determines the brightness of the reconstructed image is the diffraction efficiency, which is equal to the ratio of the light flux in the reconstructed wave to the light flux that strikes the hologram. The diffraction efficiency is determined by the kind of hologram, the properties of the recording medium, and also the conditions of recording the hologram.

Let us consider an amplitude hologram as an example. It follows from (1.117) that the amplitude of the reconstructed wave that forms the virtual image is $kta_r^2a_0$. Hence, the diffraction efficiency of the hologram is

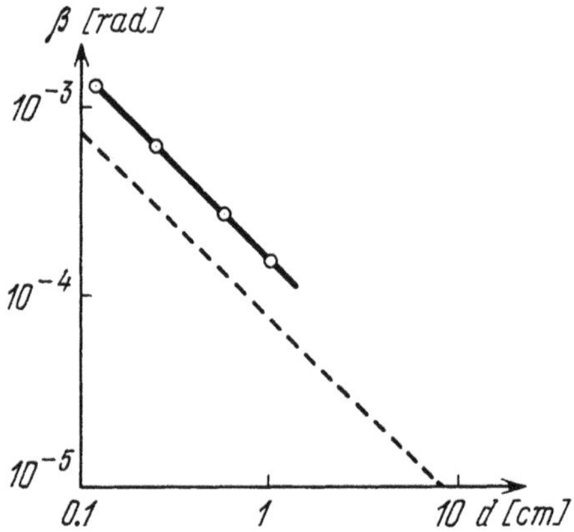

Fig.1.32. Maximum angular resolution of a hologram as a function of its size [1.49]. The dotted line shows the maximum resolution according to Rayleigh ($\beta = \lambda/d$)

$$D = \frac{(kta_r^2 a_0)^2}{a_r^2} = (kta_r a_0)^2 \quad . \tag{1.126}$$

Because the visibility of the interference pattern, in accordance with (1.45), is determined by the amplitudes of the object and reference waves

$$p = \frac{2a_r a_0}{a_r^2 + a_0^2} \quad , \tag{1.127}$$

we have $a_r a_0 = p(a_r^2 + a_0^2)/2$. Hence,

$$D = \left[\frac{kt(a_r^2 + a_0^2)p}{2}\right]^2 = \left(\frac{k\overline{H}p}{2}\right)^2 \tag{1.128}$$

where $\overline{H} = t(a_r^2 + a_0^2)$ is the mean exposure.

Let us consider a perfectly linear light-sensitive material for which the exposure dependence of the amplitude transmittance has the form shown in Fig.1.33a. Here $|k\overline{H}| = 0.5$ and with a visibility of $p = 1$ we have, according to (1.128), a diffraction efficiency of $D_{max} = 1/16 = 6.25\%$.

For a photographic material that is nonlinear to the greatest possible extent (Fig.1.33b), the maximum diffraction efficiency is somewhat higher and is 10.1% [1.36].

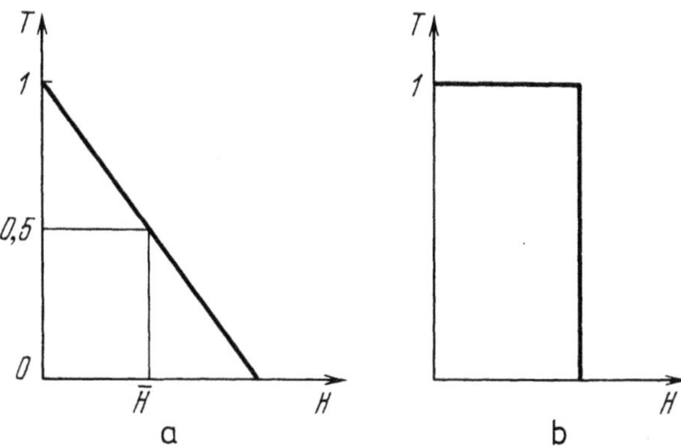

Fig.1.33. Exposure dependence of amplitude transmittance for a perfectly linear (a) negative light-sensitive material and for a material with maximum possible nonlinearity (b)

The maximum attainable diffraction efficiency of various kinds of holograms is given in Table 1.1. The experimentally achieved values of the efficiency are close to those given in the table.

Table 1.1. Maximum attainable efficiency of various kinds of holograms

Kind of hologram	Transmission		Reflection	
	amplitude	phase	amplitude	phase
Two-dimensional	6.25	33.9	6.25	100
Three-dimensional	3.7	100	7.2	100

8) If the exposure values at the maxima and minima of the interference pattern appreciably extend beyond the limits of the linear portion of the exposure dependence of the amplitude transmittance, then recording of the holograms becomes nonlinear. A linearly recorded hologram can be compared with a diffraction grating having sinusoidal distribution of the amplitude transmittance. Such gratings, as is known, do not form diffraction orders higher than the first one. In nonlinear recording, a hologram is also a periodic grating, but the distribution of the amplitude transmittance may considerably differ from a sinusoidal one. Such a grating, apart from 0th and ±1st order waves, also produces waves of higher diffraction orders. The nonlinear nature of hologram recording, however, manifests itself not

only in this aspect, but also in the distortion of the amplitudes of the reconstructed first-order waves. The influence of the nonlinear nature on the first-order image consists of amplification of the background, the appearance of haloes, distortion of the relative intensities of various points of the object, and sometimes in the appearance of false images.

The "images" formed by diffracted waves of higher orders are complicated functions (of the autoconvolution type) of the initial wave function of the object and have very little in common with the object itself [1.50]. In a number of cases, however (this relates, for instance, to image holograms of transparencies without a diffusing screen), the higher-order waves form images. The distribution of the brightness in these images, as a rule, is greatly distorted, and the phase of the k-th order wave is k times greater than that of the first-order one. This property of non-linearly recorded holograms is taken advantage of in some methods of increasing the sensitivity of holographic interferometry of transparent phase objects (see Chap. 3).

9) Volume (3D) holograms have special properties. Such holograms have a three-dimensional structure in which the surfaces of the nodes and antinodes are recorded in the form of variations of the refractive index or the reflectance of the medium. Such a pattern, when illuminated by the reference wave, functions like a three-dimensional diffraction grating (Fig. 1.34). Light that is reflected from the layers, as from mirrors, reconstructs the object wave. Indeed, the surfaces of the nodes and antinodes

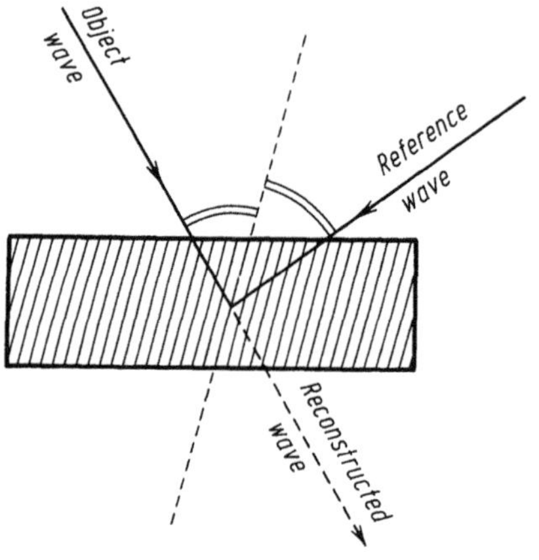

Fig.1.34. Diffraction of light on a 3D hologram

(see Sect. 1.1) are directed along the bisector of the angle formed by the object and reference beams, and this is what ensures this property of 3D holograms.

The beams reflected from different layers will amplify one another only if they have the same phase, i.e., the path difference between them must equal a whole number of wavelengths. This Lippmann-Bragg condition will be observed automatically only for the wavelength in which the hologram was recorded. This will lead to selectivity of the hologram with respect to the wavelength of the source whose light is used to reconstruct the wavefront. It therefore becomes possible to reconstruct the image with the aid of a source having a continuous spectrum (an incandescent lamp, the sun). If the hologram was exposed in the light of several spectral lines (for example blue, green, and red), then each wavelength forms its own three-dimensional pattern. The pertinent wavelengths will be separated from the continuous spectrum when the hologram is illuminated, and this will result in reconstruction not only of the structure, but also of the spectral composition of the light wave, i.e., in the obtaining of a colored image.

Three-dimensional holograms simultaneously form only one image (virtual or real depending on how they are illuminated) and do not produce zero-order waves.

1.3.5 Application of Holography

Holography is finding widespread application in the most diverse fields of science, engineering, art, medicine, etc.

Holography is used to obtain colored three-dimensional images of objects that are indistinguishable in their optical characteristics from the original. Work is actively being conducted for the creation of holographic cinematography and television [1.51-54]. Acoustical holography is being developed for purposes of technical vision in a medium opaque to electromagnetic radiation and medical diagnostics [1.55]. Methods of radio holography are used in radar and the study of radio aerials [1.56,57]. Holographic systems and methods are used for the recording and processing of information [1.58]. One of the most developed fields of holography from a practical viewpoint is holographic interferometry, to which the present book is devoted.

1.4 Holographic Interferometry

1.4.1 General Principles

The method of holographic interferometry was proposed virtually simultaneously in a number of publications that appeared in 1965 [1.59-66].

It is easy to understand the principle of holographic interferometry by considering Fig.1.25. Let us assume that a hologram, after being exposed and developed, is put in exactly the same place where it was at the moment of recording. If we now illuminate the hologram with the reference wave without removing the object, then two waves will simultaneously propagate behind the hologram: one scattered directly by the object, and another one reconstructed by the hologram. The latter wave is a replica of the one scattered by the object during exposure of the hologram. These waves are coherent and can interfere.

Should changes occur in the object (for example deformation or a change of refractive index) that result in phase distortions of the wave it scatters, this will affect the appearance of the pattern observed. Interference fringes will appear, whose shapes are determined by the changes that occur with the object. Because the interference pattern is observed simultaneously with the changes that occur in the object, this method of holographic interferometry is called real-time interferometry.

In another variant of the method, two holograms corresponding to two states of the same object are consecutively recorded on one photographic plate. The two waves that are holographic replicas of waves that existed at different moments are reconstructed simultaneously and interfere (Fig. 1.35). This variant of the method of holographic interferometry is called the double-exposure method.

Thus, the content of the method of holographic interferometry is the preparation, observation and interpretation of interference patterns formed by waves of which at least one has been recorded and reconstructed by a hologram.

1.4.2 Features of Holographic Interferometry

Both in conventional and in holographic interferometry, two (or more) light waves are compared. The interference pattern observed reveals the difference between the phase relief of the waves being compared.

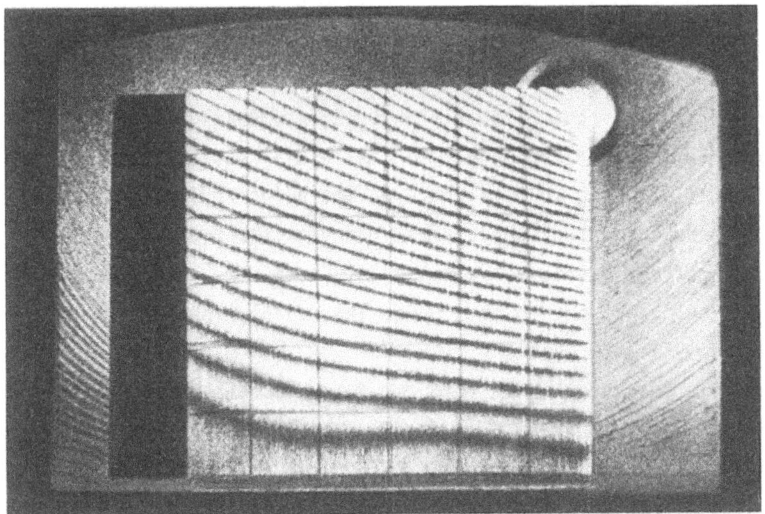

Fig.1.35. A holographic interferogram of a deformed steel plate obtained by the double-exposure method

In conventional interferometry (see Sect. 1.2), both waves being compared are formed simultaneously, but propagate along different paths. The time delay between these waves due to the difference of their optical paths must never exceed the coherence time, which even for the best single-frequency lasers does not exceed fractions of a second. In addition, the optical channels through which the waves being compared propagate must be identical, because otherwise the interference pattern will characterize not only the object being studied, but also the difference of shape of the optical components of the interference setup.

In holographic interferometry, we have to do with the interference of waves that travel along the same path, but at different moments. The appearance of the interference pattern obtained is due only to the changes that occurred in the object during the time between the exposure of the hologram and the moment of observation (or during the time between the first and the second exposures). The method of holographic interferometry is thus a differential one.

Holography makes it possible to record a light wave and to reconstruct its replica at any required moment. Consequently, holographic interferometry is not bound by the need to form the waves at the same time, and this circumstance opens up a number of interesting possibilities, which in conventional interferometry cannot be realized.

Thus, consecutive states of the same object can be compared by holographic interferometry. This automatically ensures a comparison wave that repeates the wave scattered by the object in its initial state, in all of its most minute details. Owing to this feature, the holographic interferometry method can be used to study objects of irregular shape and even rough objects that reflect diffusively. It is only necessary that the microstructure of the object should not change appreciably when the object passes from one state to another.

This is impossible in conventional interferometry because the comparison wave can reconstruct all of the details of the wave from the object only if it has a sufficiently simple form. This is why conventional interferometry studies objects that have simple forms and polished optical surfaces.

This same feature of holographic interferometry has noticeably reduced the requirements on optical components, because wavefronts that have passed through the same channel are compared instead of wavefronts that have used two different channels, as in conventional interferometry. Therefore, both interfering waves are distorted to the same extent by flaws in the optical components. Such flaws do not affect the interference pattern, because of the differential nature of the method.

The insensitivity of holographic interferometry to defects of the optical components of a setup permits interferometric investigations of objects that have virtually unlimited dimensions, whereas such investigations by conventional interference methods are difficult owing to the exceedingly great complications encountered in manufacture and the high cost of interferometers with large mirrors.

Still another feature of holographic interferometry is that if a hologram covers a large solid angle, then it can be used to reconstruct the interference pattern of light waves scattered by an object in various directions. This is of special interest for studying such spatial nonhomogeneities of a refractive index that have no axis of symmetry. In this case, to obtain a spatial distribution of the refractive index, a number of interferograms that correspond to different angles of observation must be recorded on a single hologram (see Chap. 3).

In conventional interferometry, special optical arrangements are used for simultaneously illuminating an object in different directions – the object, in essence, is simultaneously investigated in two or more interferometers.

What has just been said also relates completely to the holographic study of the deformations of objects having intricate shapes (see Chap. 4). Here the wavefront has to be reconstructed at several different angles, for unambiguous interpretation of the interference patterns.

Another unique possibility provided by holographic interferometry is that of obtaining an interference pattern formed by light waves of different frequencies (wavelengths). For this purpose, the hologram is exposed in the light of a source emitting two or more wavelengths. Such a hologram is a superposition of the holograms that correspond to the different wavelengths. When a two-wavelength hologram is illuminated with a monochromatic beam of light, replicas of the waves recorded on the hologram are reconstructed, and an interference pattern is formed that corresponds to the difference of their phase relief due both to the difference of the scale factor μ [see (1.121-124)] and to dispersion of the object being studied. Of course, an interference pattern between reconstructed waves of the same frequency is observed, although they are replicas of waves that had different frequencies. Two-wavelength methods of holographic interferometry are used to study the dispersion of plasmas (see Chap. 3), and also to investigate the relief of surfaces (see Chap. 4).

Thus, holographic interferometry has the following advantages over its conventional counterpart: it

- makes possible investigations of transparent and reflecting objects of any intricate shape and even of diffuse objects;
- does not impose strict requirements on the quality of optical components, which are so severe for conventional interferometers;
- makes possible the comparison of light waves scattered by the object at different moments of time;
- makes possible study of the interference of waves scattered by the object in different directions, within the limits of the solid angle subtended by the hologram;
- permits production of interference patterns of light waves of different wavelengths.

At the same time, we must not forget a number of significant diffisulties encountered in holographic interferometry:

— low sensitivity of hight-resolution holographic photolayers;

— high requirements on the light sources (spatial and tempoeral coherence, brightness, etc.).

1.4.3 Real-Time Method

We have already noted that the real-time method consists of observing the interference pattern of two waves, one of which is reconstructed by the hologram and corresponds to the initial state of the object, and the other is scattered by the object during observation.

The first of these waves, according to (1.117), has the following complex amplitude directly after the hologram:

$$A_1 = kta_r^2 A_{01} \quad , \tag{1.129}$$

where k is the slope of the T-H curve, t is the exposure, a_r is the amplitude of the reference wave used both for recording the hologram and for reconstructing the wavefront, and A_{01} is the complex amplitude of the wave scattered by the object during exposure of the hologram.

The second wave will have the following complex amplitude after the hologram:

$$A_2 = \beta A_{02} \quad , \tag{1.130}$$

where A_{02} is the complex amplitude of the wave scattered by the object during observation; the factor β takes into consideration weakening of the beam when it passes through the hologram.

If no changes of the object occurred during the time between recording of the hologram and the observation, i.e., $A_{01} = A_{02}$, then the two interfering waves will differ from each other only in the constant factors kta_r^2 and β. In other words, if the recording process is positive ($k > 0$), then the interfering waves in the plane of the hologram (and, consequently, in the entire space) are cophasal and will amplify each other. With negative recording of the hologram ($k < 0$), the waves will be in counterphase [$-1 = \exp(i\pi)$] and will weaken each other everywhere; when their amplitudes are equal they will completely cancel each other.

If changes of the object occur, we shall observe and interference pattern whose structure is determined by those changes. The region of localization of the interference pattern is determined by the structure of the object, the nature of the changes that occurrred in it, and also by the design of the holographic setup. These matters will be treated in Chaps. 3, 4.

The visibility of the interference pattern will be determined by the ratio of the intensities of the interfering waves. Attenuating glass filters may be introduced into the reference or object beam in the observation stage to balance those intensities. Those filters must be of high quality to prevent their distorting the phase structure of the interfering waves. Equality of the intensities of the interfering waves can also be achieved by a rational choice of the conditions of preparing the hologram. For this purpose, the following condition must be observed:

$$kta_r^2 = \beta \quad . \tag{1.131}$$

According to (1. 126), this condition can be written in the form

$$D = \beta^2\alpha \tag{1.132}$$

where D is the diffraction efficiency of the hologram, α is the ratio of the object and reference beam intensities, and β^2 is the transmittance of the hologram with respect to intensity.

For practical realization of the real-time method, the hologram after photochemical processing must be returned to exactly the place where it was exposed, i.e., the pattern of the interference field formed by the object and the reconstructing beams during the observation must be made to coincide with the pattern recorded on the hologram with an accuracy within fractions of a spatial period.

The real-time method has the appreciable merit that a single hologram obtained in the initial state of an object can be used to study with interferometric accuracy the dynamics of the process occurring with it by studying a multitude of its states one after another.

If the inclination of the reconstructing beam is changed during reconstruction of the wavefront, it is possible to obtain a system of carrier fringes of a finite width whose orientation and frequency, in accordance with the observer's wishes, will be present by the direction and the angle of inclination of the beam. As in conventional interferometry, these fringes often facilitate and make more accurate the interpretation of the interference pattern and allow the sign of the phase changes introduced by the object to be identified.

1.4.4 Double-Exposure Method

The real-time method makes it possible to compare the initial state of an object with an infinite multitude of those following. If only one of them interests us, we can use the double-exposure method, which is much simpler from the experimental standpoint.

We have already noted that this method consists in the consecutive exposure of two holograms of a single object in different states on the same light-sensitive layer. Both waves are reconstructed simultaneously, and an interferogram is thus obtained that corresponds to the changes of the object that occurred during the time between the two exposures. Now, care no longer has to be taken to return the hologram accurately to the position in which it was during the exposure. It is only necessary to ensure stability of the setup in the interval between exposures.

The configuration of the reconstructing beam may considerably differ from that of the reference beam used in recording the doubly exposed hologram. The scale distortions and displacements of the images (see Sect. 1.3) will be identical for both waves, and as a result the pattern of the interference fringes remains practically unchanged. Aberration is also identical for both waves and does not lead to a change of the structure of the interference pattern. A certain distortion of the interference pattern is possible, however, as a result of any noticeable difference of the configuration and inclination of the reconstructing beam.

If both exposures were made with identical conditions of illumination and had the same duration, then the amplitudes of both reconstructed waves would be the same, which ensures a high visibility of the interference pattern, close to unity.

In negative recording of a doubly exposed hologram, both reconstructed waves are opposite in phase with respect to the waves being recorded, and in the absence of changes in the object they do not cancel each other (as in the real-time method), but amplify each other.

The diffraction efficiency of doubly exposed holograms requires special treatment. According to (1.114), the distributions of the exposures will be as follows:

$$H_1(x,y) = t(a_{01}^2 + a_r^2 + A_{01}A_r^* + A_{01}^*A_r)$$
$$H_2(x,y) = t(a_{02}^2 + a_r^2 + A_{02}A_r^* + A_{02}^*A_r) \quad . \tag{1.133}$$

The total exposure is

$$H(x,y) = H_1(x,y) + H_2(x,y) \quad . \tag{1.134}$$

For an amplitude linearly recorded hologram, the relationship between the amplitude transmittance and the exposure is given by (1.115), and from (1.133,134) we get

$$T = T_0 + kt\{2a_r^2 + (a_{01}^2 + a_{02}^2) + (A_{01} + A_{02})A_r^* + (A_{01}^* + A_{02}^*)A_r\} \quad . \tag{1.135}$$

When a hologram having such a transmittance is illuminated with the reference wave A_r, the complex amplitudes of the waves reconstructed in the +1st order will be

$$A_{1i} = a_r^2ktA_{01} \quad \text{and} \quad A_{2i} = a_r^2ktA_{02} \quad . \tag{1.136}$$

Let us assume that the waves being recorded have the same amplitude as the reference wave and differ from each other only in phase:

$$A_{01} = a_re^{i\varphi_{01}} \quad \text{and} \quad A_{02} = a_re^{i\varphi_{02}} \quad . \tag{1.137}$$

In this case, the formula for the transmittance can be written in the form

$$T = T_0 + kt\{4a_r^2 + 2a_r^2[\cos(\varphi_{10} - \varphi_r) + \cos(\varphi_{20} - \varphi_r)]\} =$$

$$= T_0 + 4a_r^2kt\left\{1 + \cos\left(\frac{\varphi_{10} + \varphi_{20}}{2} - \varphi_r\right)\cos\left(\frac{\varphi_{10} - \varphi_{20}}{2}\right)\right\} \quad . \tag{1.138}$$

Because the factor $\cos[(\varphi_{10} + \varphi_{20})/2 - \varphi_r] \times \cos[(\varphi_{10} - \varphi_{20})/2]$ can change within the range from +1 to -1, the transmittance in different parts of the hologram will change from T_0 to $T_0 + 8a_r^2kt$.

For a perfectly linear negative material whose T-H curve is shown in Fig.1.33a, the transmittance can change from 1 to zero, i.e., $a_r^2kt = 1/8$. Thus, in opticmal recording conditions, the complex amplitudes of the reconstructed waves will be

$$A_{1i} = \frac{1}{8} a_r e^{i\varphi_{01}} \quad \text{and} \quad A_{2i} = \frac{1}{8} a_r e^{i\varphi_{02}} \quad . \tag{1.139}$$

The intensity of each of the reconstructed waves is $a_r^2/64$, i.e., one-fourth of the intensity of the wave reconstructed by a singly exposed amplitude hologram in the optimal conditions of recording.

When these waves interfere, a pattern is formed having maxima whose intensity is

$$I_{max} = \left(\frac{1}{8} a_r + \frac{1}{8} a_r\right)^2 = \frac{1}{16} a_r^2 \quad , \tag{1.140}$$

i.e., equals the intensity of the wave reconstructed by a singly exposed hologram.

In some methods of holographic interferometry, multiply exposed holograms are used (see Chaps. 4,5). Sometimes a multiple-wave interference pattern is observed, similar to that obtained in multiple-wave interferometers. For an amplitude hologram exposed m times in the optimal conditions, a similar treatment leads us to the conclusion that the intensity I_m of each of the reconstructed waves

$$I_m = \frac{I_1}{m^2} \quad . \tag{1.141}$$

The intensity at the maximum of the interference pattern observed with many exposures is

$$(I_m)_{max} = I_1 \quad . \tag{1.142}$$

In the double-exposure method, we obtain interference fringes of an infinite width. To obtain a system of fringes of a finite width, it is sufficient to change the angle between the object beam and the reference beam before the second exposure. For this purpose, a thin glass wedge may be placed in the object beam between the two exposures [1.67] (for greater details see Chaps. 2,3).

Unlike the real-time method, in the double-exposure method we cannot change the orientation and frequency of the carrier fringes in the process of reconstruction. The orientation of the fringes coincides with the direction of the edge of the wedge, whereas the frequency of the fringes is set by the angle of deviation of the object beam by the wedge. If a change of the orientation of the carrier fringes is needed for processing the interferograms, we can use special reconstruction setups that make it possible to split the waves that correspond to each of the two exposures and to change the angle between them when desired. Figure 1.36 shows such an arrangement [1.68]. The waves that correspond to the two exposures are split by the mirror prism 3 and impinge on the arms of the interferometer. The latter makes it possible to change their mutual orientation.

It is also convenient to use, for this purpose, such a variant of the double-exposure method in which the holograms are consecutively exposed on two separate plates. By changing the mutual arrangement of the two holograms in reconstruction, we can alter the frequency and orientation of the interference fringes.

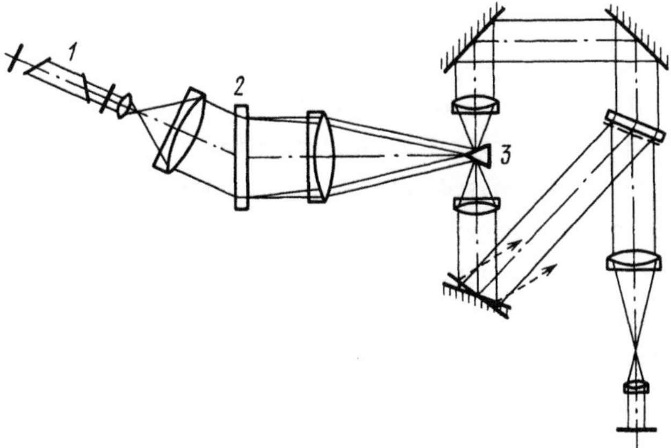

Fig.1.36. Optical reconstruction setup for changing the frequency and orientation of the interference fringes: 1 - laser; 2 - hologram; 3 - light-splitting prism

2. Experimental Techniques

2.1 Light Sources

2.1.1 Requirements for Light Sources in Holographic Interferometry

The general requirements for light sources set out in Chap. 1 are appropriate for classical interferometry. The specific nature of holography, as indicated previously, requires that special attention be given to definite characteristics of the radiation of the source. The radiation must have

1) adequate temporal coherence so that a stationary interference microstructure formed as a result of the coherent interaction of the object and the reference waves will exist in the plane of a hologram during the entire exposure;

2) sufficient spatial coherence so that the object wave scattered to a given point of the hologram from different parts of the surface being studied will create a clear interference microstructure exactly with the given element of the wavefront of the reference wave (this requirement is needed when light-scattering elements are used in a holographic setup);

3) sufficient power and the required wavelength so that the light-sensitive material used for recording a hologram will be able to respond to the action of the interference field during the exposure time.

The first two conditions can be expressed mathematically because they follow from the relationship between the visibility of the interference pattern p formed in the plane of the hologram (the visibility of the interference fringes recorded on the holographic recording material) and the degree of mutual coherence $|\underline{\mu}|$ of the object and reference waves [see (1.81)] [2.1,2],

$$p = \frac{2|\underline{\mu}|\sqrt{\alpha}}{\alpha + 1} \quad . \tag{2.1}$$

A glance at (2.1) shows that the coherent properties of the source determine the visibility of the fringe pattern recorded on the hologram and, consequently, the quality of the recorded hologram.

A consideration of the coherent properties of various sources including thermal ones shows that, to date, only lasers of certain kinds simultaneously comply with all of these essential requirements. In a number of exceptional cases, experiments have been successful with sources of other kinds, but at present they are of use only for demonstration [2.3].

A brief study of the coherent properties of radiation shows that coherence can be increased by

1) narrowing the spectral range of the emitted frequencies,
2) phasing the light trains emitted by various excited atoms, and
3) diminishing the angular dimensions of the emitter.

Lasers provide these conditions virtually simultaneously without appreciable reduction of radiated power; this is their main advantage over other light sources.

Phasing of light wave trains is possible because stimulated emission is the main mechanism for release of the excitation energy of the atoms of the active (emitting) medium. The characteristics of a stimulated-emission wave train (the frequency, phase, polarization) coincide exactly with those of the external field that acts on an atom. By properly choosing the kind of active medium and the conditions for its excitation (pumping), and also by placing the working substance in a high-Q optical resonator with strong feedback conditions, we can generate light at frequencies close to the natural frequencies of the resonator

$$\nu_n = \frac{c}{2L_r} n \quad , \tag{2.2}$$

where L_r is the length of the resonator, and n is an integer.

If several natural frequencies of the resonator can be accommodated inside halfwidth $\Delta\nu_\ell$ of the gain curve of the active medium, then several close frequencies are generated. The coherence length of such emission, in accordance with (1.84), depends on the number of emitted frequencies (they are called axial modes) and is several scores of centimeters for gas lasers. Such coherence cannot be realized in the emission of thermal sources, if the intensity of their radiation is kept sufficient for practical use in holographic interferometry.

Further increase of temporal coherence of laser radiation can be obtained by reducing the number of frequencies ν_n inside the gain curve $\Delta\nu_\ell$. This can be done, for example, by operation near the threshold of generation or by using selecting cells in the resonator. In single-frequency operation, the coherence length of laser radiation corresponds approximately to the path travelled by light in the time τ during which random changes of the phase of emission do not exceed a value comparable with π,

$$|r_1 - r_2|_0 = c\tau \approx 3 \cdot 10^5 \mathrm{cm} \quad . \tag{2.3}$$

The spatial coherence of laser radiation in light beams from 1 to 2 mm diameter is considerably greater than that from thermal sources with cross sections about 2 mm. Indeed, for a round source of radius ρ at the distance R from the region AA', provided that $|AA'|_{max} << R$ and $\rho << R$, we can get [2,1,4]

$$\underline{\mu}_{1,2} = \left[\frac{2J_1(\vartheta)}{\vartheta}\right] e^{i\psi} \quad , \tag{2.4}$$

where

$$\vartheta = \frac{2\pi}{\lambda_{av}} \rho\sqrt{p^2 + q^2} \ , \quad p = \frac{x_1 - x_2}{R} \ , \quad q = \frac{y_1 - y_2}{R} \quad ,$$

$$\psi = \frac{2\pi}{\lambda_{av}} \left[\frac{(x_1^2 + y_1^2) - (x_2^2 + y_2^2)}{2R}\right] \quad ,$$

and x_1, y_1 and x_2, y_2 are the cartesian coordinates of the points A and A'. Analyzing (2.4), we conclude that coherence completely disappears on a circle of radius AA' = 0.61 $R\lambda_{av}/\rho$, i.e., for ρ = 1 mm, R = 100 mm, and λ_{av} = 0.63 µm we have AA' = 0.04 mm. The coherence area for lasers, under the same conditions, is several orders of magnitude greater because, owing to the tight phasing of the radiation, the light beam that emerges from the entire cross section of the active element (or a considerable part thereof) has a virtually plane wavefront. Hence, the angular dimensions of the equivalent source in (2.4) sharply diminish, and the spatial coherence increases correspondingly.

Having convinced ourselves that the coherence properties of laser radiation make it possible to create a pattern of interference fringes, by interaction of the reference and object beams, in a considerable region of space, let us consider the conditions needed to record this pattern, i.e., the requirements on the energy and spectral range of radiation of various lasers.

Figure 2.1 shows the energy characteristics of lasers that determine their suitability for use in various branches of science and engineering [2.5]. The regions indicated by dashed lines in this diagram show power against pulse duration from different kinds of lasers. The constant-energy line is parallel to a straight line with a slope of 1 W/s. Of course, continuous lasers are shown by lines parallel to the duration axis, corresponding to constant power (the dashed line corresponds approximately to the maximum achieved power of continuous lasers used in holography) [2.6]. The region occupied by continuous lasers, like all the other regions in

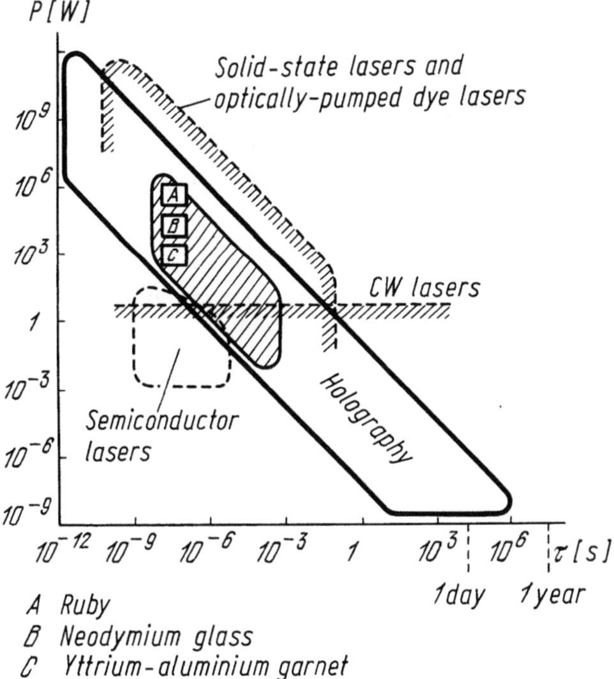

Fig.2.1. Distribution of various kinds of available lasers by power. The region of the diagram corresponding to holographic requirements is shown

this diagram, is shown very tentatively because progress in laser engineering results in constant drifting of these regions.

In view of the power sensitivity of the media used for recording a typical hologram, radiation energy of from 1 J to 1 μJ is needed; this region is shown by the closed loop in Fig.2.1. Thus, lasers characterized by points below this loop are not suitable for holography, whereas lasers whose characteristic parameters are above the loop are too powerful, but their power can, of course, be weakened by means of special attenuators. The maximum duration of the exposure of a hologram (the right-hand edge of the loop) is limited by motion, vibrations, thermal deformations, and other disturbances; this limit is quite relative and is determined by the specific conditions of an experiment. The requirements of coherence form a limit on reduction of the duration of a pulse and therefore determine the left-hand limit of the loop.

We can thus assess the region that contains the laser parameters most suitable for the purposes of holography. Inside this region is a hatched

Table 2.1. Principal laser lines

Species	Host	Wavelength, [μm]	Usual mode of oscillation
Cr^{+3}	Al_2O_3 (ruby)	0.694	pulsed, Q-switched
Nd^{+3}	glass	1.06	Ditto
Nd^{+3}	YAG	1.06	cW, repetitively pulsed
Ne	He	0.633	cW
Cd	He	0.329; 0.442	cW
CO_2	-	10.6	cW, Q-switched, repetitively pulsed
Ar^+	-	0.488 0.515	cW, pulsed Ditto
Kr^+	-	0.647	Ditto
Ga-As	-	0.840	Ditto
Rhodamine 6G	ethanol, water methanol	0.57-0.61	cW, short pulse
N_2		0.337	pulse

area that corresponds to pulsed ruby, Nd-doped glass and yttrium-aluminium garnet (Nd:glass and Nd:YAG) lasers that are in the greatest favor in holography. In view of the spectral sensitivity of holographic recording materials (see Sect. 2.2) and the simplicity of laser operation and control, lasers that radiate visible light are preferred in holography. Inspection of Table 2.1 [2.7] which gives the principal spectral lines produced by commercially available lasers shows that the following types of lasers are good for holographic interferometry: ruby, Nd^{+3}-doped (second frequency), He-Ne, He-Cd, Ar^{+}, and Kr^{+} lasers. The parameters of organic-dye and semiconductor lasers are also satisfactory, but it is too early to speak about their widespread use in holography.

2.1.2 Gas Lasers

Continuous-wave (CW) gas lasers are used in most practical interferometry based on a holographic analysis of the displacements and vibrations of surfaces, and also of the profile of surfaces. These lasers are most popular in holography at present for the following reasons:

1) Some kinds of gas lasers have been in production for a long time; they are cheaper and in greater supply than other lasers.

2) Some of them have long service life while retaining their basic output parameters.

3) Gas lasers are practically always used for reconstructing images from holograms, even if the holograms were recorded with the aid of other kinds of lasers, and also for adjusting holographic setups.

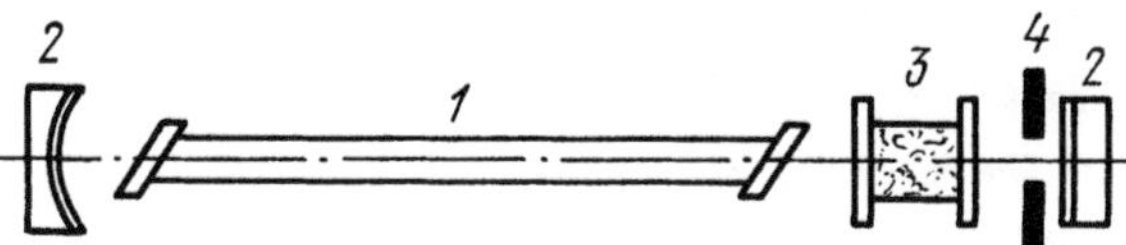

Fig.2.2. Resonator of a gas laser: 1 - active element; 2 - mirror; 3 - selective cell; 4 - diaphragm

Consequently, every holographic setup is provided with a continuous gas laser. The optical arrangement of such a laser is shown in Fig.2.2 and includes the following elements:

1) The active element — a gas-discharge tube whose output windows form a Brewster angle with the optical axis, to reduce losses. With respect to

the working-gas mixture in which a discharge is formed, the following main kinds of gas lasers are distinguished: (a) helium-neon; (b) helium-cadmium; (c) argon; (d) krypton, and (e) nitrogen. Helium-neon lasers are the ones most frequently used in holographic interferometry, with argon and helium-cadmium lasers becoming more popular.

2) Mirrors with interference coatings form an optical resonator of the confocal or, more often, of the semiconfocal kind. In the latter case, the flat mirror has a somewhat lower reflection coefficient and is used for the output window.

3) A Fabry-Perot resonator for separating one axial mode, or an absorbing cell for separating one generated line.

4) A movable diaphragm for suppressing higher transverse modes of the laser emission and increasing its spatial coherence.

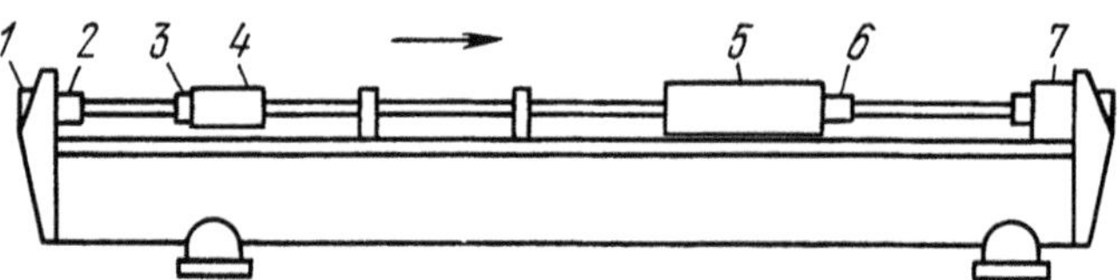

Fig.2.3. Helium-cadmium laser. The arrow shows the direction of the cadmium vapor flow

The simple optical construction of a gas laser should create no illusions as to the simplicity of design of the instrument itself.

Figure 2.3 shows the design of helium-cadmium laser that has a power of as much as 50 mW at a wavelength of 4416 Å and more than 5 mW at a wavelength of 3290 Å. The reference numbers in this figure signify: 1 - adjustable-mirror mounting system; 2 - a sylphon bellows that connects the cold Brewster window to the mirror; 3 - anode; 4 - cadmium vaporizer; 5 - cold-aluminium cathode; 6 - widened section of the tube where cadmium condenses and is removed from the discharge; 7 - moisture-proof insulator of the Brewster window, used to protect it from dirt.

A laser of this kind uses cataphoresis (the formation of a stream of cadmium ions toward the cathode) through a capillary section of the tube in which a discharge in a helium-cadmium medium is maintained. The source that continually replenishes the cadmium ions is a furnace vaporizer. The cadmium condenses and is removed at the opposite end of the tube. The temperature of the capillary-section walls is high enough to prevent condensation of the cadmium on this section, and at the same time is such that forced cooling is not needed.

Many structural elements of a laser, for example its mirrors, discharge tubes, etc., are manufactured at the limit of today's technological capabilities. Thus, to increase the service life of the active element in the more up-to-date models of helium-cadmium lasers, the direction of the cadmium stream is reversed after it is completely evaporated from the furnace – the functions of the furnace and the condenser are exchanged, and the polarity of the discharge electrodes is changed correspondingly. The use of such a system, however, makes this laser an intricate and consequently an insufficiently reliable device. For similar reasons, some of the lasers on the market are experimental models rather than instruments in the full meaning of the word – scatter of the output parameters, and also fluctuation of these parameters in the course of operation remain to be improved. We can be confident, however, that the needs of holography, which is one of the main spheres of laser application [2.8], will stimulate the appearance of new kinds of lasers, including gas lasers, that have the required characteristics.

Table 2.2 gives the basic characteristics of selected gas lasers in current production in the USSR and other countries. The data of this table show that among CW gas lasers the most acceptable for the purposes of holographic interferometry are argon lasers that produce radiated power of at least 0.2 W on each of the following lines [2.9]:

$$\lambda_1 = 5145\ \mathring{A}\ , \quad \lambda_2 = 4965\ \mathring{A}\ , \quad \lambda_3 = 4888\ \mathring{A}\ ,$$

$$\lambda_4 = 4727\ \mathring{A}\ , \quad \text{and} \quad \lambda = 4579\ \mathring{A}\quad .$$

Introduction of means for increasing the coherence length on most operating lines made it possible to record a hologram of quite long (up to one meter) objects. These results indicate that after the shortcomings of some current types of lasers on the market are eliminated (first of all the low service life and comparatively high prices), argon lasers will become more popular.

Table 2.2. Basic characteristics of selected gas lasers in current production in the USSR and foreign countries and suitable for holographic interferometry

Type of laser	Active medium	Wavelength [Å]	Output power [mW]	Operating conditions (mode)	Coherence length [m]	Remarks
LG-36	He-Ne	6238	20	Single	0.2	
LG-36A	He-Ne	Ditto	40	Single	0.2	
LG-38	He-Ne	Ditto	50	Single	0.2	Stabilized in power
LG-106	Ar	4880	1000	Multiple	Below 0.06	Can be readjusted to single mode
LG-31	He-Cd	4416	10	Single	0.3	
Spectra-Physics 125A	He-Ne	6328	50	Single	10	With Model 989 etalon
124A	He-Ne	Ditto	15	Single	0.2	
165	Ar	4579-5145	$2\text{-}5\times10^3$	Single	Below 0.06	Higher coherence achieved when using single spectral line (165, 171, Coherent Radiation)
171	Ar	Ditto	$9\text{-}18\times10^3$ (Total)	Single	Ditto	
Coherent Radiation CR185G	Ar	Ditto	18×10^3 (Total)	Single	Ditto	
CR-750K	Kr	6471 5309 5209	750 250 150	Single	0.2	1.2W of total power

2.1.3 Solid-State Lasers

Pulsed lasers using solid-state active elements are employed more than others for studying rapid processes by methods of holographic interferometry. Of solid-state laser elements used, most in favor, at present, (see Fig.2.1) are ruby, neodymium (the active element is neodymium glass) and yttrium-aluminium garnet (YAG) lasers [2.9].

The optical diagram of a solid-state laser is quite similar to that shown in Fig.2.2, except that an additional element is often put in the resonator to increase the radiated peak power by diminishing the duration of the generated pulse and to improve the coherence properties of the radiation – a shutter for Q switching. Solid-state lasers can be classified according to kind of Q switching into lasers with a passive shutter – a cell that contains a saturable dye or a saturable color filter, and lasers with synchronized Q switching by use of an electrooptical or acoustical shutter, or a rotating mirror (prism). Finally, lasers can be used in conditions of free generation, in which there is no shutter, and the radiation consists of an irregular sequence of pulses. If the coherence properties of the first two types of lasers in single-frequency mode differ only slightly from those indicated in Table 2.1 for gas lasers [2.2], then lasers with free generation (free-running lasers) are suitable for studying phase objects in setups that have full compensation of the optical phase difference of the two beams, because the coherence length of their radiation does not exceed several millimeters. Pulse-flash lamps are used as sources for pumping the active elements in solid-state lasers.

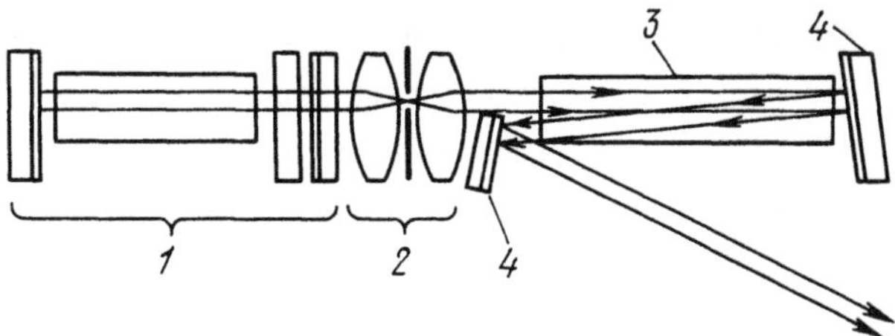

Fig.2.4. Ruby quantum amplifier: 1 - oscillator; 2 - collimator with pinhole diaphragm; 3 - active element of amplifier; 4 - mirror of amplifier

Ruby lasers are most popular among solid-state lasers in the technique of holographic experimentation. The characteristics of the typical ruby laser are as follows: wavelength 6943 Å, output energy 400 mJ in double-pulse conditions with a pulse duration of 20 ns, pulse separation from 50 to 500 μs, and a coherence length about 1 m. Notwithstanding the high amplification factor in the active medium of solid-state lasers, their power is not adequate for obtaining holograms of large-scale diffusely-scattering objects 1 m^2 and more in area; amplifiers have to be used, in which a Q-switched laser is the driving oscillator (Fig.2.4). The pulse of light from the driving laser 1 is fed through the collimator 2 with a pinhole diaphragm into the active element of the amplifier 3, whose mirrors 4 are arranged so that the beam passes several times through the active element, which is in the excited state at the time. Multistage amplification in several series-connected amplifiers is also employed. This makes it possible to increase the radiated power of the driving laser several hundreds of times without reducing the coherence of the radiation [2.10]. However, the following phenomenon must be taken into consideration in constructing holographic setups in which the reference beam is formed from a part of the radiation of the driving laser, and the object beam from the radiation of the amplifier. The maximum visibility of the interference field and, consequently, the best quality of a hologram are obtained not when the optical paths of the two beams are made equal, but when there is a definite difference between these paths. The optimal value of this difference is determined experimentally and is usually about one meter.

The cause of this phenomenon is the existence of a shift between the frequencies of radiation of the driving oscillator and the amplifier, which can be several megahertz [2.2,11].

The existence of this shift can be explained by analyzing the interaction of the running wave from the driving laser with the active medium in the amplifier. Assuming the presence of a frequency shift between the object and the reference waves (Δf), in turn, we can calculate the lag of one wave behind the other (Δt) at which the visibility of the fringes is maximum (but no longer equals unity, as it does when $\Delta f = 0$) [2.11],

$$\Delta t = \frac{\pi^2 \Delta f (\Delta\tau)^3 f_0 \delta}{1 + (\pi \Delta f \Delta\tau)^2} , \tag{2.5}$$

where f_0 is the operating frequency of the driving laser, δ is the frequency shift of the amplifier, and $\Delta\tau$ is the duration of the pulse generated.

In most methods of holographic interferometry, double or multiple exposure of the object is used to obtain an interferogram that records the changes that occur during the time between the exposures (see Chaps. 3-5). For this purpose, the different methods described below can be used.

1) If exceedingly rapid processes are being studied (for example, plasma phenomena) whose development time is measured in a few or scores of microseconds, one of the following procedures is used:

(a) double-exposure holography — the first exposure is made before the process being studied begins, the second pulse is synchronized with the beginning of the process or with any moment during it. Electrooptic and acoustooptic shutters are used for synchronization. The latter have the advantages over electrooptic shutters that a considerably lower voltage is needed to control them, and their design is simpler;

(b) if multiple-frame holography of a process is needed (cinematographic holographic interferometry), then an optical delay line is used in combination with a single-pulse laser (see Sect. 3.3), or the light-generating channels are temporally separated across the active element of the laser. One of the most interesting ways of temporal separation is the method of the moving-wave slit, whose principle is illustrated in Fig.2.5 [2.12].

The laser resonator contains an acoustic cell 1 in which the emitter generates an acoustic wave that propagates in the direction of the absorber and is modulated in time so that a short region of silence – a wave slit – exists between the two trains of running waves. Several apertures (3, 4, 5) are made in an opaque screen; together with the moving-wave slit, they open consecutively different sections of the previously excited active element. The depth of modulation of the running acoustic wave (grating) is such that the zero-order diffraction maximum is zero at the operating laser wavelength, so that apertures that do not coincide with the moving-wave slit are practically closed.

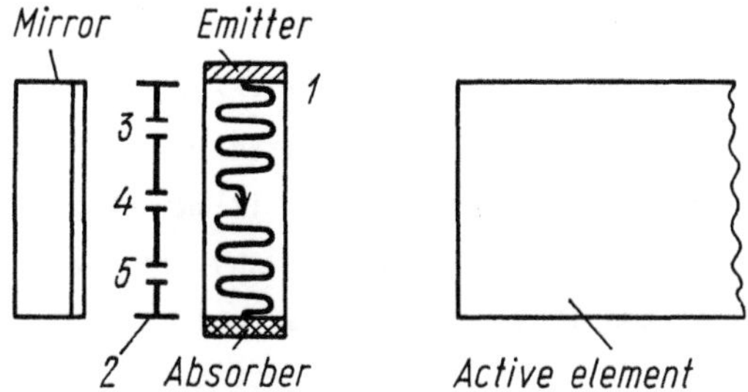

Fig.2.5. Q modulation using the moving-wave slit principle: 1 - acoustic cell; 2 - diaphragm block holder; 3-5 - diaphragm

About 5.4 μs intervals between the generated pulses can be obtained when nitrobenzene is used as the acoustic medium and the Rb active element is 10 mm in diameter and is divided into channels 2 mm in diameter. It is difficult to reduce this interval because narrowing of the generation channels is required. The interval is increased by changing the conditions of excitation of the acoustic cell, for example by two opposed emitters.

2) If the characteristic times of the process being studied are measured in milliseconds (vibrations, deformations, aerodynamic phenomena), then method 1a is used for double-exposure holography or its analogue with optomechanical Q switching of the laser (for example by use of a rotating prism).

One of the methods of cinematographic holography similar to high-speed photography is used for multiple-frame holography. It consits of scanning the information over the recording material of the hologram. There are different ways of temporal scanning (coding) of holograms, among which the following deserve attention:

(a) Use of a diffusing screen that, when illuminated with a laser ray, produces radiation, part of which is scattered diffusely, and part retains the shape of the illuminating beam (Fig.2.6) [2.13]. The radiation of the

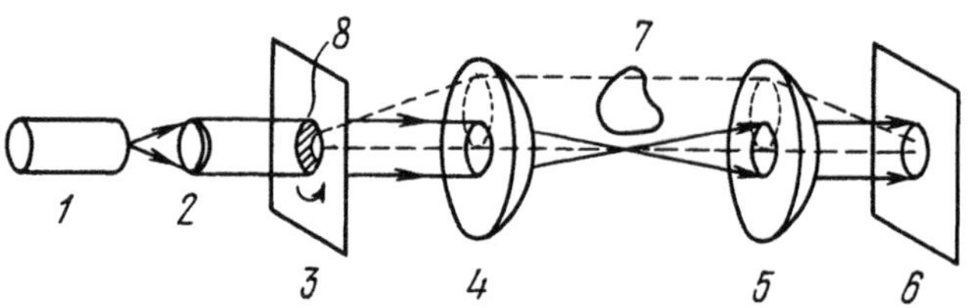

Fig.2.6. Setup for the holographic study of phase objects, with a semitransparent diffusing screen: 1 - pulsed laser; 2 - collimator; 3 - diffusing screen; 4,5 - lenses; 6 - hologram; 7 - object; 8 - rotating mask

laser 1 is expanded by collimator 2 and falls on the diffusing screen 3, after which there is a system of two lenses 4 and 5 that focus an image of the screen 3 in the plane of the hologram 6. The direction of illumination of the screeen 3 is such that the undisturbed beam is made to converge by the lens 4 away from the optical axis of the lens system, thus freeing space near the focal plane for the object 7 being studied. A rotating disk with aperture 8 is placed before the screen 3. With this arrangement, the reference beam can be directed onto the hologram at different angles that form a cone relative to the optical axis of the setup. The laser generates a sequential-pulse train. The number of these pulses in combination with the speed of rotation of the disk determines the number of frames of the cinematographic hologram.

(b) It is also possible to record temporally scanned information by masking the hologram with a rotating disc that has a small open sector. Lasers that operate in free-generation (multiple-peak) duty can be used in this setup [2.14].

3) Slow processes whose characteristic times are fractions of a second or seconds are usually studied by use of gas lasers and real-time holographic interferometry setups (see Sect. 1.4) in which the changing pattern of interference fringes is recorded on cinematographic film [2.15].

2.1.4 Dye lasers

A new class of lasers, that in many aspects have not yet been mastered, is represented by dye lasers. They transform the radiation of solid or gas continuous or pulse lasers into radiation of any wavelength of the visible spectrum [2.2,16]. The use of these lasers in the resonance holography of phase objects is very effective, notwithstanding frequently inadequate coherence of their radiation (see Sect. 3.2).

2.2 Hologram Recording Materials

2.2.1 Requirements to Recording Materials

The main branches of holography – interferometry, optical information processing, synthesis of optical elements – impose a considerable number of serious requirements that hologram recording materials must meet. These requirements are often of a specific nature and relate only to a given specific application, being of much smaller importance for other applications. For this reason, the present trend is to specialize the branches of industry that produce and study recording materials for different applications of holography.

What determines the main directions in which physical effects, to be used for transformation of two- and three-dimensional light-intensity distributions (microstructure) that illuminate holograms into distributions of optical properties of the hologram material, are being sought?

First, more-complete utilization of various optical properties that can be modified by micro-interference patterns that are formed in the region of a hologram. These properties include:

— optical transparency, usually described by the amplitude transmittance τ,

$$\tau = \frac{a_{out}}{a_{inc}} \quad , \qquad (2.6)$$

— optical thickness, determined by the thickness of the recording-material layer d and the refractive index n of the material,

$$\ell_{opt} = dn \quad , \qquad (1.7)$$

— reflectance (reflection coefficient), determined by the ratio of the intensities of the reflected and incident waves,

$$r = \frac{|a_{refl}|^2}{|a_{inc}|^2} \quad , \qquad (2.8)$$

— rotation of the plane of polarization, when polarized light passes through a layer 1 cm thick, or is reflected from it,

$$\Delta\theta = \text{ar cos}\left\{\frac{\vec{a}_{out}\vec{a}_{inc}}{|a_{out}||a_{inc}|}\right\} \quad , \qquad (2.9)$$

and a number of others.

In searching for materials in which those properties depend optimally on exposure conditions, we must take the following fact into consideration. The energy of visible-light quanta that correspond to the wavelength range of from 0.4 to 0.7 μm is very small, from 1.8 to 3.1 eV. This energy is most often inadequate to produce appreciable change of the optical properties of a medium in the region where a light quantum is absorbed. For this reason, as a rule, recording materials whose functioning is based on direct action of light on the optical properties of a solid have a low sensitivity (evaporated metal films, photochromic crystals, magnetooptic films).

The way out of this situation is to use a two-stage principle for forming the response of the recording material. In it, the action of the light field is similar to the action of a catalyst that controls the course of the recording process, or to the electric signal fed to the grid of a triode. A typical illustration of this principle is a silver halide photographic emulsion, in which a latent (invisible to the eye) image is formed under the action of light; after chemical processing, this image is transformed into a sufficiently effective change of various optical properties of the emulsion at the expense of the energy of the chemical processes. The greatest achievements have been attained by development of recording media for holography in which just such a two-stage principle of forming the response is employed. They include silver halide photographic materials, photoconductor-thermoplastic layers, and dichromated gelatin films [2.17].

The requirements of holographic interferometry on recording materials seem quite moderate compared to the requirements of other branches of holography. The spatial frequencies that a hologram has to record rarely exceed 1500 to 2000 lines per millimeter, and a quite high level of nonlinear intensity distortions is permitted – in essence, only a good visibility of the interference fringes in the reconstructed image is often all that is required. Interferometric methods are also used in holography that utilize nonlinear recording [2.18-23]. The basic requirement on recording materials is therefore maximum sensitivity to the operating wavelength of the laser, with moderate spatial resolution, noise level, and diffraction efficiency.

Such criteria for the suitability of recording materials for holographic interferometry may seem to be too mild, but they reflect how matters stand at present in developing recording materials for holography; it is barely possible to meet completely and simultaneously all of the basic requirements with photosensitive materials.

Indeed, as we shall see in Chap. 4, some of the existing methods of interpreting holographic interferograms are based on linear reconsturction of the interference pattern with a low level of hologram noise and adequate diffraction efficiency. A more urgent problem, however, is to expand the fields of use of holographic interferometry and to pass from laboratory arrangements to industrial setups operating in real industrial conditions. All this requires reduction of the time needed to expose and process a hologram. This, in turn, sharply restricts the range of materials that, at present, can be used in holographic interferometry. In essence, only two kinds of materials can be used – silver halide photographic emulsions (plates and films), and photothermoplastic devices.

2.2.2 Silver Halide Photographic Materials

Let us consider the basic characteristics of silver halide photographic emulsions for holography [Ref. 2.17, Chap. 2]. We have already mentioned that the mechanism of their action is based on the chemical transformation of the microcrystals of the silver halide that were acted upon by light. In holographic emulsions, the size of the microcrystals must be smaller than the spatial period of the intricate interference field formed when the plane (or spherical) reference wave interacts with the complex field of the object, which most frequently fluctuates statistically in space. If the scattering indicatrix of the surface of the object is sufficiently wide, i.e., the angle θ_s (Fig.2.7) formed by the extreme rays of the indicatrix is of the same order of magnitude as the angular aperture of the hologram, then a statistically fluctuating object field will exist in the region where the hologram is. This field can be described by superposition of the plane waves that form angles of convergence with the reference wave varying from θ_{max} to θ_{min}. The spatial period of the interference pattern changes over the hologram within the limits from

$$\Delta_{min} = \frac{\lambda}{2\sin(\theta_{max}/2)} \tag{2.10}$$

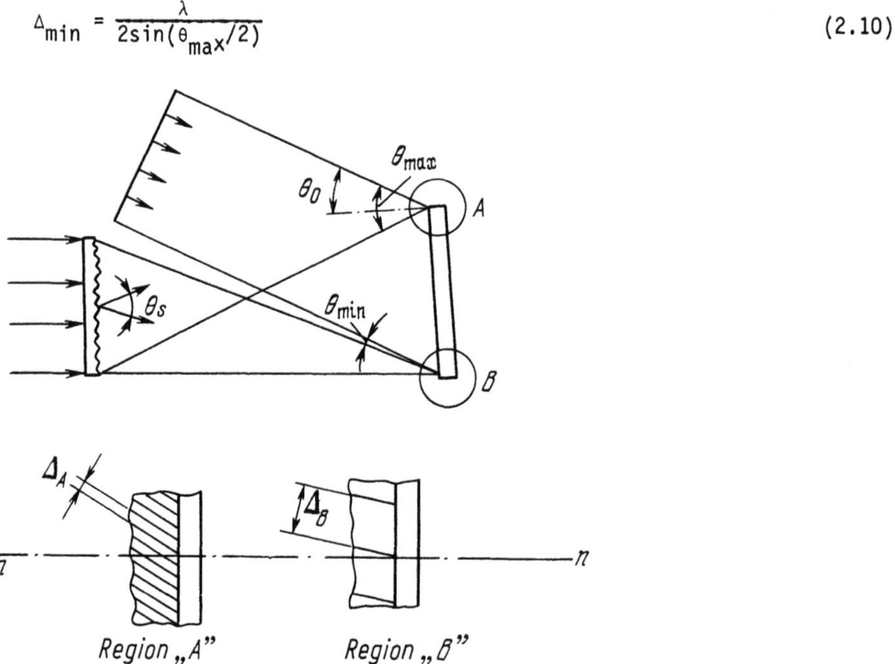

Fig.2.7. Determination of the range of spatial frequencies recorded on holographs of light-scattering objects

to

$$\Delta_{max} = \frac{\lambda}{2\sin(\theta_{min}/2)} \quad . \tag{2.11}$$

This change of the period takes place not smoothly – so that the period, for example, increases monotonically from region "B" to region "A" – but again statistically. On the average, however, the spatial frequencies of the interference pattern

$$\nu_{sp} = \frac{1}{\Delta} \tag{2.12}$$

that must be recorded on a hologram are greater in region "A" than in "B".

What happens if the high spatial frequencies are not recorded on a hologram, owing to insufficient resolution of the photographic material? Some of the components that carry information concerning the smallest details of the object being studied are lost in the wave reconstructed from such a hologram.

From the standpoint of the tasks of holographic interferometry, the main consequence of this loss consits of poorer visibility of the interference picture in the reconstructed images that show the changes of the object between exposures [2.24].

Because the object is often observed at different angles through a hologram when interferograms are interpreted (see Chap. 4), it is essential that the reconstruction of the object wavefront a_0 occur, as far as possible, without distortions over the entire aperture of a hologram (from region A to the region B, Fig.2.7), i.e., it is desirable that the following condition be observed over the entire hologram:

$$a_I = \kappa a_0 \quad , \tag{2.13}$$

where κ is a constant factor.

The intensity of the total light field that acts on a hologram at an arbitrary point $(\vec{x})$ is [see (1.32)]

$$I = |a_r + a_0|^2 = a_r^2 + a_0^2 + 2a_r a_0 \cos\Delta\varphi \quad ,$$

where the term $\Delta\varphi = \Delta\varphi(\vec{x})$ determines the spacing and the shape of the interference pattern recorded on the hologram. If the object being studied is reflecting or transparent (phase object), then the interference fringes have a regular nature. If the object scatters the incident light diffusely, then the interference pattern has a random nature, and a periodic relationship can be approximated only on separate small areas of the hologram. We can consider in this case, that the hologram records the superposition of periodic patterns whose amplitudes and phases vary statistically and whose periods are within the range

$$\Delta_{min} \leqslant \Delta \leqslant \Delta_{max} \quad .$$

Because the responses of recording materials are not linear (for a photographic material, this is illustrated by the dependence of the amplitude transmittance τ on the exposure, E, Fig.2.8), the condition of linear recording is satisfied when the variation of the exposure determined by the third term of (1.32) is not too great in comparison with the level of the exposure that corresponds to the operating point E_0. Hence, the following condition is obligatory:

$$a_0 << a_r \tag{2.14}$$

or

$$I \approx a_r^2 + 2a_r a_0 \cos\Delta\varphi \quad . \tag{2.15}$$

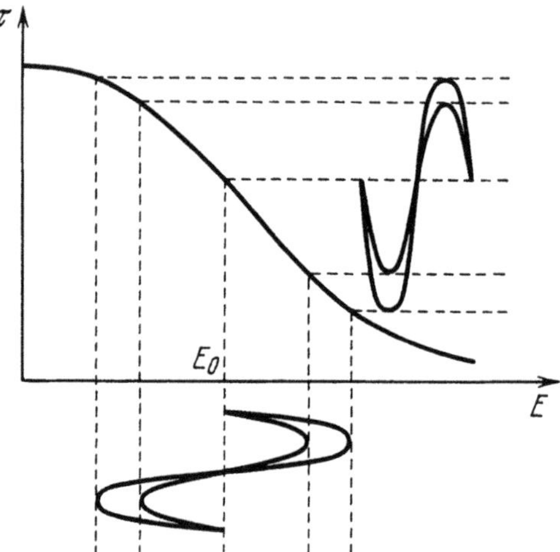

Fig.2.8. Appearance of nonlinearity in the photographic-material response when the object-beam intensity is increased

Expanding the function $\tau(E)$ into a series near the operating point E_0 in accordance with (2.14), we can disregard all of the terms in this expansion except the first two,

$$\tau(E) \approx \tau(E_0) + \frac{d\tau}{dE}\,\Delta E = \tau(E_0) + \left(\frac{d\tau}{dE}\,2a_r a_0 \cos\Delta\varphi\right)t_{exp} \quad , \tag{2.16}$$

where t_{exp} is the exposure time $(|a_r|^2 t_{exp} = E_0)$.

After processing the hologram and illuminating it with the wave $a_{inc} = a_r$, in accordance with (2.6,16) we get, directly beyond the plane of the hologram,

$$a_{out} = a_r\tau(E) = a_r\tau(E_0) + \frac{d\tau}{dE}\,(E_0)(2a_r^2 a_0 \cos\Delta\varphi)t_{exp} \quad .$$

Consequently (see, e.g. [2.24]), the amplitude of the wave that passes to the first diffraction maximum and creates the reconstructed image is

$$a_I = \frac{d\tau}{dE}\,(E_0)a_r^2 a_0 t_{exp} = \frac{d\tau}{dE}\,(E_0)E_0 a_0 \quad . \tag{2.17}$$

Comparison of (2.13) and (2.17) gives the requirement that ensures proportionality of the fields of the true image and that reconstructed from the hologram

$$\kappa = \frac{d\tau}{dE}\,E_0 \quad . \tag{2.18}$$

If we consider that the operating point (i.e., the level of exposure) E_0 is constant over the entire area of the hologram, i.e.,

$$E_0 = \mathrm{const}(\vec{x}) \quad , \tag{2.19}$$

then condition (2.18) is the well-known requirement

$$\frac{d\tau}{dE} = \mathrm{const} \quad , \tag{2.20}$$

which signifies that the variation of exposure should not extend beyond the quasilinear portion of the descending curve $\tau(E)$; this is exactly what is indicated by Fig.2.8.

Experimenters often fail to comply with condition (2.19), however, because the usual distribution of intensity in a laser beam is gaussian, and satisfaction of condition (2.19) results in great loss of power that could be used to illuminate a hologram (it is necessary to expand the reference beam greatly and use only its central part). An undesirable consequence of this is an increase of exposure time. This is why condition (2.19) is disregarded in practice, and the need to observe the following relationship follows from (2.18),

$$\frac{d\tau}{dE} = \frac{\text{const}}{E} \quad , \tag{2.21a}$$

i.e., the relationship $\tau(E)$ is required to follow a law that cannot be realized in practice,

$$\tau \propto \ln E \quad . \tag{2.21b}$$

Consequently, holographic interferometry rarely complies with the condition of linear recording and reconstruction of the object field when the hologram has a large area.

For the foregoing reasons, the reconstructed image of the object (and the interferogram) will appear brighter through some parts of a hologram than when observed through other parts. This circumstance, particularly, may make it difficult to realize systems for automatic processing of interferograms and methods of interpretation based on photometry of interference fringes (see Chap. 4).

Another important and almost inseparable property of photographic materials used in holography is that information is recorded on a hologram by various physical effects that often occur simultaneously when the hologram is processed:

1) formation of alternate dark and bright regions due to the presence of grains of metallic silver – this process determines the characteristics of amplitude holograms;

2) formation of surface relief due to shrinkage of the photographic emulsion at the places where undeveloped silver halide is dissolved and washed out, and also due to tanning of the gelatin during processing;

3) formation of regions that have different refractive indices, or more precisely, polarizability [2.25] by bleaching of developed holograms, in which metallic silver is transformed into a transparent salt whose optical coefficients differ from the characteristics of pure gelatin.

Consequently, amplitude holograms may diffract light not only because of variations of the transmittance $\Delta\tau$, but also because of variations of their geometrical thickness Δd. The amplitude of a wave at the exit surface of a hologram in which the amplitude transmittance and the thickness change simultaneously is expressed by

$$a_{out} = (\tau + \Delta\tau)a_{inc} \exp[i(\phi + \Delta\phi)] \quad ,$$

where

$$\Delta\phi = \frac{2\pi(n - 1)\Delta d}{\lambda} \quad .$$

If we know the slopes of the characteristics $\tau(E)$ and $d(E)$ obtained for the given spatial frequency near the operating point E_0, then, similar to (2.16), we can write (assuming, without limiting generality that $t_{exp} = 1$)

$$a_{out} = \left(\tau + \frac{d\tau}{dE}\,\Delta E\right)a_{inc} \exp\left[i\left(\phi + \frac{d\phi}{dE}\,\Delta E\right)\right] =$$

$$= \left(\tau + 2a_r a_0 \frac{d\tau}{dE} \cos 2\pi\nu_{sp}x\right)a_{inc} \times \exp\left[i\,\frac{d\phi}{dE}\,2a_r a_0 \cos(2\pi\nu_{sp}x)\right] \quad . \tag{2.22}$$

In (2.22), the constant phase factor $\exp(i\phi)$ has been omitted, and the spatial frequency ν_{sp} has been determined for the particular small region of the hologram within which ν_{sp} is considered constant. Equation (2.22) shows that the relationship between the output wave and the object wave becomes clearly nonlinear if the effects of the change of the optical thickness play an appreciable part in the formation of the hologram. Only when the phase modulation is weak, i.e.,

$$\frac{d\phi}{dE}\,a_r a_0 \ll 2\pi \quad , \tag{2.23}$$

can we obtain a linear response, even for a material with phase recording, by expanding the exponent in (2.22) into a series. Condition (2.23) is not advantageous for practical applications for reasons that will be treated subsequently. Imagine a hologram on which a periodic pattern has been recorded both in the form of changes of the amplitude transmittance τ_0 of a layer of unit thickness

$$\tau(x) = \tau_0 + \delta\tau \cos(2\pi\nu_{sp}x) \tag{2.24a}$$

and in the form of variations of the thickness of the photographic emulsion d (Fig.2.9):

$$d(x) = d_0 - \delta d \cos(2\pi\nu_{sp}x) \quad . \tag{2.24b}$$

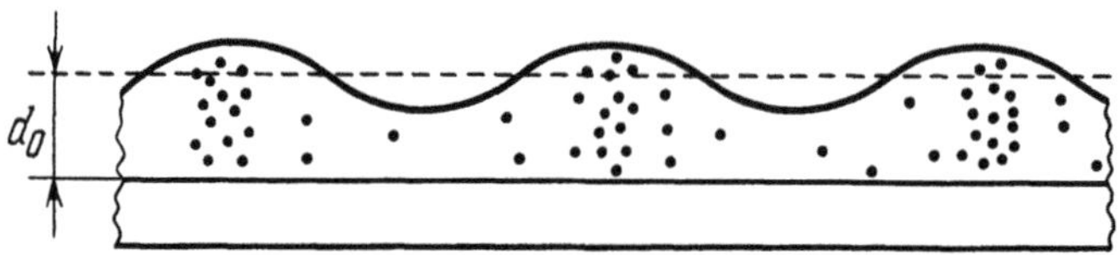

Fig.2.9. Geometrical relief in a layer of photographic material after photochemical processing

The signs of the variable terms in (2.24a,b) are opposite because the undeveloped silver halide, which is washed out as a result of fixing, reduces the thickness of the gelatin at the given spot. In the hologram whose model is shown in Fig.2.9, the incident wave is modulated in phase at the expense of variations of thickness, and in amplitude at the expense of the changes of the local amplitude transmittance $\Delta\tau$ due to variation of the exposure ΔE, and also due to thickness variations of the semitransparent layer $\tau_0\Delta d$. Calculation of the amplitude of the wave at the exit surface of such a hologram gives [2.26],

$$\begin{aligned} a_{out} &= a_{inc}\frac{\tau_0}{d_0}\left[1 - \left(\frac{\delta d}{d_0} + \frac{\delta\tau}{\tau_0}\right)\cos(2\pi\nu_{sp}x)\right] \times \\ &\quad \times \exp\left[\frac{2\pi i\delta d(n-1)}{\lambda}\cos(2\pi\nu_{sp}x)\right] = \\ &= a_{inc}\frac{\tau_0}{d_0}\left[1 - \left(\frac{\delta d}{d_0} + \frac{\delta\tau}{\tau_0}\right)\cos(2\pi\nu_{sp}x)\right] \times \\ &\quad \times \sum_{q=-\infty}^{q=\infty} \mathcal{J}_q\left[\frac{2\pi\delta d(n-1)}{\lambda}\right]\exp\left[qi\left(\frac{\pi}{2} - 2\pi\nu_{sp}x\right)\right] \quad . \end{aligned} \tag{2.25}$$

Equation (2.25) uses the expansion into a Fourier series of an exponential function whose exponent contains a harmonic function, $\tilde{J}_q$ are Bessel functions of the first kind of order q. Using the method described by GOODMAN [Ref. 1.35, Sect. 4.2], we can determine the wave of the first order of diffraction; for this purpose, we must perform a Fourier transform of the output wave and group together the waves that correspond to the focal spot closest to the axis. We thus get

$$
\begin{aligned}
a_I = a_{inc} \frac{\tau_0}{d_0} \Bigg\{ i\tilde{J}_1\left[\frac{2\pi\delta d(n-1)}{\lambda}\right] - \\
- \left(\frac{\delta\tau}{\tau_0} + \frac{\delta d}{d_0}\right) \tilde{J}'_i\left[\frac{2\pi\delta d(n-1)}{\lambda}\right]\Bigg\} = a_{inc} \frac{\tau_0}{d_0} \Bigg\{ i\tilde{J}_1\left[\frac{2\pi\delta d(n-1)}{\lambda}\right] - \\
- \frac{1}{2}\left(\frac{\delta\tau}{\tau_0} + \frac{\delta d}{d_0}\right)\left(\tilde{J}_0\left[\frac{2\pi\delta d(n-1)}{\lambda}\right] - \tilde{J}_2\left[\frac{2\pi\delta d(n-1)}{\lambda}\right]\right)\Bigg\} \quad . \qquad (2.26)
\end{aligned}
$$

Equation (2.26) shows that the first diffraction order includes two waves that differ in phase by $\pi/2$ and have amplitudes that depend on three Bessel functions whose argument is the phase-modulation index $2\pi\delta d(n - 1)/\lambda$. If phase modulation is absent, i.e., if no surface relief is formed on the hologram, then (2.26) transforms into

$$
a_I = -a_{inc} \frac{1}{2} \frac{\delta\tau}{\tau_0} \quad ,
$$

i.e., into an expression for the diffraction efficiency of an amplitude hologram [2.24].

We previously arrived at the conclusion that, in most practical holographic interferometry it is desirable to achieve the maximum diffraction efficiency possible with the photographic material that is at the experimenter's disposal[1]. Accordingly, (2.26) should be employed to find the maximum of the function $a_I(E)$, on the basis that $\delta d(E)$ and $\delta\tau(E)$ are determined from the experiment. If $\delta d(E) = 0$, i.e., if no surface relief is formed, then we get the condition $\delta\tau(E) = \max$, i.e., the requirement that

---

[1] This does not invalidate the requirements set out at the beginning of this section, to which persons who develop recording materials should adhere.

the operating point be chosen on the steepest section of the curve $\tau(E)$; this, in combination with (2.14), leads to the use of the center of the quasilinear portion of the curve as the optimal mean exposure (see Fig.2.8).

Experience shows, however, that most real holographic photographic materials form surface reliefs whose heights increase above a particular exposure level [2.27] [Fig.2.10, curve $\delta d(E)$]. Therefore, the maximum amplitude-diffraction efficiency $\sqrt{\eta} = a_I/a_{inc}$ is usually achieved not at the middle of the quasilinear part of the curve $\tau(E)$, but when the operating point is displaced in the direction of greater exposures [the curve $\sqrt{\eta_{exp}}(E)$].

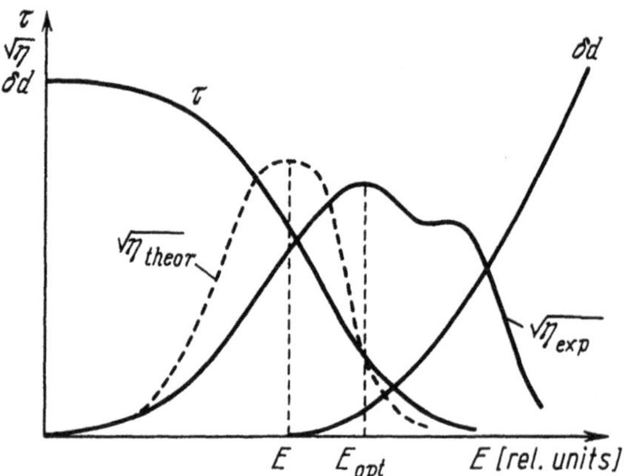

Fig.2.10. Exposure dependence of the height of the surface relief δd, the transmittance τ and the diffraction efficiency $\sqrt{\eta_{exp}}$ for a photographic material used in holography. The exposure dependence of $\sqrt{\eta_{theor}}$ for a thin amplitude hologram is also given (dashed curve)

Thus, the characteristics of real holographic photographic materials cannot be used to make holograms with low levels of nonlinear distortions if we want to achieve maximum diffraction efficiency. Therefore, investigators often accept noise and nonlinearity in the image and, in accordance with (2.22), try to increase the value of the coefficient $a_r a_0$ in the expression for the reconstructed wave by making it maximum, i.e., $a_0 \to a_r$. When the number of interference fringes on the reconstructed image is great, however, the effects described in the foregoing and also other effects that lead to nonlinear distortions begin to prevent the proper recording of the shape and number of fringes. Therefore, an acceptable relationship is [2.28]

$$a_0 \approx \frac{1}{3} a_r \quad ,$$

provided that the operating point is chosen near the middle of the quasi-linear section on the curve $\tau(E)$.

In completing our analysis of the applicability of photographic materials for holographic interferometry, we note that it was carried out for a fixed spatial frequency. Of course, in accordance with (2.11,12), the spatial frequency fluctuates over a hologram. This makes it necessary to consider the dependence of the diffraction efficiency $\sqrt{\eta}$ on two quantities — the exposure E and the spatial frequency ν (Fig.2.11). It is not a simple task to construct such surfaces $\sqrt{\eta} = \sqrt{\eta}(E,\nu)$ for real photographic materials, when account is taken of all of the effects discussed (for example, surface relief); in the following we shall give data to facilitate qualitative estimations of the forms of such surfaces for a number of materials.

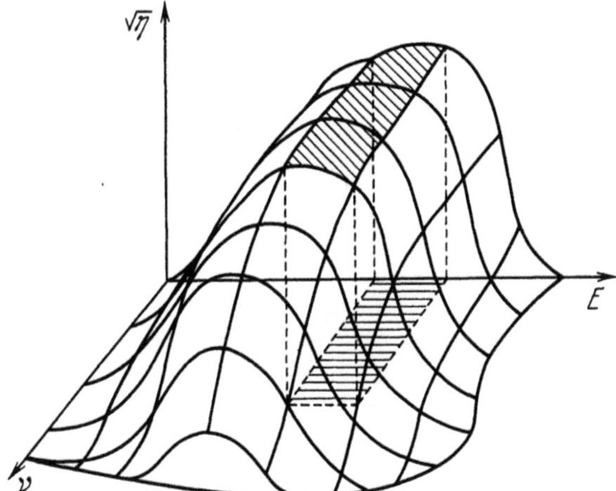

Fig.2.11. The surface $\sqrt{\eta}(E,\nu)$ that characterizes a photographic material used in holography. The region of optimal values of the arguments is hatched

Figure 2.12 shows the characteristics $\tau(E)$ for photographic materials that are most popular in holography [2.29]. When account is taken of the influence of surface relief on diffraction efficiency $\sqrt{\eta}$, it is desirable to have information on the depth of the relief Δd depending on the exposure. The diffraction efficiency $\sqrt{\eta_r}$ of plastic replicas made from holograms prepared with various levels of exposure indicates the increase of the relief for all photographic materials, beginning from a definite level E_r for each [2.26,27]. The relationship between the diffraction efficiency $\sqrt{\eta}$ and the spatial frequency ν for all photographic materials has a descending nature. This indicates poorer recording of light distributions whose spatial periods are comparable with or less than the average size of the silver halide grains (Fig.2.13). The data of Figs.2.12, 13 have been obtained with laser light of wavelength of $\lambda = 0.63$ µm.

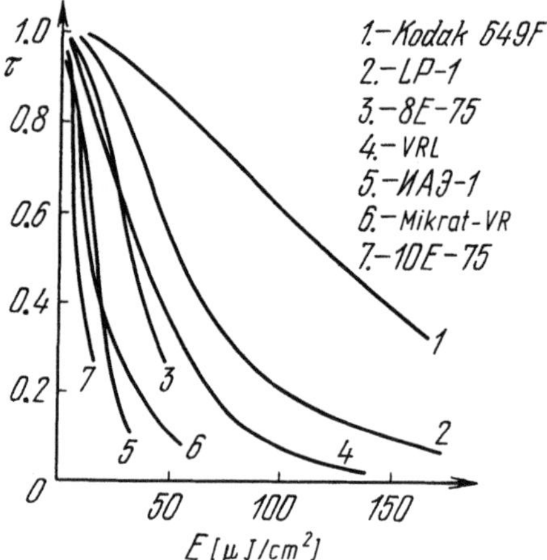

Fig.2.12. Characteristics $\tau(E)$ for various photographic materials used in holography [2.29]

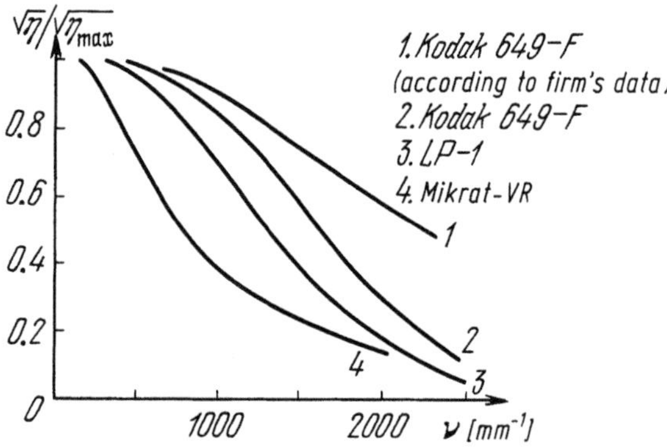

Fig.2.13. Relative diffraction efficiency against spatial frequency for various photographic materials

Table 2.3. Characteristics of selected silver halide materials for holography

Photographic material	Sensitivity [$\mu J/cm^2$]			Diffraction efficiency		Emulsion thickness h[μm]
	E_1	E_2	E_3	$\sqrt{\eta_a}$	$\sqrt{\eta_{ph}}$	
Agfa-Gevaert 8E-75	17	16	178	4.4	39	5
Kodak 649F	218	280	3200	1.2	27	15
SRBS	19	23		3.7		8
PE-1	110	80	1500	3.1	29	14
VRL	42	40	437	2.9	37	8
LOI-2	80	89	447	4	32	17

Table 2.3 contains the basic holographic characteristics of photographic materials [2.30].

The three values of sensitivity correspond to the exposure needed to obtain a transmittance of $\tau = 0.5$ (E_1), maximum diffraction efficiency $\sqrt{\eta_a}$ for amplitude holograms (E_2), and maximum diffraction efficiency $\sqrt{\eta_{ph}}$ for phase holograms (E_3), respectively.

Although silver halide photographic materials are the recording media used most frequently at present in holography, they remain quite imperfect for industrial and scientific application of methods of holographic interferometry. Their comparatively low sensitivity, their nonlinear image reconstruction, and the impossibility of repeated recording and erasure of information considerably limit the introduction of holographic methods into practice. This is why great attention is being given to investigating new recording materials for holography.

2.2.3 Photoconductor-Thermoplastic Devices

One of the most promising new materials for holographic interferometry is a photoconductor-thermoplastic [Ref. 2.17, Chap. 6]. It consists of a solid solution of abietic acid that contains a sensitizer, in an amorphous thermoplastic polymeric matrix. In another, more sensitive, variant the photoconductor-thermoplastic film is made in the form of multilayers (Fig.2.14) that consist of a glass substrate 1 that carries a transparent grounded electrode 2, a photoconductor 3, and a thermoplastic 4.

The properties of photoconductor-thermoplastic films (PTF) as applied to holography have not been studied sufficiently; hence, there is greater ambiguity in the quantitative assessment of their possibilities than is the case for photographic materials. But, because the sensitivity of multilayer PTF does not seem to be inferior to photographic materials used in holography, we will give a brief analysis of the properties of this type of photoconductor-thermoplastics.

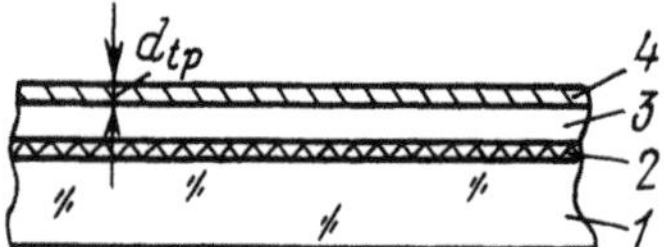

Fig.2.14. Structure of a photoconductor-thermoplastic multilayer system

The thermoplastic film 4 (see Fig.2.14) for which we will give the basic characteristics of photoconductor-thermoplastic materials functions as follows: before exposure, a uniform charge is applied to the surface of the film in darkness; and it creates a potential relative to the grounded electrode 2.

The charge may, for example, be applied by a corona discharge between a wire at a voltage of about 10 kV and a grounded electrode. Next, the PTF is exposed in the holographic setup. The internal resistance of the photoconductor is reduced where illuminated; virtually the entire potential applied initially to the photoconductor-thermoplastic system is consequently applied directly to the exposed areas of the thermoplastic. The values of the electric field at the spots where the thermoplastic was acted upon by light and at the spots where there was no light are different. When the PTF is heated to the softening temperature of the polymer, the latter is deformed by the local electric fields in proportion of the magnitudes of those fields. The thickness of the thermoplastic layer usually ranges from 0.2 to 15 μm; the latent image, fixed in the form of the electric field distribution, and the developed image, in the form of a surface relief, contain spatial frequencies that do not exceed the reciprocal of the film thickness. The response of PTF to low spatial frequencies remains unclear.

Investigation of this question by NAKHODKIN et al. [2.31] not only yielded basic spatial and temporal characteristics of the deformation of the surface of the thermoplastic, but also explained the effect of "frost deformation" i.e., chaotic distortion of the surface relief that sometimes leads to great noise in images reconstructed from photoconductor-thermoplastic films.

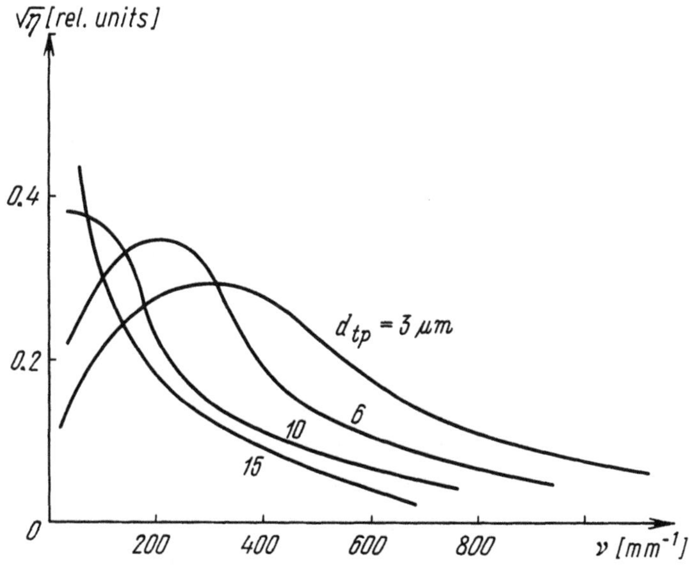

Fig.2.15. Diffraction efficiency of photoconductor-thermoplastic films of various thickness d_{th} against spatial frequency

The dependence of diffraction efficiency on spatial frequency for photoconductor-thermoplastic films has a maximum at frequencies that depend on the layer thickness (Fig.2.15); with increase of exposure, diffraction efficiency tends to saturate [2.32] (Fig.2.16). Inspection of the relevant figures shows that the characteristics of photoconductor-thermoplastic films differ greatly from those of photographic materials. It is therefore difficult to make a general comparison of the possibilities of these two kinds of recording materials.

2.2.4 New Recording Materials

The materials discussed in the foregoing are not the only ones used in holography. A constantly growing number of investigators are studying the possibility of using photochromic materials, dichromated gelatin films, photo-

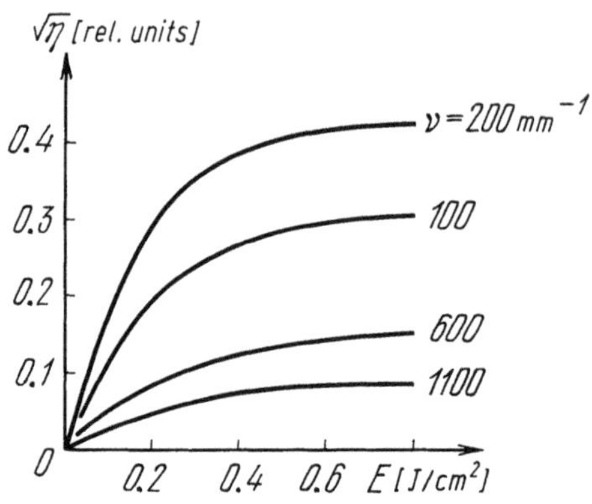

Fig.2.16. Exposure characteristics of photoconductor-thermoplastic film at different spatial frequencies

sensitive resists, chalcogenide glass, metal films, electrooptic crystals, and other substances and compounds [2.17,30,33].

The A.F. Joffe Physicotechnical Institute of the USSR Academy of Sciences has developed a new recording material FTIROS (phase-transforming interference reversible light reflector based on vanadium oxides) which to a considerable extent meets many requirements of modern holographic materials [2.34,35]. Among them are simple technology in manufacture, the possibility of repeated use both in dynamic conditions, with recording times of the order of 10^{-8}s and erasing times of about 10^{-6}s, and for permanent memory. The material can also be sensitized in the infrared wavelength range, where the demand for holographic recording materials is still greater, owing to the impossibility of using photographic materials. Here the sensitivity, diffraction efficiency, and "signal-to-noise" ratio remain quite satisfactory. The resolution of FTIROS is also quite high — at least 2000 lines per millimeter. The mechanism of FTIROS is based on a metal-semiconductor phase transition under the action of heat and attended by a sharp change of the optical constants (particularly, the reflectance). For thermal recording, a great merit of this material is its nonselectivity for wavelength. Holograms were recorded on a FTIROS layer in storage conditions [2.26] with the aid of a pulsed laser with Q modulation of the resonator at wavelengths both in the visible (λ = 0.69 μm) and in the infrared (λ = 1.06 μm) ranges.

The main shortcomings of FTIROS are its lower (in comparison with conventional photographic materials) sensitivity (about 400 μJ/cm^2) and diffraction efficiency (about 1% in visible light, and also the need to use energy to maintain a permanent memory (of the order of magnitude of several scores of mW/cm^2).

2.3 Setups

In this section, we consider the basic kinds of holographic setups for interferometric studies and their fields of application.

2.3.1 Basic Kinds of Holographic Setups

The most typical arrangement used for holographic interferometry is the classical split-beam scheme proposed by LEITH and UPATNIEKS [2.27] (Fig. 2.17a): the beam from laser 1 is expanded by collimator 2 and passes on the beam splitter 3 that forms the beam that illuminates the object 4 and the reference beam that passes directly onto the hologram 5. After the object beam is scattered by the surface 4 is also falls on the hologram 5. A virtually indispensable element of interferometric setups is the light shutter 6, which determines the exposure time of the hologram and which is synchronized in some way with the behavior of the object being studied. When stresses and strains are studied, the shutter is optomechanical and operates twice, usually before and after the application of the stress being studied (mechanical or heat, etc.).

When rapid processes are studied, a Q modulator of a pulsed laser is the shutter, or, if free-generation conditions are used, the process studied is synchronized with the pulse that triggers the pumping flashlamps (see Sect. 2.1).

If the coherence length ℓ_c of the laser is not sufficient (the setup in Fig.2.17a usually requires $\ell_c \approx$ 10-20 cm), then the path of the reference beam must be altered as shown in Fig.2.17b, in which the reference numbers have the same meaning as in Fig.2.17a; 7 and 8 signify additional mirrors. The setups in Fig.2.17a,b are used to study light-scattering objects.

A similar setup for studying phase (transparent) objects is shown in Fig.2.17c. Quite frequently, a ground glass diffuser 9 or a similar device is used to facilitate observation of the interferograms.

A block diagram that depicts the functional features of each of the elements of a holographic experiment is shown in Fig.2.17d. Here L is a

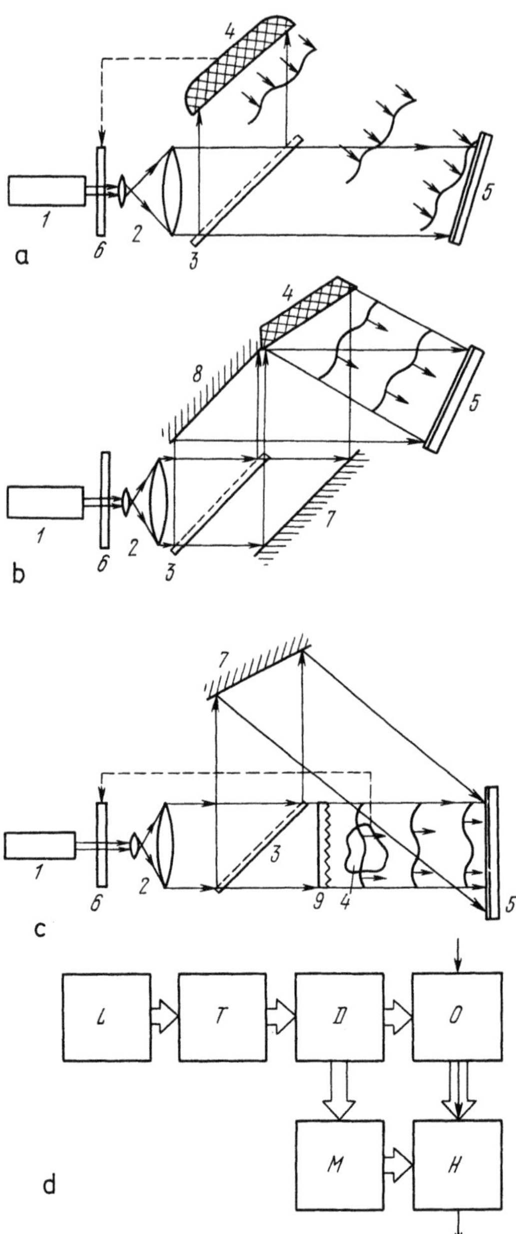

Fig.2.17. Holographic arrangements for interferometry of objects that have different light-scattering properties: (a) light-scattering objects; (b) same with restricted temporal coherence of laser radiation; (c) phase objects, with ground-glass diffusing screen; (d) block diagram of holographic interferometry; 1 - laser; 2 - collimater; 3 - beam splitter; 4 - object; 5 - hologram; 6 - shutter; 7, 8 - mirrors; 9 - diffuser

laser, T is the laser beam transforming unit (collimator, modulator), D is the beam splitter, O is the object being studied, and H is the hologram. In addition, in a number of cases, the device M for further transformation of the reference beam (focusing element, phase and/or amplitude modulator, light filter, polarizer, etc.) is included in the path of the reference beam. The wide arrows show the direction of propagation of the light, and the narrow arrows show the path of the information on the changes of the object being studied that is introduced and transferred.

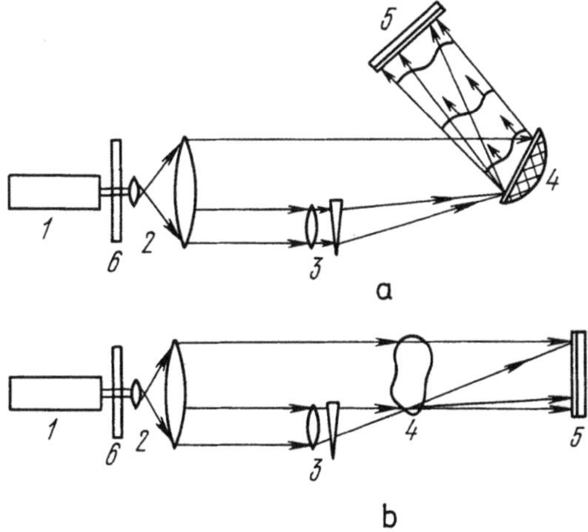

Fig.2.18. Arrangements with local reference beam: (a) for studying light-scattering objects; (b) for studying phase objects. The elements have the same reference numbers as in Fig.2.17, except that 3 is a "lens-prism" unit

The local-reference-beam method (Fig.2.18) [2.28] is used in a number of cases when this is dictated by structural considerations or the need to eliminate (compensate) great phase distortions of the object and to find relatively small changes from their background. The holographic set-up in this case differs from the principles described in the foregoing, in that the reference beam is focused on a region of the object, which for objects with light-scattering surfaces is made reflecting (Fig.2.18a). The phase shift introduced into the reference beam when the object 4 is displaced as a whole compensates the corresponding phase shift in the object beam. Thus, information concerning the change of arrangement of various points on the surface of the object, relative to the point on which the reference beam is focused, is recorded on the holographic interferogram. The local-reference-beam method is employed with especially great success in the study of vibrations; various modifications of this method have been given the general name of reference-beam phase modulation (Chap.5).

The setup shown in Fig.2.18b [2.27] is used if it is necessary to exclude the influence of changes of the transparent medium being studied on the nature of the interferogram. The reference beam is focused by the long-focus lens so that the focal plane coincides with the central plane of the object. The focal length of the lens in the unit 3 is much greater than the longitudinal dimension of the object. The prism is such that the reference beam will pass through the region occupied by the object 4 relative to which the interference fringes are being counted (most often it is the central region of the object). For example, when studying acoustic waves in a dense medium, we can thus eliminate the influence of heating of the medium and of the corresponding changes of refractive index on the interference fringes. When aerodynamic processes are studied, this method makes it possible to diminish distortions caused by deformations of the viewing windows of the chambers being studied, etc.

The setups in Fig.2.18 differ from those in Fig.2.17 in another feature – absence of beam splitters (wavefront division of the light beam is used). In the setups in Fig.2.17, amplitude division of the light beam is used – light waves that have the same wavefront shape are directed along the paths marked for the object and the reference beams, whose amplitudes are selected suitably for hologram recording.

Setups of different kinds display different sensitivities to the quite frequent phenomenon of shift of laser frequency between exposures, which affects the accuracy of experiments [2.39]. This effect is especially probable in double-exposure holography with pulsed lasers; the increase of the working temperature of the active element during the first pulse may be one of the obvious causes of shift of the emitted wavelength. Let us consider this phenomenon in greater detail by taking a very simple setup as an example in which wavefront beam splitting is used (Fig.2.19a). At the point 1, there is an equivalent source – usually this is the focus of a diverging (negative) lens beyond which the laser beam is transformed into a divergent beam and illuminates the object 2 and the mirror 3. The light from the object beam, after it is scattered by the object, and the reference beam, after it is reflected from the mirror, interfere in the plane of the hologram 4 that passes through the source S. The virtual image of the source in the mirror 3 is at the point S' at the distance p from the hologram, whereas the object is approximately at the distance q from it.

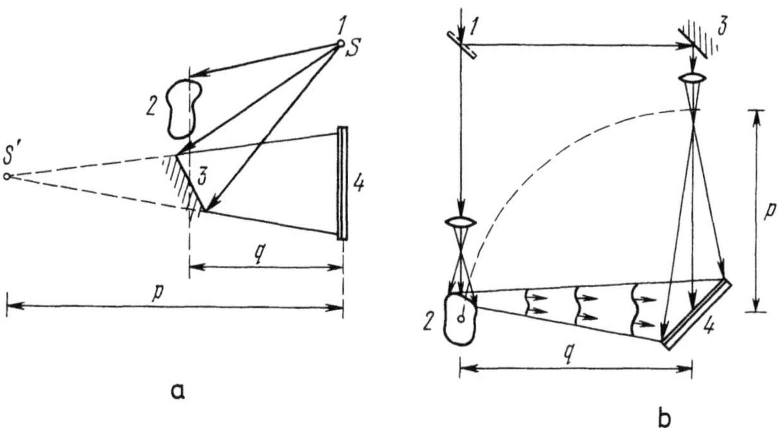

Fig.2.19. Different principles of laser beam splitting and arrangements for their embodiment: (a) wavefront division; (b) amplitude division; 1 - source; 2 - object; 3 - mirror; 4 - hologram

The positions of the fringes of the microinterference structure recorded on the hologram are determined by two simultaneously acting interferometric and holographic factors. The interferometric parameter is the difference between the optical path lengths of the reference and object beams. If all of the components of the setup are rigidly fixed in place and if the wavelength does not change, then the difference of path is the same in both exposures. A change of wavelength causes the positions of the micropattern fringes to shift if the optical lengths of the object and reference beams differ. We show in Chap.4 that the interferogram observed from a hologram can be interpreted in a number of cases as moire fringes that appear upon superposition of the two microstructures that are recorded during the first and the second exposures.

Hence it is clear that if a change of wavelength causes a shift of the microstructure during the second exposure, then the fringe pattern observed upon reconstruction will be displaced. To prevent a change of wavelength from resulting in such an effect, the optical lengths of both beams must be as nearly equal as possible, i.e., we must have

$$p \approx 2q \quad . \tag{2.27}$$

The second, holographic, parameter is the influence of a shift of wavelength on curvatures of the two wavefronts. If a hologram is recorded with one wavelength and reconstructed with another wavelength, the dimensions and an axial location of the reconstructed image are changed [2.24]. Consequently, a shift of wavelength between exposures results in two slightly displaced images in the reconstruction. In other words, fringes appear on the reconstructed image. We show in Chap.4 how this phenomenon is used to determine the relief of a surface, but here we limit ourselves to the remark that when displacements are studied uncontrolled changes of wavelength may cause considerable errors.

How can we avoid this? It is evident that if the curvatures of the partial waves from different points of the object and the curvatures of the reference wave are equal, a shift of wavelength will not cause parasite fringes, i.e., it is necessary that

$$\frac{1}{q} \approx \frac{1}{p} \quad . \tag{2.28}$$

It can be seen that (2.27,28) cannot be satisfied simultaneously, i.e., the effect of a change of wavelength between exposures cannot be eliminated in a wavefront-division setup.

On the other hand, the modified Mach-Zehnder interferometer setup shown in Fig.2.19b satisfies both conditions because the interferometric term is minimized by the equality of the two arms of the interferometer. The influence of the holographic difference is minimized by placing the focus of the lens in the reference beam at a distance p equal to the distance from the center of the hologram to the center of the object. Thus, a setup for amplitude division of a light wave is less sensitive to a change of wavelength between exposures.

Using similar reasoning, we can consider what happens when a laser (source) is displaced relative to the remainder of the setup. Such is possible in setups where the laser is not on the slab that supports the remainder of the holographic setup. It is clear from the above that a setup such as that shown in Fig.2.19b is less sensitive to such displacements and therefore can be recommended.

In general, it follows from experiments and analysis that holographic setups that use amplitude division of the light wave are considerably more convenient and trouble free than setups that use wavefront division. Laser radiation is most completely utilized in such setups; consequently, less time is needed to expose a hologram.

2.4 Experimental Aspects

The aim of this section is to analyze the conditions for obtaining high-quality holographic interferograms, and also to describe holographic supports and components designed to satisfy these conditions.

2.4.1 Premises

The requirements for premises in which to install holographic laboratories have changed during recent years. The first arrangements were installed in basements and semibasements, and the slabs that supported holographic setups weighed over a ton. Then, with the passage of time, numerous experiments showed that these requirements were too severe. Holographic setups are known that operate in the upper stories of buildings. Sources of vibrations, such as fore pumps and machine tools, are in adjacent rooms or even in the same room as the holographic setups. The single requirement that remains is that the premises be adapted for darkening and be free of dust, as far as possible. The latter requirement is ensured by using plenums and exhaust ventilation. It is best to darken a room by use of curtains placed between frames and made of a fabric that does not accumulate dust. Darkroom shutters can be used on windows to darken the room adequately [2.40]. The level of darkening required while standard photographic materials are being placed in position may be such that a person can easily find his way around the room and distinguish the components of the setup. Complete darkening is needed only during very long (x 10 min) exposures.

It is desirable to paint the laboratory walls with dull, dark paint, which reduces the possibility of scattered light striking the photographic materials and, when pulsed lasers are used, diminishes the possibility of a reflected pulse striking anyone's eyes.

2.4.2 Holographic Slab

One of the main components of a holographic arrangement is an antivibration slab or flat bench on which the optical setup is assembled and the object being studied is placed. Let us consider different variants of this unit as applied to holography by use of continuous lasers.

Among the slabs used are the following:

1) Granite blocks with a polished working surface — such slabs generally have a high degree of flatness and a high finish of their working surface, but they are costly, sensitive to blows (possible chipping) and do not completely solve the problem of adjusting the elements of the setup.

2) Cement (reinforced cement) slabs that have a sheet-metal working surfaces from 10 to 15 mm thick — such slabs are considerably cheaper than granite slabs, but a sheet metal is usually not flat enough.

3) A pile of metal sheets each of which is from 10 to 15 mm thick, with vacuum-rubber sheet spacers about 5 mm thick, placed between the metal sheets. If the working surface of such a slab is finished after assembly, it can be one of the most satisfactory kinds of holographic slab. Large slabs (over 1.2 m long) of such design, however, are not sufficiently rigid.

4) An iron casting with stiffening ribs on its bottom surface, threaded holes for jacks at the corners, and a ground working surface is most frequently used (longitudinal slots or threaded openings for fastening the elements of the setup are often made in such a slab). Standard optical-bench plates can also be used for holographic experiments.

5) Magnetic slabs of various sizes with separate switching of the magnetic field in each slab. Such slabs are bolted to a common foundation (a tee-beam or channel), and their working surfaces are levelled with the aid of gaskets.

6) A honeycomb slab combining rigidity, vibration stability, and low weight (the most progressive design). These slabs are made of steel and aluminium and are seven meters and more long. Such length is virtually impossible with any other method of manufacturing holographic slabs and plates.

2.4.3 Protection against Vibration

Vibration insulation of a holographic slab is needed because, as mentioned in the preceding sections, the positions of the microstructure fringes recorded on a hologram are determined by the phases of the beams that interfere in the plane of recording. Displacement of the phase of one of the beams by an amount of the order of π during an exposure causes blurring of the micropattern and a sharp loss of quality, up to vanishing of the reconstructed image. A reason for a change of the phase relationships of the beams, apart from changes of the optical properties and position of the object, may be displacements and vibrations of the components of the holographic setup, caused by external mechanical and acoustic vibrations.

To develop effective protection against vibration, it is essential to know the spectral composition of the induced vibrations. In one holographic laboratory, the spectra of vibrations of the building (the floor near a main wall) that appeared when a compressor was operating in a neighboring room and when persons walked in the laboratory were measured [2.41]. A glance at curve 1 in Fig.2.20 shows that the compressor induced vibrations

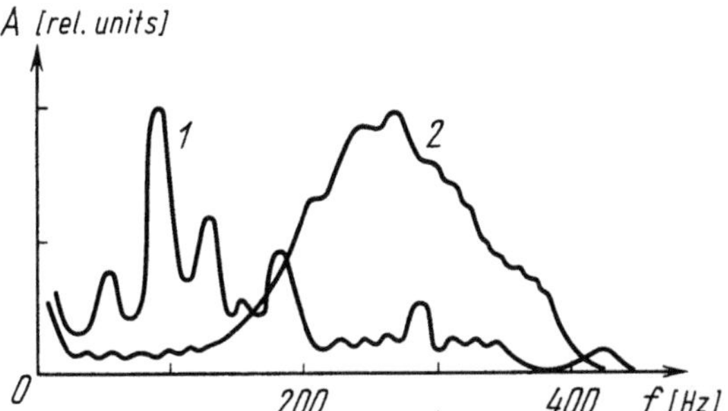

Fig.2.20. Frequency spectra of building element oscillations induced by different sources: 1 - industrial production installations; 2 - footsteps in the room

at frequencies that were a multiple of 50 Hz, whereas footsteps in the laboratory created serious vibrations in the range from 150 to 400 Hz. In addition, an increase of vibrations in the region of infra-frequencies ($f < 16$ Hz) was noted. Vibrations are also induced at those frequencies when the setup is touched accidentally.

How is it possible to assess the weakening of building vibrations in the pneumatic supports of the holographic plate? This can be done by use of the transfer coefficient, which is the ratio of the amplitude of vibrations induced in the foundation by a monofrequency vibrator to the amplitude of vibrations of the slab. This coefficient depends not only on the kind of vibration insulation, but also on the weight of the holographic slab, which was 300 kg for the measurements shown in Fig.2.21. That figure shows the frequency dependence of the transfer coefficients K for pneumatic supports (motor-vehicle tire tubes) at different pressures, and also for an aerated plastic mat that had 15 cm uncompressed. Figure 2.21 shows that reduction of pressure in the tubes decreased the transfer coefficient at frequencies of about 100 Hz, which are very serious. The transfer function for the aerated plastic mat has no resonances (curve 4) and is more advantageous for protection against vibration within a broad spectrum range.

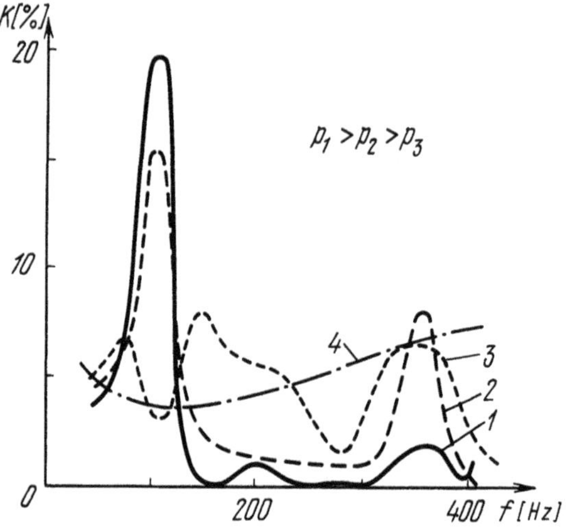

Fig.2.21. Transfer functions of vibration-damping devices: 1-3 - low-pressure tire tubes; 4 - aerated plastic mats

Figure 2.22 shows the frequency dependence of the transfer coefficient for the vibration-insulating supports used in a holographic arrangement of the firm NRC with a 400 kg holographic slab (curve 1) and a 100 kg slab (curve 2). The complete suppression of vibrations at frequencies above 10 Hz should be noted.

At present, most holographic setups are protected against vibrations by use of rule-of-thumb methods. The variety of devices used, apart from those indicated above, include motor-vehicle shock absorbers, spring suspensions, boxes with sand, and helical coils made of vacuum rubber. Multilayer gaskets made of newspapers, other paper, etc., are used for protection against tangential vibrations (horizontal displacements). We advise experimenters who construct homemade setups to use aerated plastic mats when comparatively light slabs (up to 100 kg) are employed, aerated plastic mats in combination with lengths of vacuum hose arranged vertically for protecting the mats from being crushed when slabs that weigh up to 300 kg are employed, and, finally, motor-vehicle and aircraft tire tubes and tires or tennis balls if the slabs weigh 1000 kg and more.

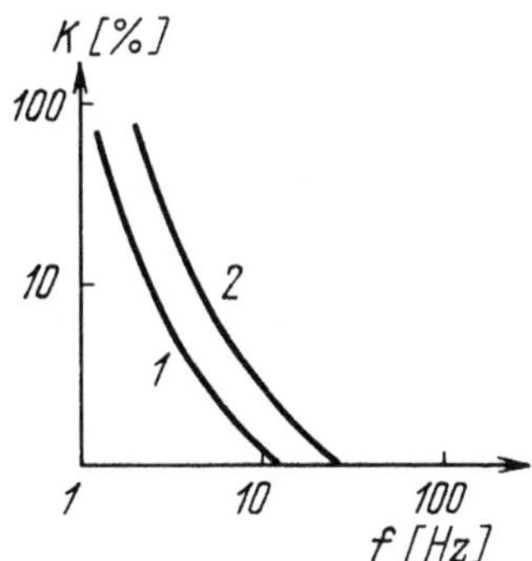

Fig.2.22. Transfer functions of special vibration-insulating supports under various loads

It is simplest to detect vibrations in a setup by use of a microscope from an optical bench placed in the vicinity of the hologram [2.42]. The magnification of the microscope is selected with a view to the angle θ of convergence of the beams (when θ is about 20 degrees, a magnification of 400 times is sufficient). If the object is transparent or a mirror image and if there are no vibrations of the setup components, a clear stationary pattern of vertical interference fringes is seen in the eyepiece. A more complicated structure is seen if the object is diffuse.

By placing a screen in the path of each of the beams in turn and watching the disappearance of the fringe pattern, we can see that the latter is indeed formed as a result of interference of the reference and object beams. The immobility of the fringe pattern is easily confirmed by observing the fringes relative to the line of sight, or the fringes relative to the edge of the field of vision. If the fringes shift, the mechanical stability of separate elements of the setup must be controlled.

2.4.4 Setup Elements. Pinhole Diaphragm and Collimator

In quantitative measurements, the use of collimated radiation (a plane wavefront) is often dictated by considerations concerning the convenience of interpreting the interferograms. As we show in Chap.4, it is good to maintain a constant angle between the beam that illuminates an object and its surface.

The design of the collimator differs from the traditional design, by use of a pinhole diaphragm as the focus of the first lens. The effect of this diaphragm is spatial filtration of light waves that are diffracted by dust particles and inhomogeneities in the optical path preceding the output lens of the collimator. Because the increase of the diameter of the beam in the collimator is of the order of 100, the particles and inhomogeneities are clearly seen as diffraction zones. They disturb the spatial homogeneity of distribution of the intensity in the beam and change the conditions of exposure at various points of a hologram.

Fig.2.23. A "microscope objective-pinhole diaphragm" system for laser-beam spatial filtering in a collimator: 1 - diaphragm; 2 - microscipe objective; 3,4 - microadjusters

A pinhole diaphragm 1 (Fig.2.23) from 20 to 30 μm in diameter is usually made from aluminium, nickel or other foil from 10 to 12 μm thick and is placed in the focus of the first lens of the collimator 2 (most often a microscope objective with a magnification of about 40 times). To ensure accurate coincidence of the diaphragm with the focal point of the objective, it is convenient to place the latter on a microadjuster 3 movable in the direction of the collimator axis, and to fasten the diaphragm on a microadjuster 4 movable in two perpendicular directions (a bench with orthogonal axes of motion).

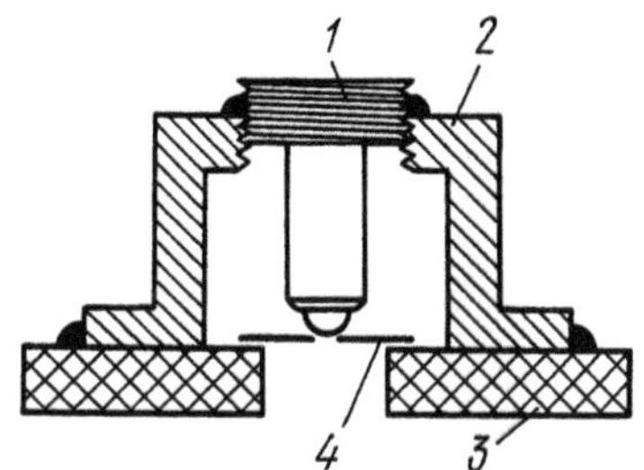

Fig.2.24. "Microscope objective-pinhole diaphragm" system with preliminary adjustment: 1 - microscope objective; 2 - threaded sleeve; 3 - ferrite ring; 4 - diaphragm

If microadjusters are not available, the "microscope objective-pinhole diaphragm" system can be made by preliminary adjustment (Fig.2.24). The system consists of the following elements: a microscope objective (with a magnification from 40 to 20 times) 1, a steel threaded sleeve 2, a ferrite ring 3, and a foil disk with a pinhole diaphragm 4. The disk 4 is glued to the ring 3 so that the diaphragm is approximately at the center of the ring. Next, the objective 1 is screwed into the sleeve 2, and the latter is held to the ferrite ring by magnetic attraction.

The system is adjusted as follows: the ferrite ring is placed horizontal, the unit is illuminated with an undiverged laser beam, from above, and a paper screen is placed below the unit on which the distribution of the intensity of the light that passes through the diaphragm is observed. First, a weak spot of light of small size is observed on the screen, which is positioned at the center of the latter by horizontal movement of the threaded sleeve containing the objective relative to the ferrite ring. After this, the objective is gradually screwed in, while the position of the spot is watched and corrected by corresponding movements of the sleeve. When the pinhole diaphragm is thus made to coincide with the focal spot of the objective, a sharp increase of the size and brightness of the light spot is observed on the screen. Further screwing in of the objective detracts from the brightness of the spot. After optimum adjustment of the arrangement of the system, epoxy resin should be poured over the upper section of the objective thread and the joint of the ferrite ring and the sleeve. The system is fastened to a vertical rocking table in the holographic setup, which makes it possible to ensure coincidence of the axis of the system and the direction of the laser beam without any difficulty.

In conclusion, we give data that show how the diameter of the diaphragm should correspond to the magnification of the lens used:

Magnification [times] ...	2	5	10	20	30	40	60
Diameter [m]	50	25	25	15	10	10	5 .

The distortions of the wavefront in collimators of serial holographic setups that incorporate pinhole diaphragms do not exceed $\pm 0.1\lambda$ within the range from $\lambda = 0.4 - 1.0\ \mu m$.

2.4.5 Beam Splitter

The optimal ratio between the amplitudes of the object and the reference waves, determined by the characteristics of a beam splitter is selected in view of the light scattering properties of the object and the distance from the object to the hologram. If we consider Fig.2.17a, then it becomes clear that, the less the light scattered from the object toward the hologram and the greater the distance to the latter, the greater is the portion of the light wave that must be directed by the beam splitter along the path of the object wave in order to record the interference fringes with maximum modulation on the hologram [2.38], i.e.,

$$a_0/a_r \approx 0.3 \quad .$$

Deviations from this requirement, and very considerable ones, occur in holographic investigations of oscillating objects. This will be considered in greater detail in Chap.5.

Usually, plane-parallel plates and wedge-shaped glass plates are used as beam splitters. One of their faces is coated with a dielectric or metallic reflecting layer that has a transmittance of about 0.3 to 0.7. It is quite obvious, however, that when the objects being studied are changed, the possibility must be provided to readjust the beam splitter, i.e., to smoothly change continuously the ratio of the amplitudes of the waves directed along the object- and the reference-beam paths. Use of a new kind of optical element – holographic diffraction gratings and lenses – makes it possible to solve the problem of readjusting the beam splitter for setups with amplitude division of the light wave [2.43]. Figure 2.25a shows a setup with a beam splitter that consists of a holographic diffraction grating with a small aperture, and Fig.2.25b with a large grating as the beam splitter. As shown by COLLIER et al. [2.2], the diffraction efficiency of a thick or volume hologram, including a grating, is a function of the angle of illumination. By changing the angle, we can change the ratio between the amplitudes of the zero and first diffraction orders (the negative first order is not used in the setup – the dash line in Fig.2.25a).

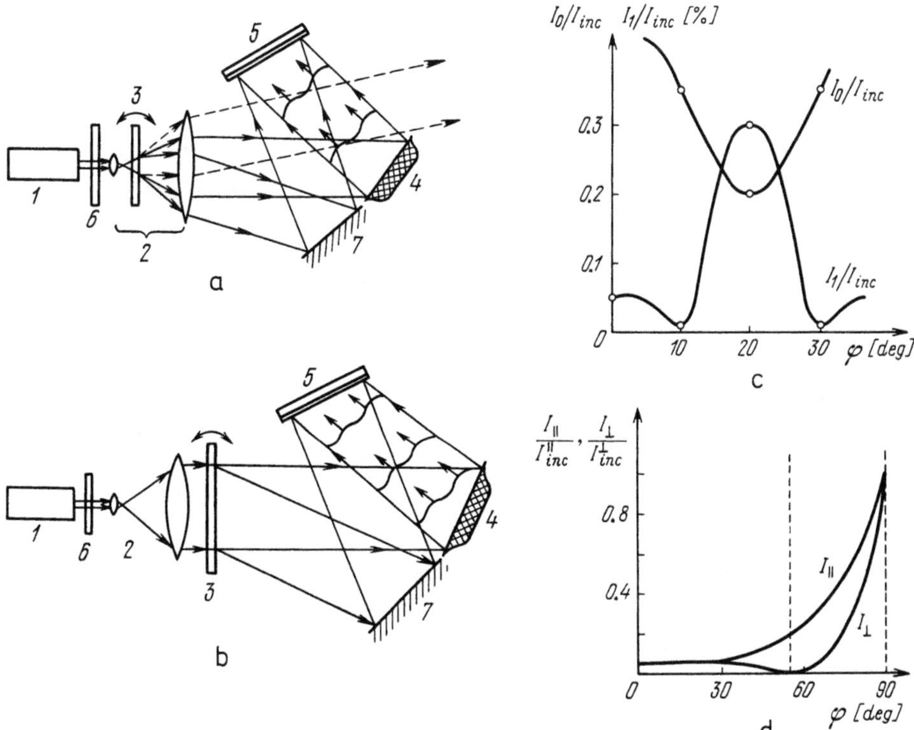

Fig.2.25. Arrangements and characteristics of readjustable beam splitters: (a) holographic grating as a beam splitter inside a collimator; (b) holographic beam splitter outside a collimator; 1 - laser; 2 - collimator; 3 - holographic grating; 4 - object; 5 - hologram; 6 - shutter; 7 - mirror; (c) ratio of intensities of direct and diffracted waves caused by change of beam-splitter orientation; (d) dependence of Fresnel reflectance on angle of incidence, for various polarizations

Therefore, by placing inside the collimator, or at its outlet, a thick-hologram diffraction grating 3 whose period is such as to separate the object and reference beams, we obtain a setup that is very convenient for studying a broad range of objects (Fig.2.25a,b, respectively). Depending on the size and the light-scattering properties of the object, the angular position of the grating is such that the ratio of the amplitudes of the object and reference waves will be within the limits from about a_0/a_r = 0.3 to 0.8.

Figure 2.25c shows how the ratio of the intensities of the two waves used for recording a hologram changes when the grating is turned through a certain angle relative to the incident beam. Double-refracting systems, for instance Wollaston prisms and crystals of calcite or other double-refracting substances, are also used to change continuously the ratio of the beam intensities. The beams that emerge from such systems are polarized in mutually perpendicular planes. To make them interfere, they are passed through a polarizer oriented at an angle of 45 degrees to the planes of polarization of both of the beams. Half of the light energy is thereby lost. The planes of polarization can be made parallel without any loss of light by use of elements that rotate the plane of polarization, for example by passing the light through quartz-crystal plates cut at right angles to their optical axis. The plane of polarization is rotated through 90° by means of a quartz plate 4.8 mm thick ($\lambda \approx 6328$ Å).

A partially diffusing screen can also be used for amplitude division. The diffused radiation that emerges from it illuminates the object, whereas the part that remains unscattered is used for the reference beam [2.13,44].

With a weak reference beam and a correspondingly great relative intensity of the object wave, the interference of the light waves emitted by different points of the object becomes noticeable. Every point of the object can be considered as a reference source with respect to its other points. Upon reconstruction, the image of the reference source will be surrounded by a wide halo caused by diffraction of light superimposed on the interference pattern. The halo may be superimposed on the reconstructed image and spoil it. If the reference beam illuminates the photographic plate several times as much as that produced by the object beam, then the superimposed interference pattern is usually not noticeable on the hologram.

In a well-constructed setup, the desired ratio of the illuminations of the hologram by the reference and the object beams is achieved by choosing the proper beam splitter instead of using filters. The energy radiated by a laser is utilized to the fullest extent. It should be borne in mind that, generally, the energy losses in the object-beam path are much greater than in the reference beam (from 10 to 1000 times). It is, therefore, often advantageous to use a simple glass wedge without a coating as the beam splitting mirror, employing one of the beams it reflects as the reference one. The ratio of the light fluxes in the beams can be varied within broad limits by inclining the beam-splitting wedge, because the Fresnel reflection coefficient depends on the angle of incidence (Fig.2.25d).

To change the ratio of intensities of the beams, it is still more convenient to use the relationships between the reflected and transmitted light energy and the orientation of the polarization plane. If light strikes a glass plate ($n = 1.52$) at an angle of $56^{\circ}41'$ (the Brewster angle), then by rotating the plane of polarization of the incident light through 90°, the ratio of the beam intensities can be changed from five to infinity. At an angle of incidence of 70°, the limits of the change in the intensity ratio are from 2 to 18. Quartz wedges cut at right angles to their optical axis are used for smooth and controlled rotation of the plane of polarization. The rotatory power of quartz is about 19°/mm for $\lambda = 6328$ Å.

A polarization beam splitter with a variable ratio of the beam intensities is described by BROWN [2.45]; it consists of a Rochon prism with a half-wave plate placed in front of it and rotatable about its optical axis.

2.4.6 Light-Scattering Screen

A light-scattering screen or its equivalents are used in holographic interferometry of phase objects. Information on ways of using light-scattering screens and experimental fine points associated with them are discussed in Sect. 3.1. Here we shall give the basic characteristics of various screens prepared by the abrasive grinding of glass plates [2.46,47].

Figure 2.26 shows scattering indicatrixes for glass finished with various abrasives. The dashed curves correspond to measurements when the light was incident on the ground surface of the scattering screen and the solid curves are indicatrixes obtained with the more customary conditions — the smooth surface was illuminated. The latter way of illumination is more convenient if only because the beam reflected from the smooth front surface can be used as the reference beam [2.44].

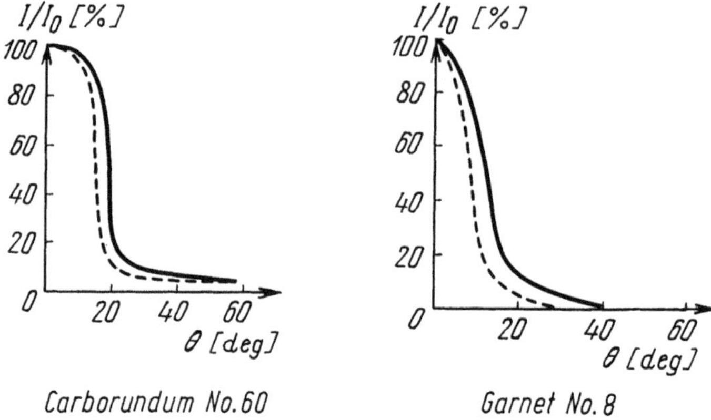

Fig.2.26. Scattering indicatrixes of ground glass processed with different abrasives [2.46]

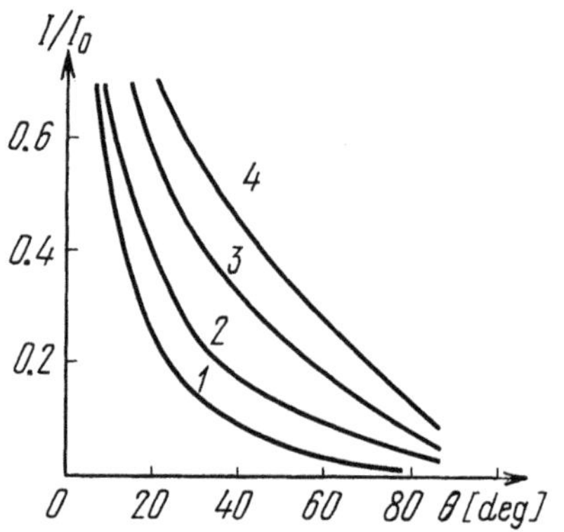

Fig.2.27. Indicatrixes of various light-scattering screens [2.47]: 1 - ordinary ground glass; 2 - glass with two ground surfaces; 3 - thin layer of milk glass; 4 - thicker milk glass

Figure 2.27 shows how different light-scattering screens can be used to achieve a greater field of view by extending the indicatirx. The angular distribution of the scattered light can also be changed if we change the manner of illumination of the screen (Fig.2.28). This figure also shows why it is convenient to use a holographic diffraction grating that yields a set of narrowly directed beams of high intensity, instead of a ground-glass screen. The use of a holographic grating appreciably facilitates the interpretation of holograms obtained with such a discrete scattering screen

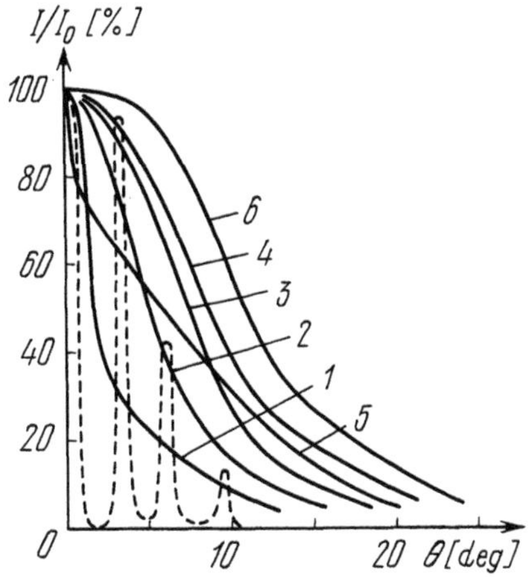

Fig.2.28. Scattering indicatrixes for different conditions of illumination: 1 - average grain ≈ 5 μm, illumination by plane wave; 2,3 - same as 1, illumination through lenses f = -100 mm and + 10 mm; 4 - average grain ≈ 20 μm, illumination by plane wave; 5,6 = same as 4, f = -100 mm and +10 mm. The dashed curve is for a holographic grating

because when the aperture of the optical system is reduced a speckle pattern (Sect. 3.1) does not appear on the reconstructed image. The ratios between the intensities of different orders of diffraction can be equalized if methods of preparing holograms for image multiplication (gratings with an intricate profile) [2.48] are used to create the light-scattering screen. The technology of fabricating such gratings with the use of photolithography is quite complicated. Therefore, a different way can be proposed that is available to every experimenter; it consists of double printing a hologram in a projection setup (Fig.2.29) [2.49].

In the first step (Fig.2.29a), a pattern of the interference of two plane waves is recorded on a hologram. The exposure must be such that the record will not be linear, i.e., the developed hologram should have a transmittance of the order of magnitude of several per cent. This hologram is next bleached and placed at the input of the setup shown in Fig.2.29b. The secondary hologram obtained is also bleached. Its reconstruction is a convenient substitute for a ground-glass screen, because the waves diffracted on it are directed within a wide range of discrete and multiple diffraction angles (Fig.2.29c).

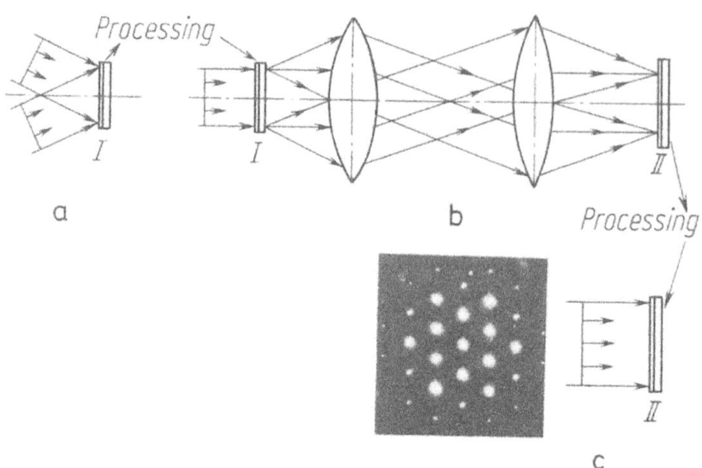

Fig.2.29. Holographic preparation of a light-scattering screen with discrete diffraction angles: (a) preparation of the primary hologram; (b) arrangement for preparing the secondary hologram; (c) illumination of the secondary hologram

2.4.7 Hologram Fasteners

A hologram is fastened in a setup so as to ensure its mechanical stability during the exposure, and also, when necessary, to put it accurately in place and to process it directly in the setup. Provisions for the latter are needed only in real-time holographic interferometry (Sect. 1.4).

There is no practical difficulty in providing for stability of a hologram during one or several exposures.

For real-time studies, however, it is necessary to develop the hologram after the first exposure and to return it to its original location with an accuracy as close as $\lambda/10$. Or, the hologram can be processed directly in the holographic setup, without disturbing it.

The first way requires an immersion hologram holder that ensures accurate replacement of the hologram. Immersion is needed to prevent parasite fringes that appear if there is any shrinkage of the emulsion in drying. An immersion holder consists of a metal vessel with glass windows 1 and 2 and two microadjusters 3 and 4 (Fig.2.30). A hologram, fastened in a special stainless steel frame 5 is tightly inserted into a holder filled with water and is kept there for about 15 minutes before exposure. This is essential for preliminary swelling and for sensitization of the emulsion. Next, the hologram is exposed, removed in its frame from the holder, and the water is replaced by the solutions required for photographic processing. After the processing is completed, pure water is substituted for the solutions, the holder with the processed hologram in it is replaced in the setup. Certain changes of optical thickness that occur owing to swelling of the emulsion or changes of location owing to inexact replacement of the frame in the holder manifest themselves in that the image observed when the object is viewed the hologram is covered with a small number of interference fringes. They can usually be eliminated quite easily by use of the microadjusters.

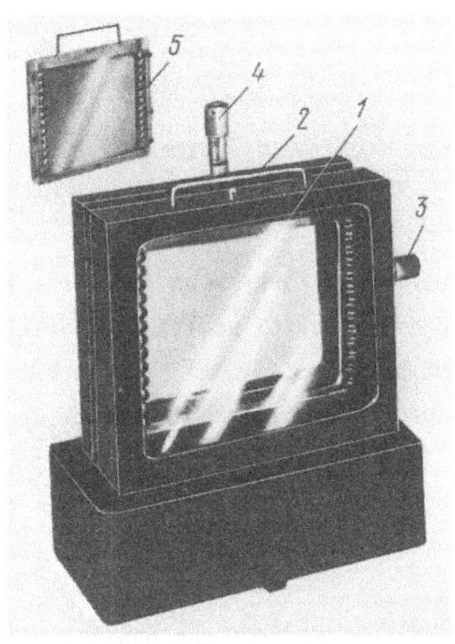

Fig.2.30. Device for developing a hologram in a holographic arrangement: 1, 2 - glass windows; 3, 4 - microadjusters; 5 - frame with hologram

Several variants of processing holders have been proposed for processing hylograms directly in a holographic setup. One of them is similar to the holder shown in Fig.2.30, but has provisions for changing the solutions.

A different type of unit is based on the use of a total-internal-reflection prism 5 (Fig.2.31) [2.50]. The hologram 1 is placed, with its emulsion downward, under the spring clamps 2 in the round tray 3 that is cemented to the top face of the prism 5. The solutions are poured in from above. No air bubbles may remain under the hologram; for this reason the sides 6 of the prism are held so that the hologram is slightly inclined relative to the horizontal. Distilled water is poured into the space between the hologram and the top face of the prism from 10 to 15 minutes before the exposure. The exposure and the following reconstruction of the image from the hologram are performed in this state. The solutions are drained off via the tube 4.

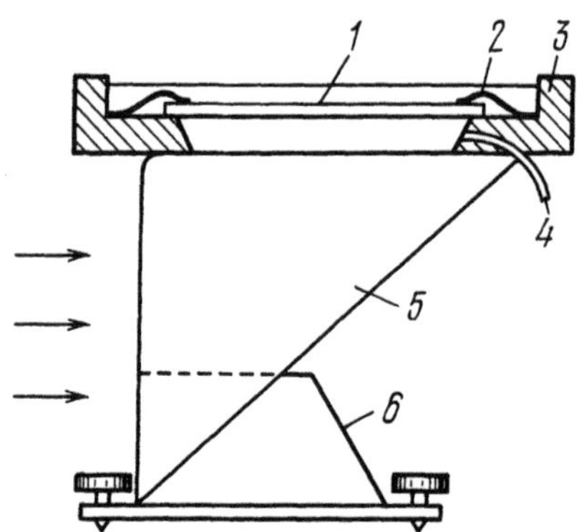

Fig.2.31. Development unit based on a glass prism: 1 - hologram; 2 - clamp; 3 - tray; 4 - drain tube; 5 - prism; 6 - side faces of prism base

Experiments have shown that this holographic setup makes it possible to obtain high-quality undistorted interferograms with lifetime of several days on standard plates "Mikrat VRL".

Many of the holographic setups on the market are provided, apart from the elements listed above and clamps for fastening them, with shutters, including ones that automatically close after a given quantity of light has passed through them, with adjustable beam splitters of various kinds, etc.

We have considered the basic elements of setups for holographic interferometry in such detail to facilitate fabrication of them by the reader, independently so far as possible.

In conclusion, we shall describe briefly some holographic setups, test devices, and cameras.

1) The UIG-2 laser test device, produced in the USSR (Fig.2.32) is intended for study of objects and processes by means of optical holography and holographic interferometry. It can record holograms of stationary objects, obtain holographic interferograms of processes in transparent and opaque objects, for example in hydraulic and air flows, in flames and plasma; it can be of great use in deformation and vibration analysis, and so on.

The unit consists of a massive iron slab mounted on an antivibration pneumatic suspension, a gas laser LG-36 (or LG-36A), a collimator, and a set of holders, tables, and optical elements for arranging holographic setups. Provision is made for processing holograms directly in a holographic setup, which permits real-time holographic measurements.

Fig.2.32. The UIG-2 laser test unit

2) The Gaertner holographic systems (USA) are among the simplest setup for holographic experiments. They need a minimum number of components to record holograms and can be used for interferometric research. All of the components of these systems are secured to a flat optical bench with vibration damping by use of magnetic bases. A typical setup includes a low-power He-Ne laser (rated at a few milliwatts), a beam splitter, pin-hole collimater, plate holder and mirrors.

3) Rottenkolber Holo-System GmbH (FRG) produce a considerable variety of holographic setups. They include:

(a) Holographic test systems LT-1000H and LT-500H, suitable for holographic deformation and vibration analysis. These systems are mounted on a rather heavy (1350 kg) damped T-slot table, optical elements such as plane mirrors, beam splitters, lenses and pin-holes and their housings, a plate holder and shutter. A Spectra-Physics argon laser is used as a light source. Photothermoplastic 35-mm film PT-1000 or PT 1000-Hs, in special PT-Holo-instant-picture cameras can be supplied to simplify investigations and to adapt them to environmental conditions.

(b) The holographic interferometer HIF 150-2 for air flow studies in wind tunnels, plasma, heat-transfer research, and other interferometric studies having to do with transparent media. The observation-field diameter is 120 mm, so that the HIF 150-2 instrument successfully competes with classical Mach-Zehnder interferometers that are much more expensive. This setup uses a He-Ne laser at about 15 mW. The latter can be replaced with a more powerful or pulsed laser for special purposes.

(c) The holographic interference camera HIK 1000-FE (Fig.2.33). This is a more universal setup, which can be used to study reflecting objects in addition to transparent objects. The standard camera is equipped with an argon 700-mW laser so that a field of 1000 mm can be illuminated. It can be used with the PT holographic instant-picture accessory mentioned above and with a video-camera and monitor to simplify the interpretation of interferograms.

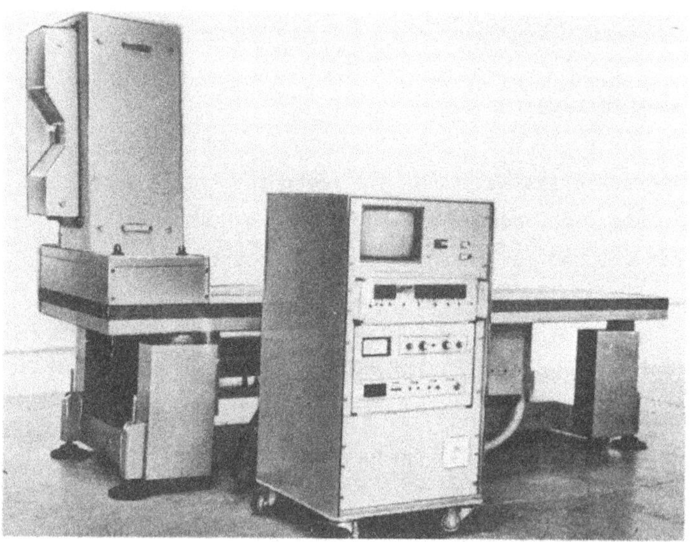

Fig.2.33. Holographic interference camera HIK 1000-FE

(d) The double-pulse hologram camera PHK 1 (Fig.2.34). This mobile unit has been designed especially to cope with vibration problems. The ADP ruby laser adopted in this setup was described on an earlier page. Use of this kind of light source allows the PHK 1 camera to be employed for vibration analysis of vehicle and ship structural members, turbine component studies, etc.

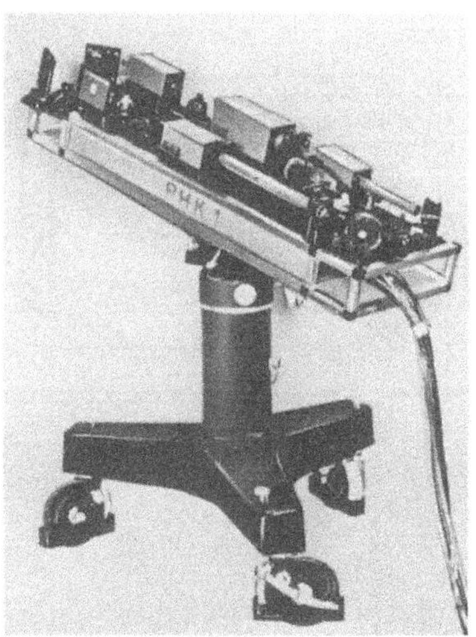

Fig.2.34. Double-pulse hologram camera PHK 1

The above Soviet and foreign-made arrangements are intended both for experimental work and for studying a broad range of objects. Most arrangements in the future will probably be produced for specific purposes instead of for universal application; this will sharply reduce their cost and diminish their size. An example is the vibrometric system for studying industrial products (Fig.2.35) [2.51]. It consists of an opaque rigid metal chamber 1 with sound-absorbing walls. The article 2 being studied is attached to the front end of the chamber, and the holder for the hologram 3 is at its back end. The flexible light guides 4 from the laser enter and are secured in the chamber walls. Microlenses that provide illumination of the article and the reference beam are fastened on the ends of the light guides. The system can operate in illuminated space and requires no special protection against vibration.

Another example of a specialized holographic setup is the Holo Tire Testing Unit HRT 56 manufactured by the Rottenkolber Holo-System (Fig.2.36). It has been developed for testing tires in industrial conditions and is protected against the influence of dust and vibration. The diagnostic system used in this setup is essentially fault free. The testing procedure is nondestructive, but permits rather heavy loads to be applied to tires being examined and service-life tests of them.

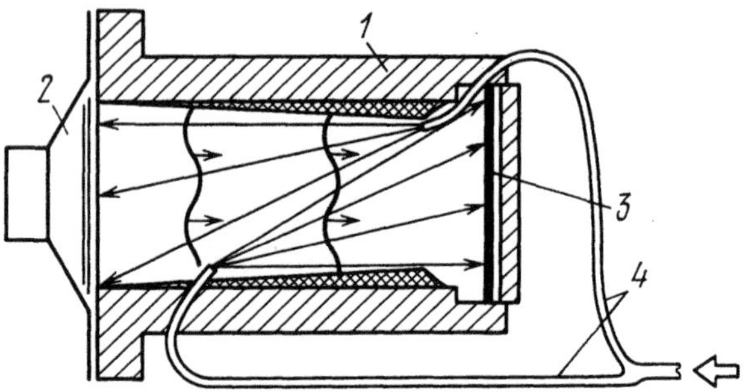

Fig.2.35. Specialized installation for holographic vibrometry [2.51]: 1 - chamber; 2 - object; 3 - hologram; 4 - light guides

Fig.2.36. Holographic tire-testing unit HRT 56

3. Investigation of Transparent Phase Inhomogeneities

3.1 Features of Holographic Interferometry of Transparent Objects

An object being studied produces amplitude and phase changes in a light wave. They change the interference pattern. The amplitude changes affect the visibility of the interference fringes; the phase changes affect their shapes and mutual arrangement. Although, in principle, the appearance of the interference pattern can be used to determine both kinds of changes in a light wave, photometric methods are customarily employed to study amplitude changes, and interference methods for phase distortions.

These distortions may be produced when a light wave is reflected from the surface of an object, or when a light wave passes through a transparent object. In the former case, the light wave carries information on the shape of the reflecting surface, in the latter on the spatial distribution of the refractive index.

In both cases, the phase of the object wave with respect to the reference wave is the quantity measured; in this sense, there is no difference of principle between objects of the two kinds. Nevertheless, the nature of typical phase distortions of the interference setups, the localization of the interference fringes, and also the specific features of the translation from the interference pattern to the parameters of the object make it expedient to treat transport phase inhomogeneities – objects studied by passing light through them – in a separate chapter.

Objects of this kind can be divided into two basic groups. The first group includes slight phase inhomogeneities such as flames, gas streams, plasma, and shock waves that produce comparatively small path differences, from fractions of a wavelength to several scores of wavelengths. The second group of objects includes transparent optical components such as lenses, prisms, and plates that produce great path differences, of the order of 10^3-10^4 wavelengths. Although this division is conditional, the investigation of objects of each of these groups nevertheless has a number of distinctive features determined by the structure of the interference pattern and its localization.

3.1.1 Methods for Visualization of Phase Inhomogeneities and the Relationship Between the Spatial Distribution of the Refractive Index and the Quantity Being Mesu red

Let us consider a narrow light ray AB (Fig.3.1) that, in the absence of the inhomogeneity, reached the screen at the point B. As a result of passing through an optically inhomogeneous medium, i.e., a medium whose thickness and/or refractive index depend on the coordinates, the light ray AB deviates from its initial direction through the angle ε and strikes the screen at the point B' at a distance of Δx from the point B. In addition, when passing thr ough the phase inhomogeneity, the ray lags in time. Each of these quantities can be determined by use of the relevant optical methods (schlieren, shadow, and interference).

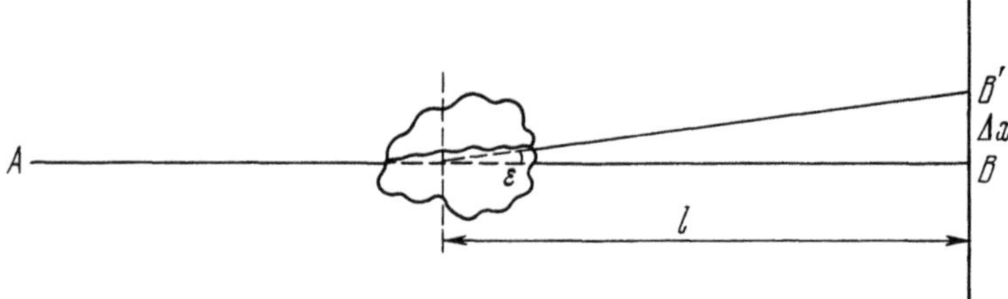

Fig.3.1. Passage of light through a transparent phase object

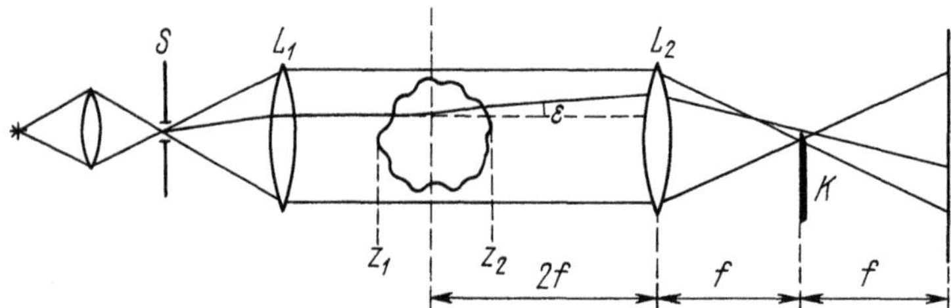

Fig.3.2. Schlieren system

In the schlieren method, shown schematically in Fig.3.2, what is measured is how the illumination is distributed in the image of the object. This distribution is proportional to the angular deviation ε. In the absence of inhomogeneity, the light that forms the image of the slit S is completely blocked by the knife edge K. Deflection of the light by the inhomogeneity causes a displacement of the image of the slit and a change of the illumination in the image of the relevant portion of the object. It can be shown [3.1,2] that the deflection of a ray, when it passes through the phase inhomogeneity, is determined by the gradient of the refractive index in the direction perpendicular to the direction of propagation of the ray,

$$\varepsilon \approx \frac{1}{n} \cdot \int_{z_1}^{z_2} \nabla_{\perp} n(x,y,z) dz \quad . \tag{3.1}$$

Thus, in the schlieren method, the integral of the gradient of the refractive index over the entire thickness of the layer is measured.

Whereas in the schlieren method a visualizing diaphragm (the knife edge K in Fig.3.2) and the optical system L_2 are placed between the inhomogeneity and the screen, in the shadow method no optical devices are placed between them[4].

In the shadow method, the measured quantity is the redistribution of the illumination on the screen, which results from the different displacements Δx of the rays that pass through different portions of the optical inhomogeneity. If the deflections of the rays when passing through the inhomogeneous medium are small, then the relative changes of the illumination on the screen in the first approximation are [3.1]

$$\frac{\Delta I}{I} \approx \ell \int_{z_1}^{z_2} \left(\frac{\partial^2}{\partial x^2} + \frac{\partial^2}{\partial y^2} \right) n(x,y,z) dz \quad , \tag{3.2}$$

where ℓ is the distance between the inhomogeneity and the screen.

[4] Here we shall keep to the classification proposed by LADENBURG [3.1]. A number of other authors (for example, VASILEV [3.2]) apply the name "shadow" to methods in which inhomogeneities are made visible by use of diaphragms of various kinds, which redistribute the illumination in the plane of the image that is optically conjugate to the plane of the object, and the angular deflection is measured. They call the method described by us as the shadow method the "luminous-point method".

Equation (3.2) shows that the change of the illumination is proportional to the second derivative of the refractive index integrated along the ray through the object.

In the interference method (both in the conventional and the holographic one), what is measured is the phase shift between the object wave and the reference wave due to the temporal lag of the ray when it passes through the inhomogeneity. Figure 3.3 shows the path of a ray through the inhomogeneity and in the absence of it. The objective L projects the cross section of the object MM' onto the screen P. The interference pattern formed by the rays ABC'O (the ray that has passed through the system in the absence of the inhomogeneity) and ACDO (the ray that has passed through the inhomogeneity) is observed at the point O on the screen. In conventional interferometry, these rays pass simultaneously along different branches of the interferometer, whereas in holographic interferometry they are separated in time instead of in space.

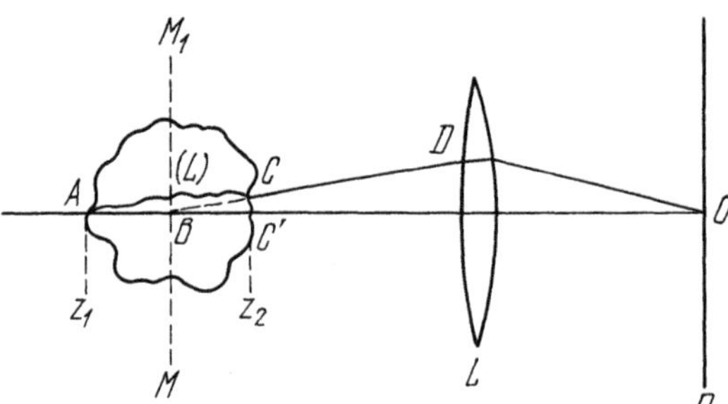

Fig.3.3. Calculation of phase shift introduced into a light wave by a phase object

The relevant phase shift, computed with account taken of the tautochronism of the lens L, is

$$\Delta\varphi = \frac{2\pi}{\lambda}\left[\int_{(S)} n(x,y,z)d\ell - (AB + BC)n_0\right] , \qquad (3.3)$$

where the integral is taken over the trajectory of the ray (S) from A to C, and n_0 is the refractive index in the absence of the inhomogeneity.

In the overwhelming majority of investigations that use interferometry of transparent inhomogeneities, conditions are such that insignificant angular deflections and displacements of a ray occur within the inhomogeneities. This permits replacement of integration along the exact trajectory of the rays with integration along the ray path followed when the inhomogeneity is absent. In this case, we have

$$\Delta\varphi = \frac{2\pi}{\lambda} \int_{z_1}^{z_2} [n(x,y,z) - n_0]dz \quad . \tag{3.4}$$

An analysis of the conditions in which it is possible to disregard the deviations and displacements of the rays, and also the errors that result as a consequence, can be found in [3.1,3-6].

The purpose of optical investigation of phase inhomogeneity is to study the spatial distribution of refractive index (or its derivative). The latter is directly associated with such characteristics of the object as the spatial distribution of the density of a gas, the concentration of atoms, electrons, and temperature. The integral quantities measured along the line of observation [see formulas (3.1,2,4)] are used to calculate the local values of the refractive index or its derivatives. We shall consider how this problem is solved at the end of the present section.

Thus, the interference, schlieren, and shadow methods give information on different quantities: the refractive index and its first and second derivatives, respectively. We could find the absolute values of the refractive index from the first and second derivatives by integration, but such calculations are usually not sufficiently accurate.

Therefore, the interference, schlieren, and shadow methods are not competitors, but supplement one another. The interference method is the most accurate. It permits us to measure directly the absolute values of changes of refractive index. The shadow method is the most sensitive to abrupt changes of refractive index and can be used for visualization of shock waves, turbulent streams, etc. A significant advantage of the shadow method is the simplicity of the experimental setup. The schlieren method occupies an intermediate place with respect to experimental possibilities and complexity.

Holography makes it possible to record the light wave that has passed through a transparent inhomogeneity. The thus-recorded wave can be studied further, not only by the interference method, but also by the schlieren and shadow methods. In this way the maximum amount of information can be obtained about the inhomogeneity. The possibility of sequentially studying the same wave by use of different methods that supplement one another, and also the freedom of choice and subsequent modification of the frequency and directions of the fringes in interference studies [3.7] or of the

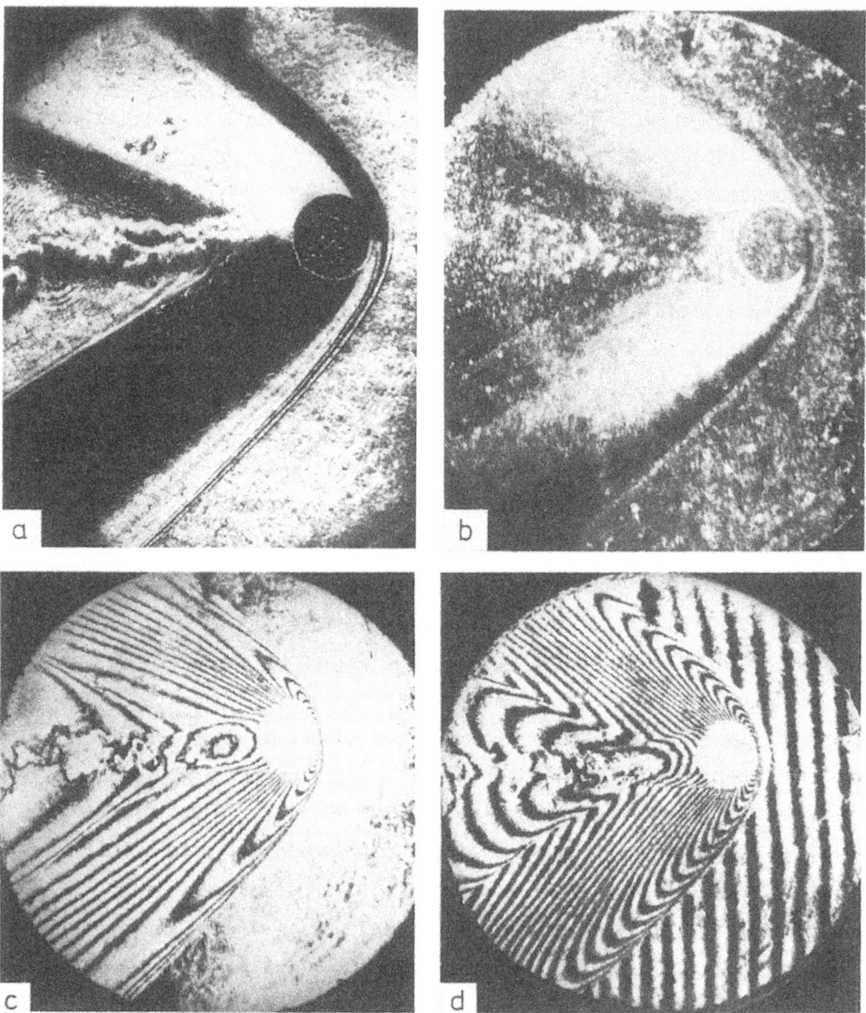

Fig.3.4. Schlieren photographs (a and b correspond to different orientation of the knife edge and slit) and interferograms (c - in fringes of finite, and d - of infinite width) obtained by use of the same hologram. The object is the air that flows around a steel ball in free flight with Mach number M = 1.8

direction of the knife edge in the schlieren system [3.8] appreciably increase the information content of investigations and improve the reliability of the results obtained.

Figure 3.4 shows, as an example, schlieren and interference patterns of the shock wave near a bullet, obtained from the same hologram [3.9].

3.1.2 Setups Without Diffusing Screens

The simplest setup for studying transparent phase inhomogeneities is one without a diffusing screen, in which a plane coherent light wave is passed through the object (Fig.3.5).

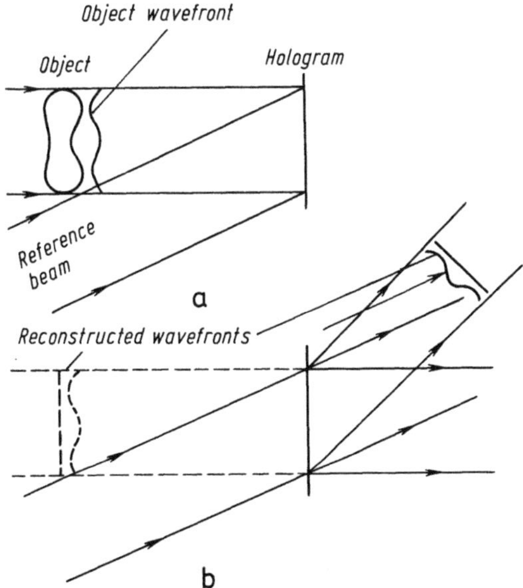

Fig.3.5. Arrangement for obtaining a hologram of a transparent object without a diffusing screen (a) and reconstruction of the wavefronts (b)

Assume that the double-exposure method is used to obtain a holographic interferogram, and that the object is absent during one of the exposures. Consequently, upon reconstruction, we shall observe the interference of the wave distorted by the object with the plane comparison wave. The interference pattern in this case will be observed in both the positive first and the negative first diffraction orders, at any distances from the hologram. It is simplest, however, to compare directly the interference pattern with the local parameters of the transparent phase inhomogeneity. This is done by observing the interference pattern in the plane of the virtual or real image of the object, where the relationship between the phase shifts of the waves and the refractive index of the object is given by the integral (3.4). The fringes observed indicate equal optical thickness of the phase inhomogeneity.

The phase relief of the waves reconstructed in the first positive and first negative orders will be reversed (the corresponding waves, as we have shown earlier, are complex conjugate). This will not affect the form of the interference fringes, however (see Fig.1.15).

The merits of a setup without a diffusing screen are its simplicity, complete utilization of the light, the easy interpretation of the interference pattern (a collimated beam of light is passed through the object), and absence of a speckle structure of the image. In addition, setups without a diffusing screen can use multiple-mode lasers as light sources, because it is only necessary to superimpose mode structure in the object and reference beams (if the disturbances of the wavefront by the object are not too great). The intensity of illumination of both the object and the hologram, however, is not uniform in holography without a diffusing screen.

When a hologram is obtained without a diffusing screen, light rays that pass through different portions of the object reach only the corresponding portions of the hologram. A fragment of such a hologram reconstructs only the corresponding part of the object, and not the entire object.

Among the shortcomings of setups without a diffusing screen is the fact that the reconstructed image of the object is seen on the background of a luminescent point, which is not a satisfactory visual condition. (in setups with a diffusing screen, the object is seen on the background of a more-or-less uniformly luminescent screen.) In addition, the interferogram obtained without a diffusing screen corresponds to passing of light through the object at only one fixed angle; it is not possible to determine the local parameters of objects that have no axes of symmetry.

3.1.3 Setups with a Diffusing Screen

Use of a diffusing screen in the object beam (Fig.3.6) eliminates many of the aforementioned shortcomings. In such setups, every part of the diffusing screen illuminates the entire area of the hologram, and, conversely, light scattered by all points of the screen impinges on every portion of the hologram. As a result, every part of the hologram carries information concerning light waves that have passed through every part of the object, at all different angles. This makes it possible to determine the local parameters of objects that have no axes of symmetry. This same circumstance, however, causes a number of difficulties in observation and interpretation of interference patterns.

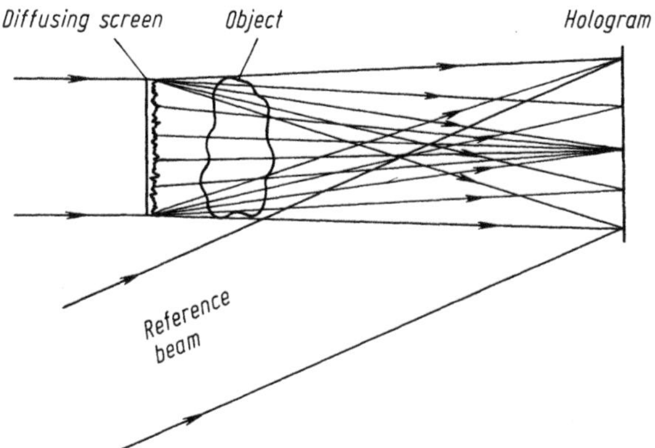

Fig.3.6. Arrangement with a diffusing screen

The wavefront recorded on a hologram, either in the absence or in the presence of an object has an intricate microstructure, due to the interference of light waves that reach any given point of the hologram from different points of the diffusing screen. The phases of the waves emitted by even closely neighboring points of the diffusing screen differ chaotically. This results in a chaotic change, not only of the phase, but also of the amplitude of the object wave in the plane of the hologram (Fig. 3.7). When such a wave passes through the object, an additional phase difference is introduced into it that slowly changes from point to point. This phase difference is due to the spatial distribution of the optical thickness of the object.

Thus, two waves are reconstructed with an identical phase microstructure, but owing to the fact that one of the waves has passed through the object, their phase macrostructure differs (Fig.3.8).

The low-frequency interference pattern that characterizes the properties of the object will not be observed everywhere, but only where the cor-

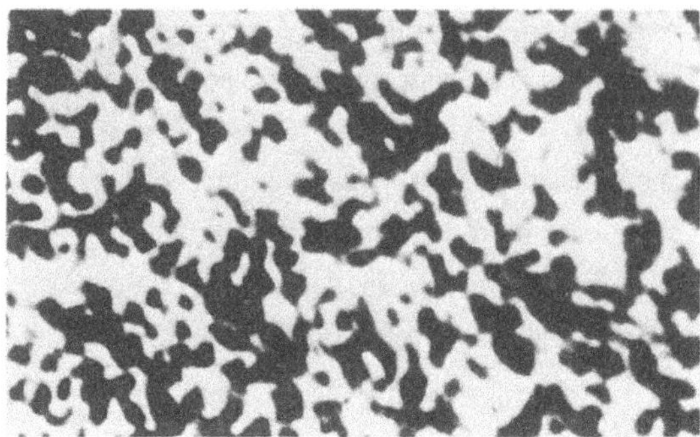

Fig.3.7. Microstructure of a laser beam after it passes through a diffusing screen

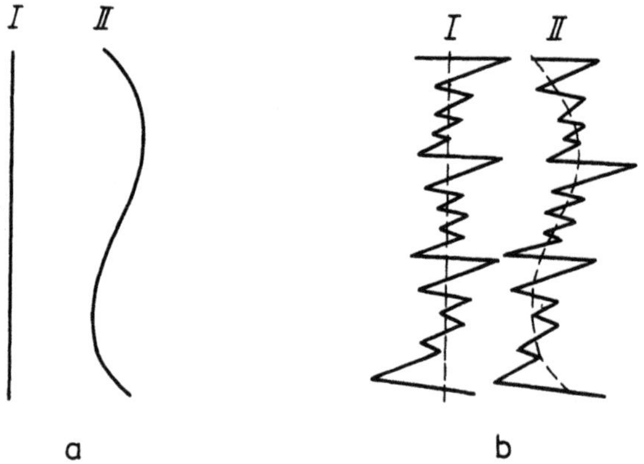

Fig.3.8. Phase reliefs of the comparison wave (I) and the wave that has passed through the object (II): (a) hologram obtained without a diffusing screen; (b) hologram obtained with a screen

responding sections of two interfering wavefronts, i.e., sections with an identical microstructure, are superposed. Transverse displacement of these wavefronts leads to reduced visibility of the fringes. The latter thus become localized in space.

Every interference fringe is the combination of the points of superposed wavefronts at which their phases differ by a constant value, i.e., in the general case, they are fringes of equal path difference. This difference is determined by the optical thickness of the object in the direction of light passage; i.e., it depends on both the spatial distribution of the refractive index of the object and on the angle of incidence of the object ray.

There are limiting cases, however, in which interference fringes due to only one of these parameters can be observed. The first case is when light passes through the object at one and only one angle, i.e., in a parallel beam. The changes of the path difference between beams over the field of view are due only to local variations of the refractive index of the object. These are the so-called fringes of equal thickness. It is this case that is usually realized in setups for the holographic interferometry of phase objects without a diffusing screen.

Another limiting case occurs when the object is homogeneous in thickness and refractive index (a plane-parallel plate), but light passes through it in all possible directions (within a certain solid angle). In this case, the path difference is the same for all rays that pass through the object at the same angle. We have equally inclined fringes, in the form of concentric rings observed in the focal plane of the objective. A particular case of such rings is the interference pattern formed by a Fabry-Perot interferometer. Such rings can also be obtained in holographic setups, particularly with a diffusing screen, which ensures that light passes through an object in all possible directions [3.10].

Because fringes of equal thickness must be obtained to determine the local parameters of transparent inhomogeneities, setups for reconstructing a wavefront by use of holograms obtained with a diffusing screen must confine the light from the object to a definite direction. This can be done by use of the setup depicted in Fig.3.9a in which a diaphragm is placed in the focal plane of a lens that projects an image of the object onto a screen. The diaphragm selects a parallel beam that corresponds to a definite direction from all of the rays reconstructed by the hologram. By moving the diaphragm, we can use the same doubly exposed hologram to obtain interference patterns that correspond to passage of light through the object in different directions.

If the scattering indicatrix of the diffusing screen is wide enough and the optical thickness of the object changes noticeably when the direction of the light passing through it is changed, then increase of the opening in the diaphragm reduces the visibility of the fringes until they completely vanish. This occurs in the investigation of optical components, where a change of the direction of the light that passes through an object

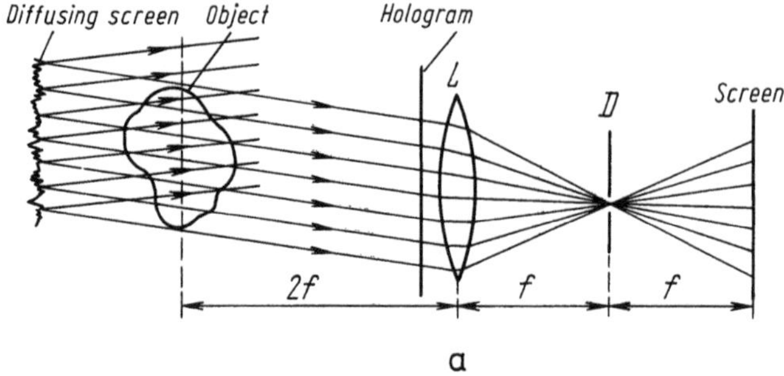

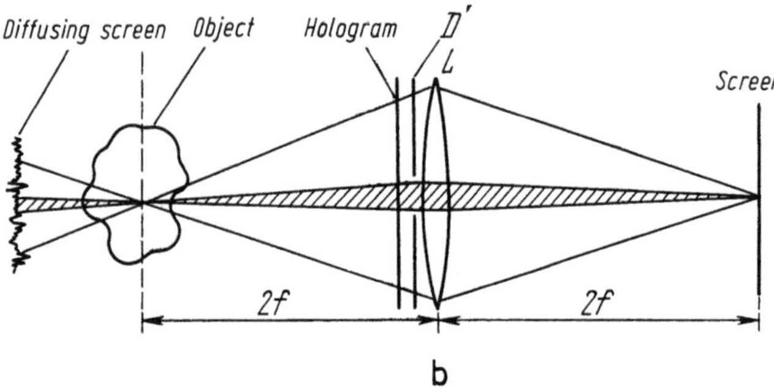

Fig.3.9. Obtaining an interference pattern corresponding to a specified direction of light passing through an object by use of a pinhole diaphragm (a) and by use of an objective with a small aperture (b)

by fractions of a degree may result in a change in the path difference by a whole wavelength (Fig.3.10a,b).

Another way of observing interference patterns by use of holograms obtained with a diffusing screen is to use an objective with a small aperture (see Fig.3.9b). In this case, the diaphragm D' picks out, from the entire cone of rays that pass through a point of the object, a narrow beam within which the path difference introduced by the object does not change significantly. Figure 3.9b shows that the light will pass through different portions of the object in different directions. Hence the fringes observed in such a setup, generally speaking, are not fringes of equal thickness, and the change of direction of observation must be taken into consideration in their interpretation.

The same pattern will be observed when a doubly exposed hologram is viewed with the naked eye.

It must be noted, however, that reduction of the aperture of the objective, of the dimensions of the hologram, or of the dimensions of the

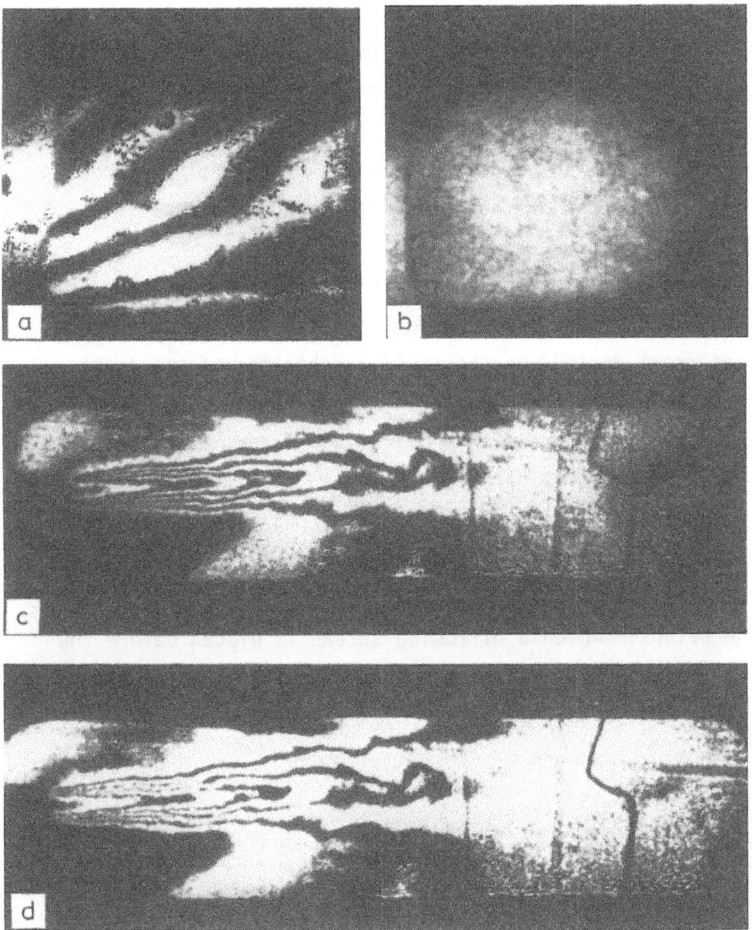

Fig.3.10. Interferograms of phase objects: (a, b) a glass plate, using small and large apertures, respectively (arrangement of Fig.3.9b); (c, d) a gas stream (arrangement of Fig.3.9b) using small and large apertures, respectively

diaphragm D (see Fig.3.9a) selects a very narrow range of spatial frequencies in the spatial spectrum of waves scattered by the diffusing screen. Near the direction parallel to that of the wave that illuminates the diffusing screen, this frequency range is in the lowest-frequency region, i.e., it corresponds to waves diffracted by the largest inhomgeneities. The result is the appearance of a speckle structure that hinders study of the pattern of interference fringes.

In the study of slight phase inhomogeneities that have total optical thickness equal to only a few wavelengths, a small change of the angle of light passage through the object does not cause an appreciable change of the interference pattern. It can therefore be observed with even quite a large opening in the diaphragm D (see Fig.3.9a), or entirely without it when the scattering indicatrix of the diffusing screen is narrow (see Fig. 3.10c,d) [3.11].

When a setup with a diffusing screen is used to prepare holograms, different portions of the incident object wave interfere with the reference wave. This imposes great demands on the spatial coherence of the light used to prepare the holograms (see Sect. 2.1). This is why single-mode lasers are usually used in setups with a diffusing screen. If the diffusing screen is projected onto the hologram, however, then it is possible, as when setups without a diffusing screen are used, to superimpose the structures of the object and reference beams accurately [3.12]. This allows use of a multiple-mode laser.

Apart from the setup in which a diffusing screen is placed before the object, the setup shown in Fig.3.11, where the diffusing screen follows the object, is also used [3.13-15].

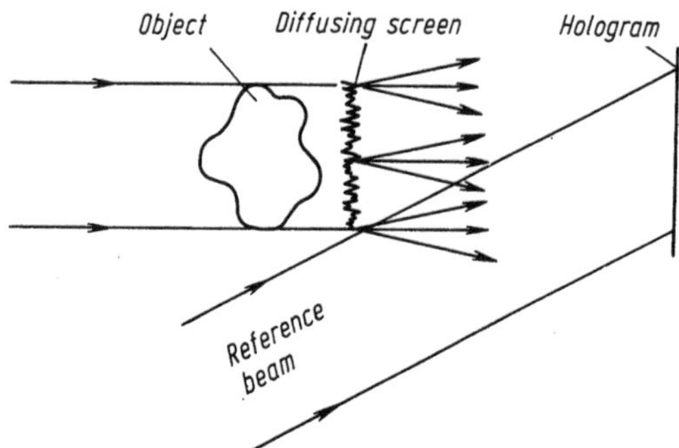

Fig.3.11. Recording a hologram of a transparent object placed before the diffusing screen

This setup does not have one of the advantages of setups with a diffusing screen — the possibility of obtaining interference patterns that correspond to different directions of observation. For an axially symmetrical object, however, all of the needed information can be obtained with one direction of observation. The other merits of a setup with a diffusing screen – distribution of the information concerning each point of the object over the entire hologram and the possibility of observing the interference pattern directly with the eye, on the background of the illuminated diffusing screen – are retained.

This setup also has a merit of setups without a diffusing screen, i.e., to obtain contrast fringes, there is no need to diaphragm the hologram or to use special reconstruction setups (see Fig.3.9). The interference patterns obtained in this setup are the projection of the interferogram of the object onto the diffusing screen and, as in the setup without a diffusing screen, are fringes of equal thickness.

It is sometimes convenient to replace a ground-glass diffusing screen with a phase diffraction grating that makes it possible to pass light through an object in several discrete directions [3.11,16] (see Sect. 2.4). For example, VEST and SWEENEY [3.16] used a holographic grating with 50 lines per mm, which was produced bright beams of three diffraction orders at each side of the normal, which gave seven interferograms covering an angle of vision of about 11 degrees (every 1.8 degrees) with the aid of a single hologram. The interferograms reconstructed in this way do not have speckle because diffusely scattering elements are absent in the setup.

3.1.4 Production of Interferograms with Fringes of Finite Width. The Wedge Method

If the orientation of the object beam and the reference beam remain unchanged during the time between the first and second exposures (or between the moment of exposure and that of observation in the real-time method), then the only difference between the two waves reconstructed by the hologram is due to the changes of the object during the time between the two exposures. The interferogram obtained is a pattern of fringes of infinite width (Fig.3.12a).

To obtain an interferogram with fringes of finite width, the angle between the object and the reference beams must be changed during the time between the two exposures. This can be done by slightly turning the beam splitters or mirrors that direct the light beams onto the hologram. The most convenient way to change the angle between the object and the reference beams, however, is to put a thin glass wedge into one of them.

When a light beam passes through a thin wedge with an angle γ at its apex, it is deflected through an angle δ equal to

$$\delta \approx (n - n_0)\gamma \quad , \tag{3.5}$$

where n is the refractive index of the wedge, and n_0 is that of air.

Thus, introduction of a wedge into the object or reference beam during one of the exposures produces a change δ of the angle between the object beam and the reference beam and a corresponding change of the spatial frequency of the hologram pattern.

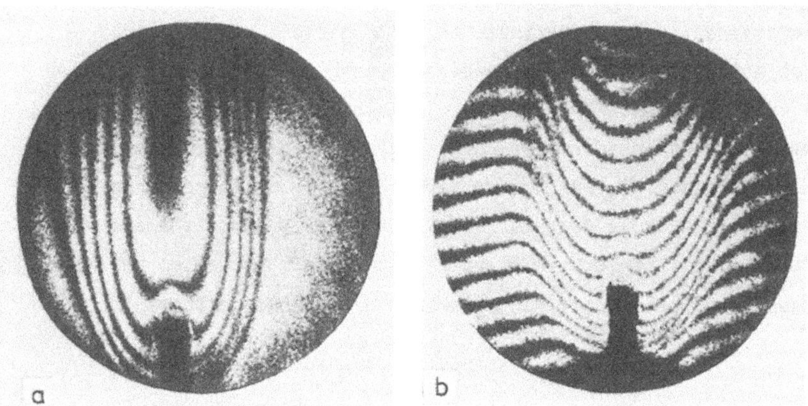

Fig.3.12. Holographic interferograms of an alcohol-lamp flame in fringes of infinite (a) and finite (b) width

When such a hologram is illuminated by the reference beam, two object waves are reconstructed that deviate at the angle δ from each other. Interference between them produces a system of carrier fringes whose frequency, in accordance with (1.41) is

$$\nu = \frac{2\sin(\delta/2)}{\lambda} \approx \frac{\gamma(n - n_0)}{\lambda} \quad . \tag{3.6}$$

The distortions of the object wave caused by the object cause displacements and distortions of the carrier fringes (Fig.3.12b).

The position of the surface of interference fringes localization depends on the position of the wedge in the holographic setup. We shall show in the following that it is most convenient to put the wedge in the object beam very close to the object.

A thin glass wedge was first used to produce holographic interferograms with fringes of finite width in [3.17]. JAHODA et al. [3.18] used a hollow wedge for the same purpose. The wedge could be evacuated and filled with various gases in the interval between two exposures. Carrier fringes of the desired frequency could be obtained by varying the pressure of the gas in the wedge. The frequency of the carrier fringes can be changed with a glass wedge with refraction angle γ placed permanently in the object beam of a holographic setup; between the two exposures, it is turned about the axis of the beam through a small angle α [3.19]. The frequency of the fringes is

$$\nu = \frac{2\gamma(n - n_0)}{\lambda} \sin\frac{\alpha}{2} \approx \frac{\gamma\alpha(n - n_0)}{\lambda} \quad . \tag{3.7}$$

The fringes are perpendicular to the bisector of the angle formed by the dihedral edge of the wedge in its initial and final states. By changing the angle of rotation of the wedge α, we can vary the frequency of the carrier fringes.

3.1.5 Localization of Interference Pattern

Two light waves are recorded on a doubly exposed hologram. Both waves are reconstructed simultaneously when the hologram is illuminated with the reference beam (Fig.3.13). Each of the rays of the reference beam that arrives at the point A of the hologram is divided at this point into two rays that correspond to the two waves recorded on the hologram.

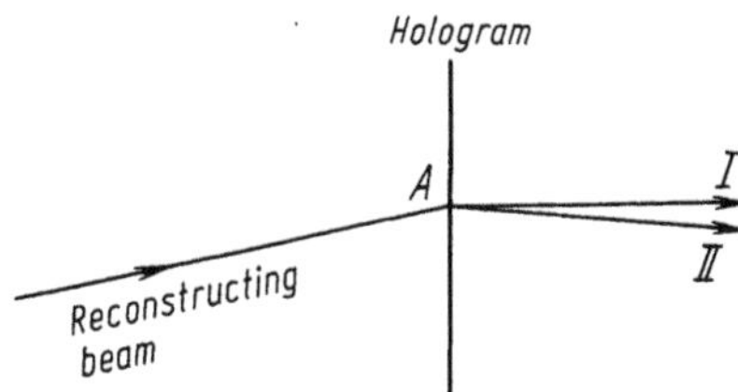

Fig.3.13. Reconstruction of the waves of a doubly exposed hologram

If we understand the localization surface (as in conventional interferometry) to mean the region where the light rays obtained by amplitude division from one initial ray converge, then the surface of localization of the interference fringes coincides with the surface of the hologram.

The depth of the localization zone understood in this sense depends on the spatial coherence of the reconstructing radiation. Indeed, on doubly exposed holograms recorded without a diffusing screen, we can observe interference fringes localized on the surface of the hologram even when it is illuminated by a conventional incandescent lamp.

If (as is customary practice) a gas laser whose radiation has a high degree of spatial coherence is used as the reconstructing source, then the waves reconstructed by the hologram are coherent and form a sufficiently high-contrast interference pattern in the entire region where the reconstructed beams overlap. The form of this interference pattern at different spots is different, however, and depends appreciably on the shape of the wavefronts recorded on the hologram and the angle between them. For example, if both wavefronts recorded on a hologram during the two exposures are plane, then in any region where the reconstructed waves overlap, a system of equidistant planes of nodes and planes of antinodes will be formed normal to the bisector of the angle between the normals to the fronts of the interfering waves (Fig.3.14).

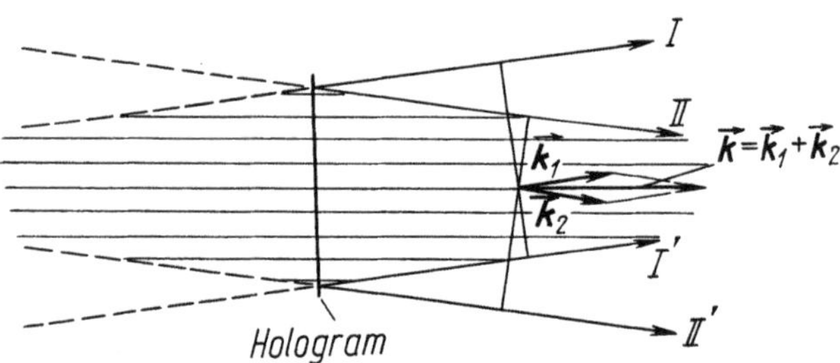

Fig.3.14. Interference of two plane waves reconstructed by a doubly exposed hologram

If the wavefronts recorded on a hologram have an intricate high-frequency structure (for example, if a diffusing screen was used in recording the hologram), then a low-frequency interference pattern that characterizes the distortions introduced into one of the waves by the phase object being studied will be observed only in the region where the corresponding sections of the reconstructed wavefronts are encountered (i.e., sections that have the same microstructure). This region (near the line AB in Fig.3.15) is called the <u>region of localization of the interference fringes</u> in holographic interferometry.

In the remaining region, noncorresponding sections of the wavefronts will be superposed on one another. The interference pattern will have a high spatial frequency and a very intricate structure that is due mainly to the structure of the diffusing screen and not of the phase object.

The position of the region of localization depends appreciably on whether the holographic interferogram has been obtained in fringes of infinite or finite width, i.e., it depends on whether or not the angle between the object and the reference waves was changed during the interval between the first and the second exposures.

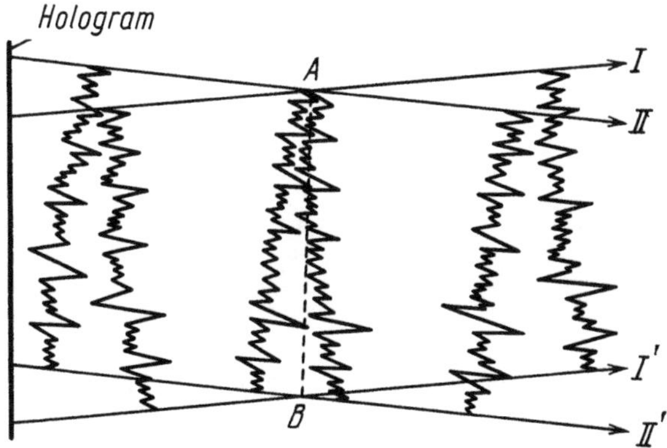

Fig.3.15. Superposition of two reconstructed wavefronts with an intricate microstructure

If the position of the diffusing screen and the angle between the object and the reference beams remain constant, then the difference between the two waves recorded on the hologram is associated with only the changes that occur in the object during that time. Let us separate from the object wave (Fig.3.16) the ray that passes through the point A of the diffusing screen and the point B of the object. In both exposures, the direction of the ray is the same up to the object. After passing through the object, the ray AB deviates from its initial position; owing to the changes that occur in the object, the angle of deflection is differnt in the two exposures; and the ray arrives at different points on the hologram (C and C'). Thus, the object in this case plays the part of a beam splitter that divides the single initial ray AB into two corresponding rays, BC and BC'.

It is obvious that when the hologram is illuminated by the initial reference beam, both rays BC and BC' are reconstructed; the point of their intersection D, as a rule, is near the virtual image of the object. Thus, when a holographic interferogram of a phase object consists of fringes of infinite width, the interference pattern is localized very close to the virtual image. For a thin wedge-shaped phase object, the surface of localization coincides with the plane of the image of the object. (This is true not only for the virtual, but also for the real image of an object formed by diffracted waves of the -1 order.)

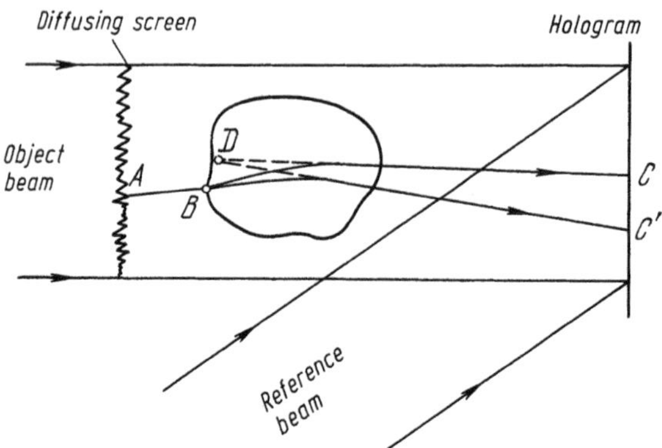

Fig.3.16. Recording of a holographic interferogram in fringes of infinite width

The problem of localization is much more complicated when the angle between the object and reference beams is changed between the two exposures in order to obtain fringes of a finite width. Let us assume that this angle is changed by altering the angle of incidence of the reference beam onto the hologram (Fig.3.17). If no changes occur in the object during the time between the two exposures, both object waves recorded on the hologram are identical. If such a hologram is illuminated by a reference wave whose direction coincides with that during the first exposure, then one of the reconstructed waves (which corresponds to the first exposure) forms an image of the diffusing screen at the location where it was when the hologram was recorded. The other reconstructed wave (which corresponds to the second exposure) propagates at the angle α to the initial one in accordance with the change of the inclination of the reconstructing beam relative to the corresponding reference beam.

Thus, two images of the diffusing screen are reconstructed that are displaced relative to each other. The corresponding rays intersect in the plane of the hologram; the latter is therefore the surface of localization of the interference pattern. In the plane of the image of the object, on the other hand, the interference fringes have low visibility or are not seen at all. In this case, it is possible to observe a sharp image of the object and a high-quality interference pattern only when the plane of the image coincides with that of the hologram (i.e., in an image hologram).

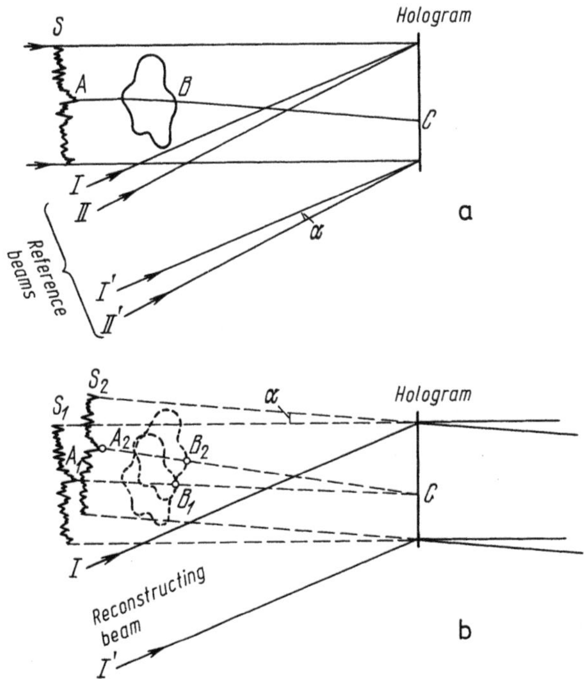

Fig.3.17. Inclination of the reference beam between exposures: (a) recording of a double-exposure hologram; (b) reconstruction of the wavefronts

If no diffusing screen is used when the hologram is recorded, then, as already noted, interference fringes are observed over the entire region of superposition of the two reconstructing waves, including the plane of the image. The two images of the object, however, are displaced relative to each other; the interference pattern obtained is called a shearing interferogram (it is assumed that the state of the object remained unchanged between the two posures). If the object was present in the object beam only during one exposure, then a conventional interferogram with fringes of finite width is observed in the plane of the image. As a rule, however, the quality of the fringes in such an interferogram is lower than in the region of localization. The reason is that noncorresponding wavefront sections participate in the formation of the interference pattern, and distortions of them due to imperfections of the optical components of the holographic setup are not excluded.

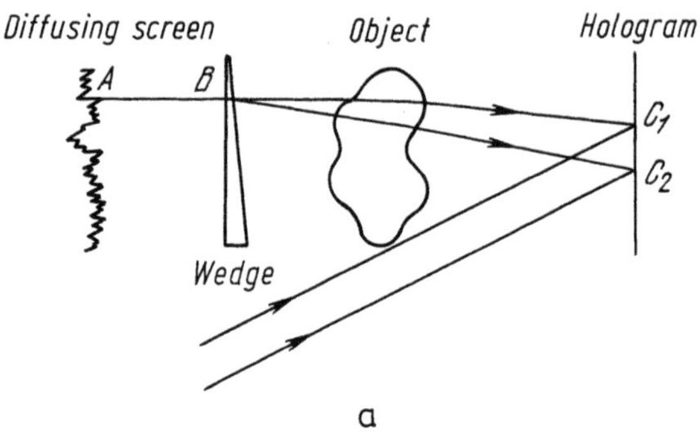

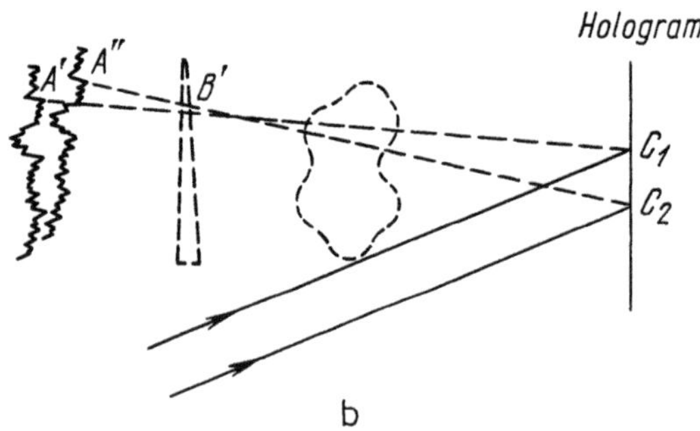

Fig.3.18. Inclination of the object beam between exposures: (a) recording of the hologram; (b) reconstruction of the wavefronts

Let us now consider the case when the direction of the object beam is changed between the exposures, for example by introducing a thin glass wedge into it (Fig.3.18). If the state of the object remains unchanged, then the wedge is the only element that causes any difference between the two waves recorded on a hologram, i.e., it plays the part of a beam splitter (compare Fig.3.16 in which the object plays the part of the beam splitter). The interference pattern is therefore localized in the plane of the wedge. If the state of the object changes during the time between the exposures (but the corresponding deviations of the object wave are small in comparison with its deviation due to the introduction of the wedge), then the position of the plane of localization also changes insignificantly. Thus, to observe a high-quality interference pattern in the plane of the image, the wedge must be arranged as close as possible to the object.

If the wedge cannot be placed very close to the object, then two wedges can be used, one placed before and the other after the object (Fig.3.19). Depending on the arrangement of the wedges and their apex angles, the interference pattern can be localized in any region, including a particular section of the object. It should be remembered, however, that introduction of a second wedge between the object and the hologram results in displacement of the images of the object that corresponds to the two exposures. Hence, two wedges should be employed only when the object is present in the object beam during only one of the exposures.

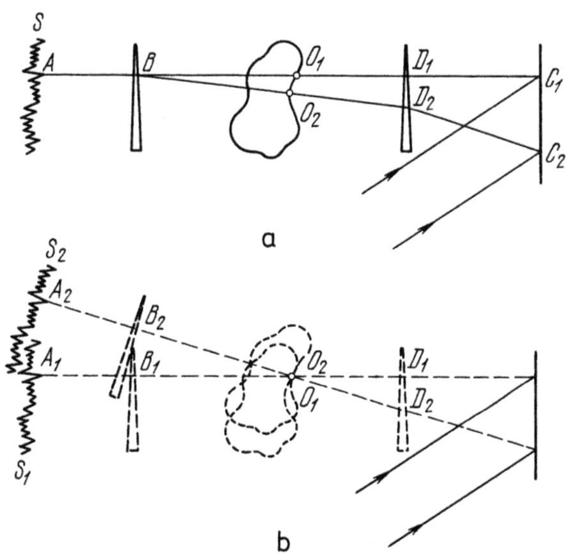

Fig.3.19. Inclination of the object beam using two wedges: (a) recording of the hologram; (b) reconstruction of the wavefronts

Fringe localization also has a tremendous significance for the holographic interferometry of deformable objects that scatter light (see Chap. 4). The following approach is often used in this case: the interference of light that arrives at the point of observation from corresponding elements of the microstructure for two states of the scattering surface is considered. The region of localization of the fringes is considered to be the region where the phase difference of the rays that arrive at the point from all pairs of corresponding points on the object (within the limits of the detector aperture) is constant [3.20-22].

The localization of fringes can also be considered from the same viewpoint in holographic interferometry of transparent objects with setups that use a diffusing screen [3.23]. In that case, the points of the diffusing screen are stationary with respect to the initial light beam, unlike the points of a deformable object; this is a simplifying circumstance.

A specific feature of investigating phase objects by use of a diffusing screen is that the purpose of the investigation is not to determine the nature of the displacement of the scattering surface, as when deformations are studied, but to investigate the phase distortions of the object wave when it passes through the object. Although, in the first case, the experimenter cannot arbitrarily change the region of localization (it is determined by the nature of the deformations of the object), in the second case (the investigation of phase objects) he should make the surface of localization of the interference pattern coincide with the image of the object. This should be done because the correspondence between the coordinates of the object and the interferogram is unambiguous in only that plane, and the relationship between the displacements of the interference fringes and the optical thickness of the object is simplest [see (3.4)].

3.1.6 Calculation of the Spatial Distribution of the Refractive Index from an Interferogram

The three-dimensional nature of the images produced by a hologram can used to advantage in investigations of the spatial distributions of dust particles, fog droplets, tracks of fast particles formed in bubble chambers, etc. When the reconstructed image is studied with an optical system that has a small depth of focus, the locations of particles at different depths in an object can be determined, for one depth after another [3.24-26]. Although particles outside the optical depth of field create a diffused background that decreases the visibility of the image [3.27], they do not prevent observation of the particles in the layer being studied.

By analogy, it might seem that to determine the spatial distribution of the refractive index in a transparent object, it would also be sufficient to focus the telescope consecutively to different fringes at depths of the image reconstructed by the doubly exposed hologram. Unfortuantely, this is not true! The displacements of the fringes on an interferogram are due to the change of phase of the object wave when it passes through the entire thickness of the object. Consequently, no matter on what plane in the reconstructed image of the object we focus the telescope, the displacements of the fringes are determined by the integral of the refractive index of the object over the entire line of observation [see (3.4)].

3.1.7 The Two-Dimensional Case

Of course the relationship between the displacements of the interference fringes and the refractive index is simplest when the object is homogeneous along the line of observation, i.e., the refractive index depends only on the coordinates x and y and does not depend on z (two-dimensional distribution). Then (3.4) has the form

$$\Delta\varphi(x,y) = \frac{2\pi\ell}{\lambda}[n(x,y) - n_0] \quad , \tag{3.8}$$

where $\ell = z_2 - z_1$ is the thickness of the layer being studied.

Converting from the phase advance $\Delta\varphi$ to the number of fringes k by which the interference pattern is displaced when an inhomogeneity is introduced, we have

$$k(x,y) = \frac{\Delta\varphi}{2\pi} = \frac{\ell}{\lambda}[n(x,y) - n_0] \quad . \tag{3.9}$$

Hence

$$n(x,y) = \frac{k(x,y)\lambda}{\ell} + n_0 \quad . \tag{3.10}$$

The problem thus consists of measuring the quantity k on an interferogram as a function of the coordinates x and y.

As we have noted previously, for fringes of infinite width the interpretation of the interference pattern is ambiguous — the sign of the change of k from one fringe to another cannot be determined from the form of the interference pattern. Thus, several possible relationships k(x) correspond to the interference pattern shown schematically in Fig.3.20 (some of them are shown at the bottom of the figure). The proper distribution may often be selected on the basis of physical considerations. If this is impossible, then an interferogram with fringes of finite width must be obtained.

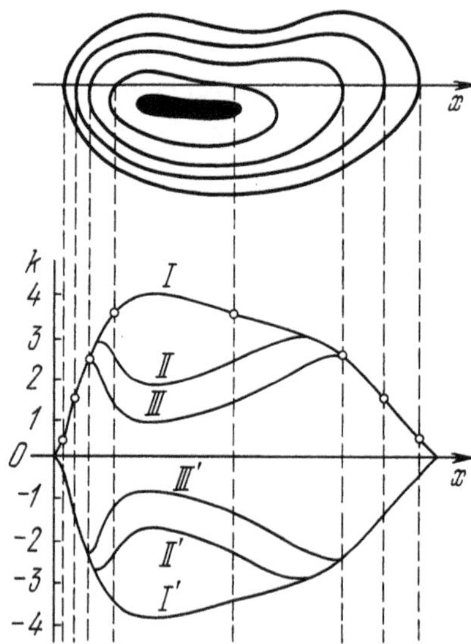

Fig.3.20. Interferogram with fringes of infinite width and a possible distribution of refractive index along the x axis

The fringes are counted from a point in the field of vision at which we know the refractive index. For example, for the interferogram shown in Fig.3.20 we assumed that the refractive index outside of the zone of disturbance is n_0. In real-time interference investigation of a transparent inhomogeneity, the fringes may be counted from the initial moment (before the change being investigated begins). Sometimes, when watching the development of the interference pattern, we can distinguish a type I distribution from type II or III (Fig.3.20). The uncertainty of the choice between I and I', II and II', and so on, remains as previously. For fringes of finite width, the sign of the change of k from fringe to fringe is determined by the direction of rotation of the object wave between the two exposures. If this rotation is produced by putting a wedge into the object beam, then k increases in the direction of the gradient of the wedge thickness.

A typical form of interference pattern for fringes of finite width is shown in Fig.3.21. In measurements of such a pattern, for every point of the field it is necessary to determine the number of the fringe at that point and to compare it with the number of the fringe that would pass through that point in the absence of the inhomogeneity. For example, for point A of the interference field shown in Fig.3.21, k = 3 because fringe No. 6 is at that point (in the absence of a disturbance fringe No. 3 would have passed through it).

Fractions of fringes are usually found by linear interpolation.

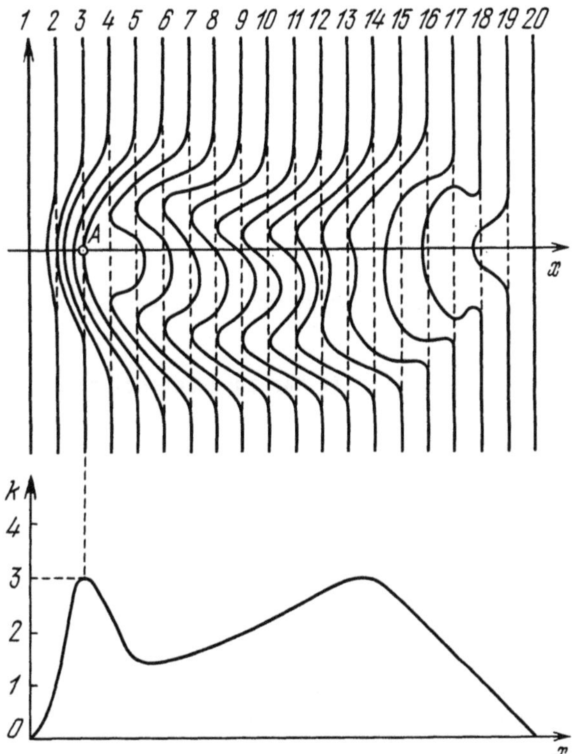

Fig.3.21. Interferogram with fringes of finite width and distribution of refractive index along the x axis

3.1.8 The Axisymmetric Case

In the particular case of an axisymmetric phase object, there are no appreciable difficulties in determining the distribution of refractive index from the measured fringe displacements.

Let us consider an object that is symmetrical relative to the x axis (Fig.3.22). The beam passes through the object in the direction z. In this case, the refractive index is the function of the two coordinates x and $r = \sqrt{y^2 + z^2}$, and (3.4) has the form

$$k(x,y) = \frac{\Delta\varphi(x,y)}{2\pi} = \frac{2}{\lambda} \int_{|y|}^{R} [n(x,r) - n_0] \frac{r dr}{\sqrt{r^2 - y^2}} \quad . \tag{3.11}$$

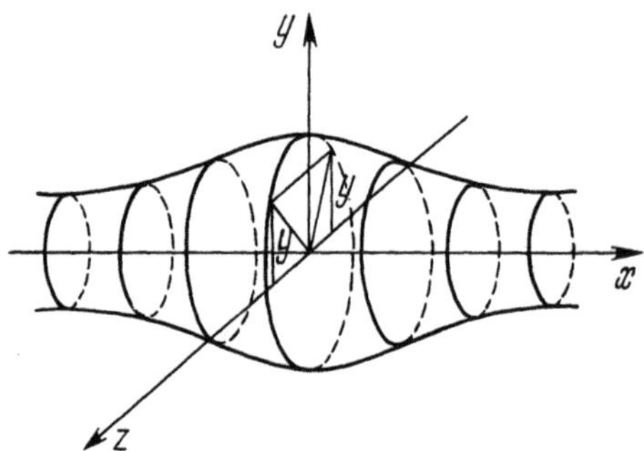

Fig.3.22. Axisymmetric phase object

Equation (3.11) is an Abel integral equation. It has an exact solution

$$n(x,r) - n_0 = -\frac{\lambda}{\pi} \int_{r}^{R} \frac{dk(x,y)/dy}{\sqrt{y^2 - r^2}} dy \quad . \tag{3.12}$$

Because the distribution k(x,y), however, is given not analytically, but in the form of experimental curves, numerical methods are generally used to solve (3.11) [3.28-35]. The axisymmetric distribution is divided, as a rule, into m concentric annular zones, and integration in (3.11) or (3.12) is replaced with summation. Thus, the task of finding the radial distribution n(r) consists of solving a system of algebraic equations of the form

$$k_\ell = \frac{1}{\lambda} \sum_{i=\ell}^{m} A_{\ell i} \Delta n_i \quad , \tag{3.13}$$

or

$$\Delta n_i = \lambda \sum_{\ell=i}^{m} C_{\ell i} k_\ell \quad , \tag{3.14}$$

where $\Delta n_i = n_i(r) - n_0$.

The different methods vary in the way of approximating the values of $\Delta n(r)$ and $k(y)$ within the limits of a zone. Stepwise, linear, or parabolic approximations are used.

Replacement of an integral with a finite sum always introduces a systematic error that depends on the kind of approximation and the number of zones formed. It is also essential to take into account the stability of the solution to chance errors in measuring the quantity $k(y)$.

Questions of the accuracy of determining $\Delta n(r)$ according to the measured integral values of $k(y)$ are discussed by many authors [3.29, 33-38]. The Abel integral equation (3.11) has the single solution (3.12) only when the function $k(y)$ is known exactly. If, however, $k(y)$ has been measured with a definite error, then a broad class of functions that differ appreciably from one another satisfy (3.11). In this sense, the solution of the Abel equation belongs to the type of so-called "conditionally correct" solutions [3.29,40]. When the integral (3.11) is replaced with the system of equations (3.13) or (3.14), the lack of correctness of the initial integral equation manifests itself in that the accuracy of the obtained values of n_i is considerably lower than that of the measured values of k_ℓ and diminishes with an increasing number of equations. The magnitude of the errors $\delta(\Delta n_i)$ depends on the method of calculation, the errors of the initial data δk_ℓ, the form of the function $\Delta n(r)$, and the number of zones m into which the distribution was divided.

If an experimental plot of $k(y)$ has a great number of points, then the accuracy of the smooth curve drawn through them is appreciably higher than the accuracy of the discrete values of the function k_ℓ. It is therefore customary in finding the radial distributions to use, not the values of k_ℓ directly determined in an experiment, but the values determined from the smoothed curve. Sometimes the smoothing is included in the calculation procedure, and sometimes it is carried out beforehand.

The simplest and clearest method of numerical solution of the Abel equation is stepwise approximation, which was described by PEARCE [3.28]. In this method, the object is divided into concentric annular zones of equal increment $\Delta r = R/m$, where m is the number of zones (Fig.3.23). Within each of these zones, the required quantity $\Delta n(r)$ is considered to be constant. The object is also divided into a number of zones by lines parallel to the z-axis. When observing along the straight line passing through the center of the ℓ-th zone, the quantity being measured can be represented as the sum of the local values of n_i from $i = \ell$ to $i = m$ with definite coefficients $a_{\ell i}$:

$$k_\ell = \frac{R}{\lambda m} \sum_{i=\ell}^{m} a_{\ell i} \Delta n_i \quad . \tag{3.15}$$

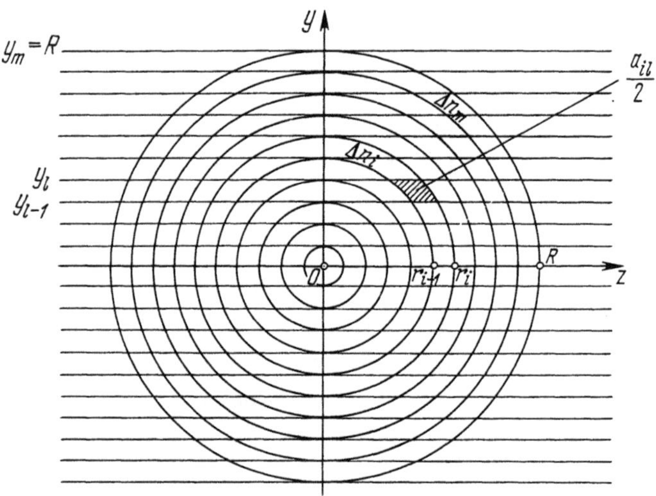

Fig.3.23. Calculation of radial distribution by the Pearce method

The values of $a_{\ell i}$ are proportional to the areas of the elements bounded by two arcs of radii r_i and r_{i-1} and two straight lines with coordinates y_ℓ and $y_{\ell-1}$. The values of $a_{\ell i}$ were tabulated by PEARCE [3.28]. The task of finding Δn_i consists of solving the system of linear equations

$$\left.\begin{aligned} a_{11}\Delta n_1 + a_{12}\Delta n_2 + \dots + a_{1m}\Delta n_m &= \frac{\lambda m}{R} k_1 \quad , \\ a_{22}\Delta n_2 + \dots + a_{2m}\Delta n_m &= \frac{\lambda m}{R} k_2 \quad , \\ \dots\dots\dots\dots\dots\dots\dots\dots & \\ a_{mm}\Delta n_m &= \frac{\lambda m}{R} k_m \quad . \end{aligned}\right\} \tag{3.16}$$

The solution should be begun from the last equation. The value of Δn_m obtained is introduced into equation No. m-1, and Δn_{m-1} is found, and so on.

The accuracy of the Pearce method was compared with other methods of the numerical solution of (3.11) [3.41]. It was shown that when the appropriate smoothing procedure is used, the accuracy of the Pearce method is not inferior to that of other considerably more intricate methods.

3.1.9 The Three-Dimensional Case

In the cases of two-dimensional and axisymmetric distributions, considered above, it is sufficient to have only one interferogram, obtained with light passed through the object in one direction, to determine the spatial distribution of the refractive index. In the general case of a three-dimensional distribution, light has to be passed through the object in many directions to determine the local values of the refractive index. In conventional interferometry, this requires simultaneous investigation of the object in several interferometers oriented at an angle to one another, or, when stationary objects are being studied, sequential recording of a number of interferograms with different orientation of the object.

Holography makes it possible to record waves that have passed through an object in many directions within the solid angle subtended by the hologram. In this respect, it is very convenient for interferometric investigation of three-dimensional distributions.

A number of authors have described the investigations of spatial distributions of refractive index of three-dimensional objects [3.42-49]. The most detailed and consistent treatment of various methods has been given by SWEENEY and VEST [3.48].

Let us consider the three-dimensional field of refractive index $n(x,y,z)$. We shall determine the optical-path difference for rays transmitted in the direction S_i as

$$\phi_i = \int_{S_i} \Delta n(x,y,z) dS_i \quad , \tag{3.17}$$

where $\Delta n(x,y,z) = n(x,y,z) - n_0$. The quantity ϕ_i is related to the phase shift $\Delta\varphi_i$ and to the displacement of the interference fringes k_i by

$$\phi_i = \frac{\lambda}{2\pi} \Delta\varphi_i = \lambda k_i \quad .$$

Let us consider the intersection of the field $\Delta n(x,y,z)$ with a plane perpendicular to the z axis. The plane contains a transmitted ray. In this case, z will be a parameter, and the three-dimensional case will be reduced to a two-dimensional one.

Disregarding, as previously, the refraction of the transmitted rays in the phase inhomogeneity being studied and transforming to polar coordinates in the plane x, y, we have

$$\int_S \Delta n(x,y)dS = \phi(\rho,\theta) \quad , \tag{3.18}$$

where ρ is the least distance from the transmitted ray to the origin of coordinates, and θ is the angle formed by this ray with the x axis (Fig. 3.24).

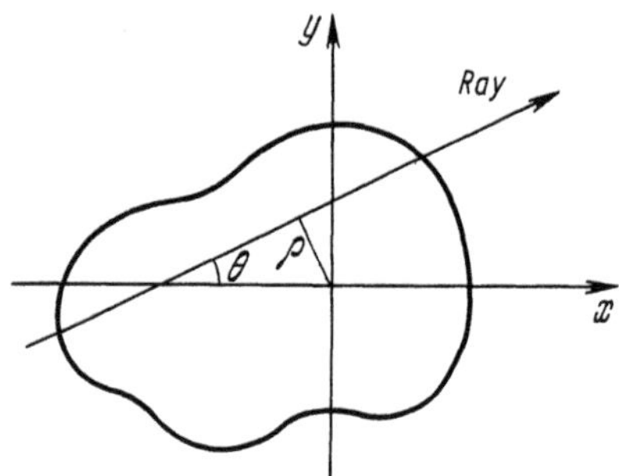

Fig.3.24. Calculation of the three-dimensional distribution of refractive index

It can be shown [3.48] that the integral (3.18) has an exact solution

$$\Delta n(r,\alpha) = \frac{1}{2\pi^2} \int_{-\pi/2}^{+\pi/2} d\theta \int_{-\infty}^{\infty} \frac{\partial\phi(\rho,\theta)/\partial\rho}{r\sin(\alpha - \theta) - \rho} d\rho \quad , \tag{3.19}$$

where r and α are the polar coordinates of the point at which the refractive index is being determined.

This direct-inversion equation makes it possible to determine the field of refractive index of an object directly from data on the path differences measured by transillumination of the object at different angles out to 180°.

Experimental data on the optical-path differences, however, are given not in the form of a continuous function $\phi(\rho,\theta)$, but in the form of a discrete set of quantities measured in interferograms. SWEENEY and VEST [3.48] give variants of the numerical solution of (3.19) for discrete input data. The analysis of those authors shows that the method of direct inversion is the best if it is possible to transilluminate an object out to an angle of 180°.

As a rule, it is not possible to transilluminate an object within such a wide solid angle in a real holographic experiment. In this case, other methods of processing the experimental data can be used that are based on the expansion of the field $\Delta n(x,y)$ into a series.

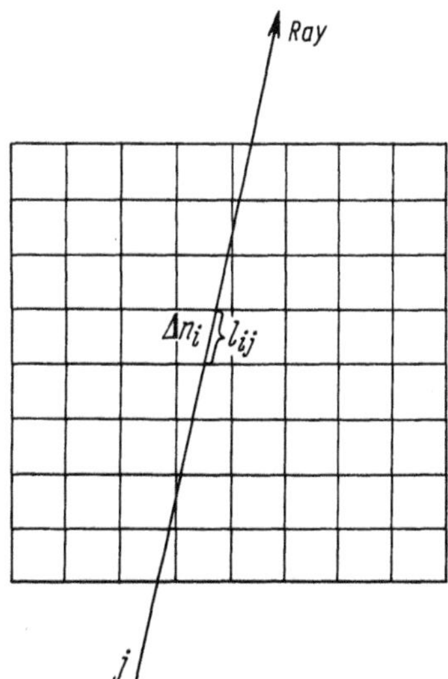

Fig.3.25. Calculation of three-dimensional distribution of refractive index by the method of stepwise approximation

Let us consider, as an example, the very illustrative method of stepwise approximation (similar to Pearce's method for axisymmetric objects) [3.48,49]. In this method, the field $\Delta n(x,y)$ is divided into a number of rectangular elements (Fig.3.25); within the limits of each such element, the quantity $\Delta n(x,y)$ is considered to be constant and equal to Δn_i. If the length ℓ_{ij} of the j-th ray is within the limits of the i-th element, then its relative contribution to the optical path difference is $\Delta n_i \ell_{ij}$, and the total path difference for the j-th ray will be

$$\phi_j = \sum_i \ell_{ij} \Delta n_i \quad , \tag{3.20}$$

summation being carried out over all of the elements through which the j-th ray passes. The coefficients ℓ_{ij} can be determined from geometrical considerations [3.48].

If the number of independent equations of type (3.20) equals the number of elements into which the field is divided, then, in principle, the system (3.20) can be solved for the quantities Δn_i. However, the stepwise approximation of the smooth function introduces appreciable errors, and the experimental values of ϕ_j are measured with limited accuracy. Consequently, the equations may be inconsistent. In addition, it is difficult to choose directions of the transmitted rays that give an independent system of equations. Therefore, in practice, the number of transmitted rays must be greater than the number of elements of the refractive index field.

The task thus consists of solving an inconsistent, overdetermined system of algebraic equations. The set of the quantities Δn_i that are an approximate solution of such a system can be found from the condition of achieving a minimum of the quantity

$$\sum_j |\phi_j - \sum_i \ell_{ij} \Delta n_i|^2 \quad .$$

The errors of measurements and the errors of approximation are all averaged when such systems are solved. A detailed analysis of this method is given in [3.49].

Apart from the stepwise method, other kinds of approximation of the required refractive index field can be used. For example, SWEENEY and VEST [3.47,48] described the expansion of the function $\Delta n(x,y)$ into a series according to the sinc functions[5].

The same authors [3.48] compared different methods of determining three-dimensional distributions of refractive index. They treated matters associated with the accuracy of the calculation methods, the required rates of sampling the initial data, and also their required redundancy. They showed that the required redundancy increases as the angle within which light is transmitted through the object is decreased.

---

[5] By a sinc function is meant a function of the form $(\sin \pi x)/\pi x$.

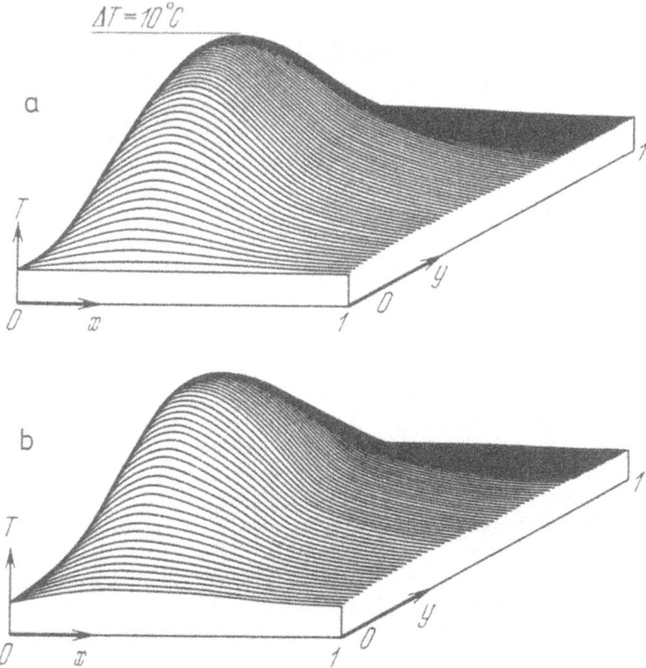

Fig.3.26. Preset (a) and reconstructed (b) temperature distributions [3.48]

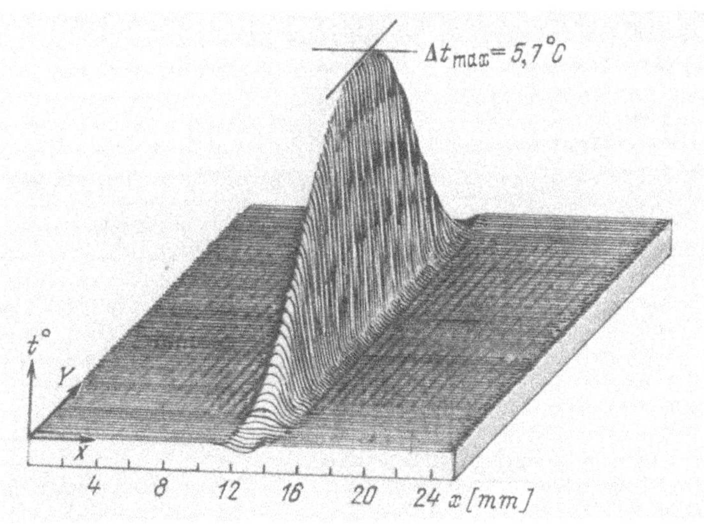

Fig.3.27. Distribution of air temperature over a heated horizontal wire, reconstructed from interferometric data [3.47]

SWEENEY and VEST also described numerical experiments that model interference measurements of the refractive index field associated with the distribution of temperature prescribed by a definite function. Figure 3.26 shows, as an example, the prescribed temperature distribution (a) and that reconstructed by the method of stepwise approximation (b). Figure 3.27 shows the experimental results.

SWEENEY and VEST [3.48] analyzed the accuracy of various methods without taking into consideration the finite error of the initial data. The stability of the solutions to the errors of an experiment, however, is of an appreciable significance (the same as for axisymmetric distributions reconstructed by numerical solution of the Abel equation).

3.2 Sensitivity of Holographic Interferometry and Methods of Changing It

Displacement of interference fringes is the quantity that is directly measured in the interferometry of phase objects – gas streams, plasma, flames, etc. The magnitude of this displacement is used to determine the phase shift introduced by the object. The latter, in turn, is associated with such of its characteristics as the distribution of temperature, concentration of atoms, electrons, etc. At this stage, however, we shall not deal with the nature of these diverse relations, but shall limit ourselves to considering the sensitivity of the method of holographic interferometry for the determination of the phase shift.

The number of fringes k by which the interference pattern is displaced is proportional to the phase advance $\Delta\varphi$ of the incident wave when it passes once through the object:

$$k = C\frac{\Delta\varphi}{2\pi} = \frac{C}{\lambda}\int_{z_1}^{z_2}(n - n_0)dz \quad . \tag{3.21}$$

Here C is a constant of proportionality that can characterize the <u>sensitivity</u> of the method. For twin-wave interferometry (both conventional and holographic), C = 1 [see (3.9)].

Thus, C is a coefficient of sensitivity of the method, which is the ratio of the sensitivity of a given method to that of twin-wave interferometry.

There is another definition of sensitivity of a method of measurement, in which the minimum value of a quantity measurable by the given method is called its sensitivity (see for example [3.50]). As applied to interferometry, this sensitivity, which we shall call the limiting sensitivity, is characterized by the minimum detectable value of $\Delta\varphi_{min}$.

The sensitivities defined in these ways are related to each other; their relationship is determined by the minimum-detectable displacement of the interference fringes k_{min}. From (3.21)

$$\Delta\varphi_{min} = 2\pi \frac{k_{min}}{C} \quad . \tag{3.22}$$

Thus, $\Delta\varphi_{min}$ can be reduced either by increasing the sensitivity coefficient C or by reducing the value of k_{min}, i.e., by improving the accuracy of measurement of the interference-fringe displacement.

In various practical problems, the sensitivity of conventional twin-wave holographic interferometry may be either too low or too great. For example, in gas-dynamical flows at low pressures, plasmas with a low electron concentration, etc., the fringe displacements on an interferogram may be smaller than k_{min}. On the other hand, if the gradient of the optical-path difference within the confines of the object being studied are too great, the interference fringes will be too close together on the hologram, and it will be impossible to trace or count them.

The sensitivity of interference measurements can be changed in various ways. Some of them can be applied in both conventional and holographic interferometry, whereas others are features of only holographic interferometry.

3.2.1 Sensitivity of Twin-Wave Interferometry

We have already mentioned that for twin-wave interferometry the sensitivity coefficient C = 1 and $\Delta\varphi_{min} = 2\pi k_{min}$. The quantity k_{min} is determined, on the one hand, by the method used to record the displacements of the interference fringes, and on the other by the quality of the fringes, namely, their visibility and noise level.

When the position of an interference fringe on an interferogram of medium quality is determined visually, it is generally assumed that $k_{min} = 0.1$, which corresponds to $\Delta\varphi_{min} = \pi/5$.

A greater accuracy can be achieved by photometric measurement of the fringes. The distribution of intensity in a twin-wave interference pattern has the form

$$\frac{I}{I_0} = \frac{1 + p \cos\Delta\varphi}{2} \quad . \tag{3.23}$$

Assuming that the visibility of the interference pattern p = 1, we have

$$\frac{I}{I_0} = \cos^2 \frac{\Delta\varphi}{2} \quad . \tag{3.24}$$

It follows from (3.24) that when $\Delta\varphi = \pi/5$, we have $I/I_0 = 0.9$, i.e., an error of 10 per cent in measuring the intensity corresponds to the value of $\Delta\varphi_{min} = \pi/5$.

Assuming that $I/I_0 = 0.99$, which corresponds to an accuracy of photometric measurement of 1 per cent, we get $\Delta\varphi_{min} = \pi/15(k_{min} = 1/30)$.

The above values which characterize the limiting sensitivity of twin-wave interferometry, do not take into consideration the influence on accuracy of reduced visibility of the interference pattern, noise, speckle structure of the image and other factors.

3.2.2. Multipass Setups

The sensitivity coefficient C can be increased by passing the beam many times through the object. This method is used both in conventional and in holographic interferometry.

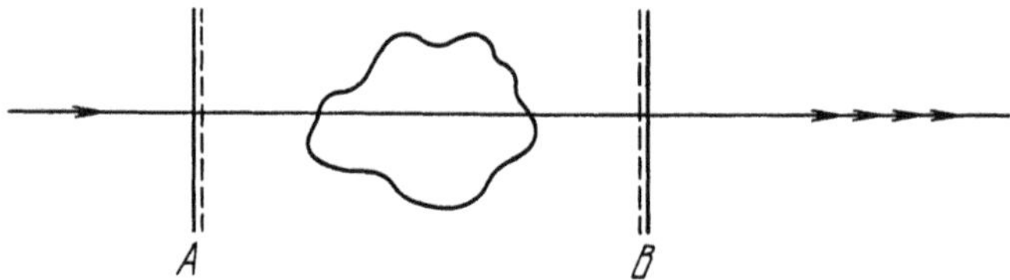

Fig.3.28. Multipass interference method

The inhomogeneity being studied is placed between two semitransparent mirrors A and B (Fig.3.28). A number of light beams that have passed through the object 1, 3, 5, etc. times are obtained at the output.

To separate the beam that corresponds to a given number of passes, in conventional interferometry, the mirrors A and B are inclined relative to each other through the small angle α. The consecutive beams will be inclined relative to one another by the angle 2α and are easily separated by a spatial filter (a lens with a diaphragm at its focal point). However, displacement of the beam each time it passes through the object is inevitable; and this detracts from the spatial resolution of the method.

Holographic interferometry makes it possible to circumvent this difficulty [3.50]. The mirrors A and B are arranged strictly parallel to each other and at right angles to the incident beam. If the distance between the mirrors exceeds half the coherent length of the laser radiation then, of all the beams that pass through the object a different number of times, only the one whose optical-path length differs from that of the reference beam by less than the coherent length of the laser radiation will participate in the formation of the hologram. Thus, the sensitivity of the method is multiplied by the number of times that the beam that forms the hologram passes through the inhomogeneity being studied.

The multipass holographic method was used to increase the sensitivity in studying the shock waves for ballistic trails at a reduced pressure [3.51]. Figure 3.29 shows the results obtained.

The multipass holographic method was also used by TSURUTA and ITOH [3.52] to increase the sensitivity in testing the mirrors of a Fabry-Perot interferometer.

3.2.3 Interference Patterns Formed by Conjugate Waves

The circumstance that waves of the first positive and first negative orders are conjugate, i.e., have opposite phase reliefs, can be used to increase the sensitivity of holographic interferometry. Superposition of two such conjugate waves produces an interference pattern in which the fringe displacements are double those that would be obtained if one of those waves interfered with a plane reference wave, i.e., the sensitivity coefficient, in this case, equals 2.

Such a method of increasing the sensitivity was proposed by DE and SEVIGNY [3.53]. The hologram is formed by interference of the object beam and two reference beams symmetrical to the object wave (Fig.3.30). When the hologram is illuminated by one of the reference beams, two conjugate

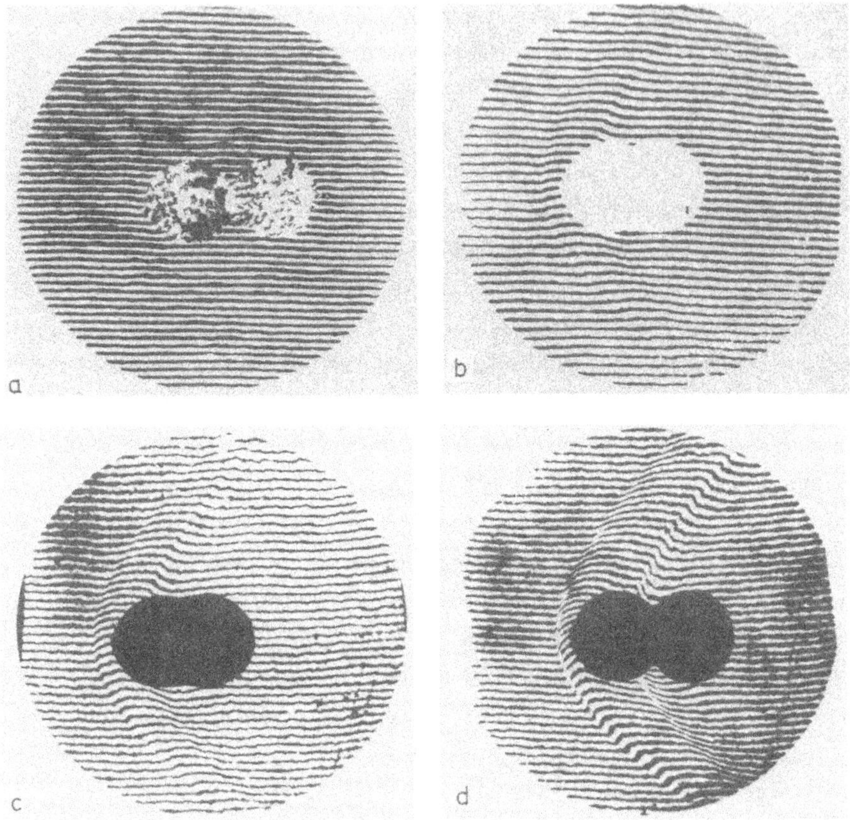

Fig.3.29. Interferograms of a shock wave at a gas pressure of 6 mmHg obtained in single (a), fivefold (b, c) and elevenfold (d) passage of the object beam through the object. The Interferograms were formed by interference of waves of the 0-th and +1st orders (a, b), and of the +1st and -1st orders (c, d)

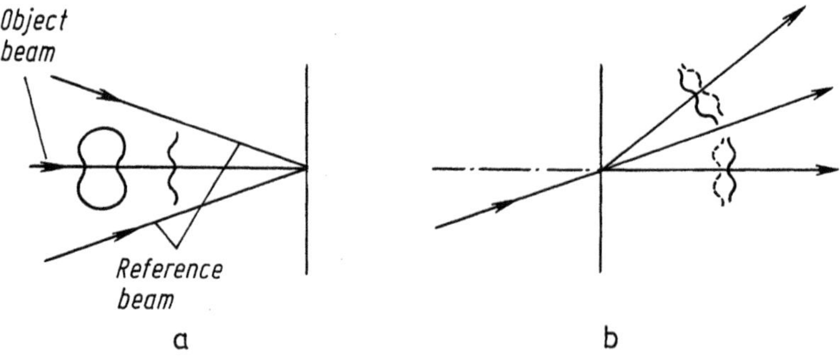

Fig.3.30. Interference of conjugate wavefronts

waves propagate in the direction of the normal to the hologram. They interfere and form a pattern of fringes of infinite width (unlocalized fringes). If the angles between the object beam and each of the reference beams were not strictly the same when the hologram was recorded, fringes of finite width will appear in the reconstruction stage. Their spatial frequency will be determined by the degree of asymmetry of the reference beams when they strike the hologram. Another result will be lateral displacement of the reconstructed images of the object, which vanishes only in image holograms.

A merit of the method is the possibility of obtaining an interference hologram with a single exposure. A shortcoming of the method is that the distortions introduced by imperfections of the optical elements of the setup are not compensated, but, on the contrary, are doubled, like the phase distortions being studied.

This method can be combined with the multipass method of increasing the sensitivity [3.50,51] (see Fig.3.29c,d).

3.2.4 Three-Wave Interferometry

The method of three-wave interferometry proposed by MUSTAPHIN and SELEZNEV [3.54], like that of DE and SEVIGNY [3.53], makes is possible to obtain a double sensitivity coefficient C. Unlike the latter method, however, it compensates the distortions introduced by imperfections of the optical elements.

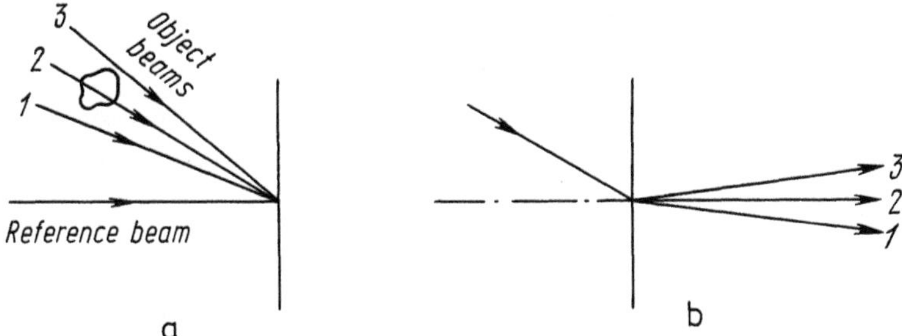

Fig.3.31. Three-wave interference method: (a) recording of a triple-exposure hologram; (b) reconstruction of the wavefronts

The method consists of triple exposure of a hologram; the inhomogeneity being studied is introduced into the object beam only during one of the exposures (2, Fig.3.31a). Beam 2 is directed approximately along the bisector of the angle between beams 1 and 3. Such a hologram reconstructs three wavefronts (Fig.3.31b), of which one is distorted by the object; the other two are reference wavefronts. All three reconstructed wavefronts are distorted to the same extent by imperfections of the optical elements.

The three-wave interference pattern obtained can be represented as the coherent superposition of three interferograms formed by the beams 1 and 2, 2 and 3, and 1 and 3. The first two patterns have an approximately identical frequency and oppositely directed fringe displacements. Each of these patterns is distorted to an equal extent by imperfections of the optical elements. When they are superimposed, a moire pattern is formed, with double fringe displacements and a carrier frequency corresponding to the difference of angles between the beams (1, 2) and (2, 3). The imperfections of the optical elements are cancelled out. Fringes of infinite width are observed when these angles are equal. A third pattern is formed by the beams 1 and 3; it is not disturbed by the inhomogeneity. The fringes of this system are superimposed onto the useful pattern; unfortunately, they reduce its visibility.

3.2.5 Two-Wavelength Methods of Changing the Sensitivity

The sensitivity of holographic interferometry can be increased or reduced by using a two-wavelength source to expose the hologram. If such a hologram is recorded, the reference beams of wavelengths λ_1 and λ_2 are directed at different angles to the surface of the hologram so that the spatial frequencies ν_1 and ν_2 [see (1.41)] corresponding to λ_1 and λ_2 are approximately the same, then the waves reconstructed by each of the recorded holograms will propagate in one direction and will interfere.

The single-exposure two-wavelength method of holographic interferometry was first proposed and employed for studying the relief of reflecting objects [3.55] (see Chap. 4). Use of a similar procedure was later proposed for study of transparent inhomogeneities [3.56].

The kind of pattern obtained by interference of reconstructed wavefronts, corresponding to the two wavelengths in the radiation, depends on how much the shapes of these wavefronts differ at the spot where the interference pattern is observed. These differences are determined, first, by the fact that the phase shift introduced into the wave by the object is inversely proportional to the wavelength,

$$\Delta\varphi = \frac{2\pi\Delta}{\lambda} \quad . \tag{3.25}$$

Here Δ is the optical-path difference introduced by the inhomogeneity. In addition, the dimensions and positions in space of the reconstructed images depend on the ratio of the wavelengths λ_1 and λ_2 of the light sources used in recording the hologram and on the wavelength λ_3 used in its reconstruction. This ratio is different for each of the reconstructed waves ($\lambda_1/\lambda_3 \neq \lambda_2/\lambda_3$).

When transparent inhomogeneities are studied, the optical-path difference

$$\Delta = \int_{z_1}^{z_2} (n - n_0)dz \tag{3.26}$$

may also depend on the wavelength, owing to the difference of the refractive index of the medium being studied for two wavelengths, i.e., owing to dispersion of the object. It is obvious that when the reliefs of reflecting surfaces are studied, this factor is of no significance.

Let us consider a setup for recording a two-wavelength hologram and reconstructing the wavefronts by use of the method proposed by WEIGL et al. [3.56]. We shall assume, for simplicity, that the object beam, in the absence of the inhomogeneity was normal to the photographic plate (Fig.3.32a), and that the reference beams reached the hologram at angles of α_1 and α_2. Hence, the spatial frequencies are

$$\nu_1 = \frac{\sin\alpha_1}{\lambda_1} \quad \text{and} \quad \nu_2 = \frac{\sin\alpha_2}{\lambda_2} \quad . \tag{3.27}$$

If the angles α_1 and α_2 comply with the condition that

$$\frac{\sin\alpha_1}{\lambda_1} = \frac{\sin\alpha_2}{\lambda_2} \tag{3.28}$$

then $\nu_1 = \nu_2 \equiv \nu$.

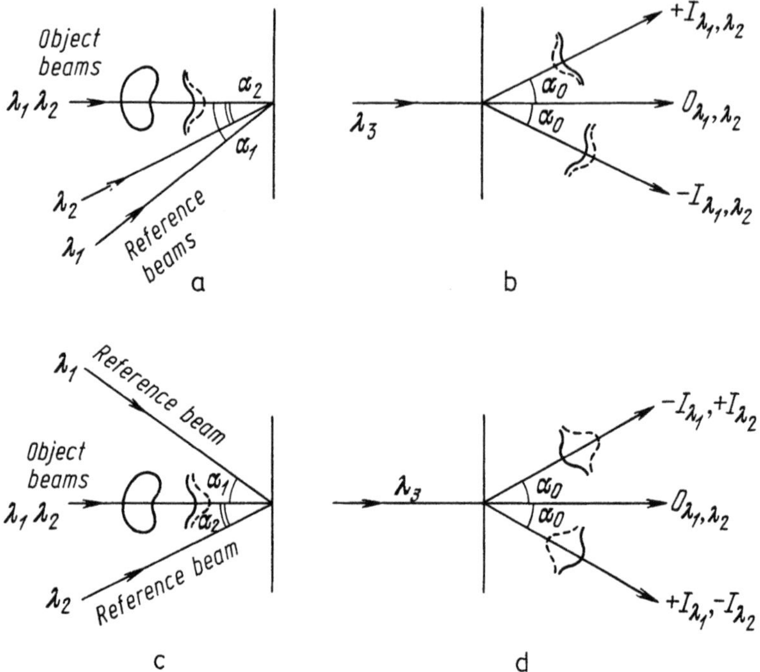

Fig.3.32. Diagram of two-wavelength interferometry with a reduced sensitivity (a, b) and with an increased sensitivity (c, d)

If we illuminate such a hologram with a beam of light of wavelength λ_3 normal to the surface of the hologram (Fig.3.32b), then the waves will be reconstructed that travel at angles $\pm\alpha_0$ that comply with the condition

$$\sin\alpha_0 = \nu\lambda_3 \quad . \tag{3.29}$$

Let us now consider the distortions introduced into the wave by a phase inhomogeneity. The phase advance for light with the wavelengths λ_1 and λ_2, according to (3,25,26), is

$$\Delta\varphi_1 = \frac{2\pi}{\lambda_1}\int_{z_1}^{z_2}(n_1 - n_0)dz, \qquad \Delta\varphi_2 = \frac{2\pi}{\lambda_2}\int_{z_1}^{z_2}(n_2 - n_0)dz \quad . \tag{3.30}$$

The displacement of the interference fringes on such an interferogram will evidently be

$$k_{2\lambda} = \frac{\Delta x_1 - \Delta x_2}{2\pi} , \tag{3.31}$$

where the subscript 2λ signifies "two-wavelength".

If we disregard the dispersion of the medium and assume that $n_1 = n_2 \equiv \equiv n$, then

$$k_{2\lambda} = \frac{\lambda_2 - \lambda_1}{\lambda_1\lambda_2} \int\limits_{z_1}^{z_2} (n - n_0)dz \quad . \tag{3.32}$$

Comparing this expression with (3.21) for $\lambda = \lambda_1$, we get $C = (\lambda_2 - \lambda_1)/\lambda_2$, where $\lambda_2 > \lambda_1$. Thus, by choosing the wavelengths λ_1 and λ_2 sufficiently close to each other, we can achieve a multifold reduction of the sensitivity of the method, in comparison with conventional twin-wave interferometry.

If, when a hologram is recorded, the reference beams of wavelengths λ_1 and λ_2 are directed at angles α_1 and α_2 that comply with condition (3.28), but at opposite sides of the normal (see Fig.3.32c), then the waves of the positive first order for λ_1 and the negative first oder for λ_2 will interfere in the reconstruction stage (Fig.3.32d). Equation (3.32) will thus become

$$k'_{2\lambda} = \frac{\lambda_1 + \lambda_2}{\lambda_1\lambda_2} \int\limits_{z_1}^{z_2} (n - n_0)dz \quad , \tag{3.33}$$

which corresponds to an increase of sensitivity, to $(\lambda_1 + \lambda_2)/\lambda_2$ times that for λ_1.

The greatest increase of sensitivity (two times) is achieved when $\lambda_1 = \lambda_2$, but this variant of two-wavelength interferometry does not differ from the method of DE and SEVIGNY [3.53], described above.

WEIGL used a setup with a diffraction grating (Fig.3.33) to utilize the two-wavelength method [3.57]. He placed in the focal plane of the lens L a spatial filter that separated the beams of the zero (λ_1, λ_2) and first

(λ_1, λ_2) diffraction orders to obtain reduced sensitivity. Alternatively, a spatial filter that separated the beams of zero (λ_1, λ_2), positive first (λ_1) and negative first (λ_2) orders could be used to increase the sensitivity. Condition (3.28) is automatically obeyed, because the spatial frequency of the hologram pattern is the same for both wavelengths and equals the frequency of the diffraction grating used. A similar arrangement was used earlier by LEITH and UPATNIEKS [3.58] to record achromatized holograms.

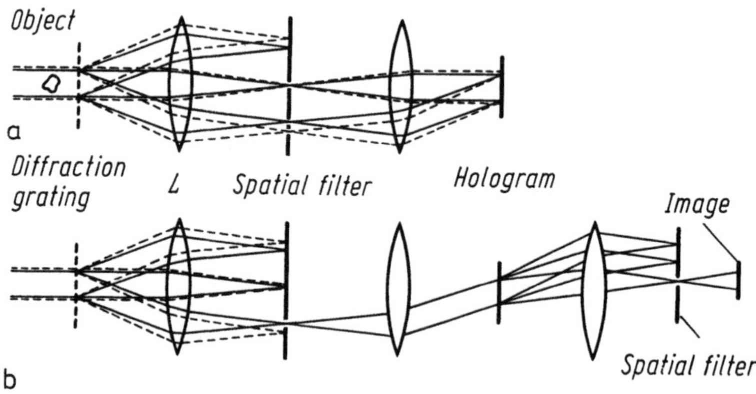

Fig.3.33. Arrangement for recording a two-wavelength hologram (a) and reconstruction of the wavefronts (b) by use of diffraction gratings (variant for reducing the sensitivity)

3.2.6 The Use of Nonlinear Effects

As we have already noted in Chap. 1, if a hologram is recorded on a nonlinear recording material then, apart from waves of the positive and negative first orders, higher-order waves will be reconstructed whose phase relief is m times that of the initial wave, where m is the number of the order.

Assume that two waves are incident on the surface of a photographic plate: the object wave $u = a(x,y)\exp[i\varphi(x,y)]$ and the reference wave $u_0 = a_0 \exp(i\varphi_0)$. We shall consider the amplitude a_0 and the phase φ_0 of the reference wave to be constant, which corresponds to a plane reference wave normal to the surface of the hologram (Fig.3.34a). We can write the phase of the object wave in the form

$$\varphi(x,y) = \frac{2\pi x}{\lambda} \sin\alpha + \Delta\varphi(x,y) \quad , \tag{3.34}$$

where the first term corresponds to a linear change of phase along the hologram, due to the inclination of the object wave, and $\Delta\varphi(x,y)$ is the phase advance when the object wave passes through the object.

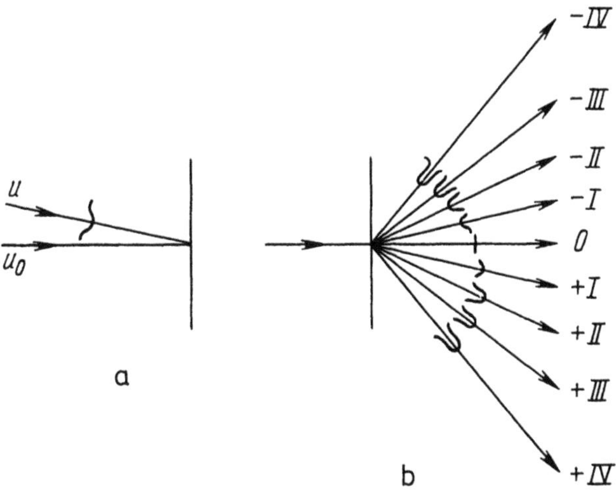

Fig.3.34. Diagram of hologram recording (a) and reconstruction of the wavefronts by a nonlinearly recorded hologram (b)

The distribution of the illumination on the hologram at $\varphi_0 = 0$ will be

$$E = a_0^2 + a^2 + a_0 a[e^{i\varphi} + e^{-i\varphi}] \quad . \tag{3.35}$$

We assumed on an earlier page that the amplitude transmittance of a hologram is a linear function of the exposure [see (1.115)]. The transmittance of a real plate is related to the exposure by an expression appreciably differing from (1.115). The transmittance is a complex function of exposure

$$\underline{T}(H) = T(H)e^{i\theta(H)} \quad . \tag{3.36}$$

The presence of the phase multiplier is associated with the fact that not only the blackening, but also the thickness and refractive index of the developed emulsion depend on the exposure (see Sect. 2.2). The nonlinear effects observed for phase holograms are due exactly to this multiplier. Here we shall limit ourselves to consideration of nonlinear effects only for purely amplitude holograms, i.e., we shall assume that $\theta(H) = 0$.

Examination of (3.35) shows that the illumination of a hologram is a periodic function of φ. Because the phase of the object wave usually changes along the hologram much more rapidly than its amplitude, the illumination and, consequently, the exposure in the plane of the hologram change sinusoidally. In linear recording, the transmittance of a hologram is also a sinusoidal function of the coordinates. When we pass beyond the linear region of the curve of T against H, the transmittance of a hologram remains a periodic function of φ, but the form of this function may differ appreciably from a sinusoid (Fig.3.35). This periodic function can be expanded into a Fourier series according to the cosines of φ [3.59]:

$$T = \frac{c_0}{2} + c_1 \cos\varphi + c_2 \cos 2\varphi + c_3 \cos 3\varphi + \dots$$

$$\dots = \tau_0 + \tau_1(e^{i\varphi} + e^{-i\varphi}) + \tau_2(e^{2i\varphi} + e^{-2i\varphi}) + $$

$$+ \tau_3(e^{3i\varphi} + e^{-3i\varphi}) \quad , \qquad (3.37)$$

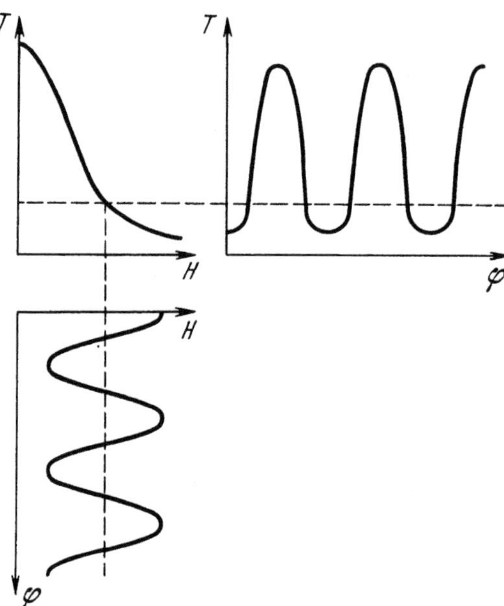

Fig.3.35. Distortion of sinusoidal distribution of exposure caused by the nonlinear nature of the curve T = T(H)

where the Fourier coefficients c_0, c_1, etc. are given by the formula

$$c_n = 2\tau_n = \frac{1}{\pi}\int_0^{2\pi} T[H(\varphi)]\cos(n\varphi)d\varphi \quad . \tag{3.38}$$

In the reconstruction of light waves by passing a plane monochromatic wave through a hologram, we carry out Fourier analysis of its spatial structure. Diffracted waves of not only the positive or negative first orders are formed, but also higher orders (see Fig.3.34b). The angles at which they propagate correspond to the harmonics fundamental frequency of the spatial structure of the hologram, whereas their intensities are proportional to the coefficients c_n.

Let us consider, for example, the reconstructed second-order wave that corresponds to the component of the amplitude transmittance $\tau_2\exp(2i\varphi)$. The phase of this wave is

$$2\varphi = \frac{4\pi x}{\lambda}\sin\alpha + 2\Delta\varphi(x,y) \quad , \tag{3.39}$$

which corresponds to its propagation at an angle whose sine is 2 times $\sin\alpha$. In addition, this wave has a phase relief twice as great as that of the initial wave. A similar treatment can be carried out for higher orders of diffraction.

An m-fold increase of the phase of the reconstructed wave in the m-th order makes it possible to use these waves to increase the sensitivity of holographic interferometry [3.60-62]. For this purpose, a nonlinearly recorded hologram with amplitude transmittance (3.37) is illuminated with two plane waves (Fig.3.36a), one of which strikes the hologram normally, and the other at the angle β whose sine is m times the sine of the angle α (i.e., the angle at which the object wave struck the hologram in recording,

$$v_1 = a_0 \quad \text{and} \quad v_2 = a_0\exp\left(i\frac{2\pi x}{\lambda}\sin\beta\right) = a_0\exp\left(i\frac{2\pi m x}{\lambda}\sin\alpha\right) \quad . \tag{3.40}$$

Consequently, two waves will propagate in the direction of the normal to the hologram – a plane wave of the zero order for v_1, i.e., a wave of complex amplitude $a_0\tau_0$, and a wave of the m-th order for v_2, i.e., $a_0\tau_m\exp(-im\varphi)\exp(\frac{i2\pi m x}{\lambda}\sin\alpha) = a_0\tau_m\exp(-im\Delta\varphi)$.

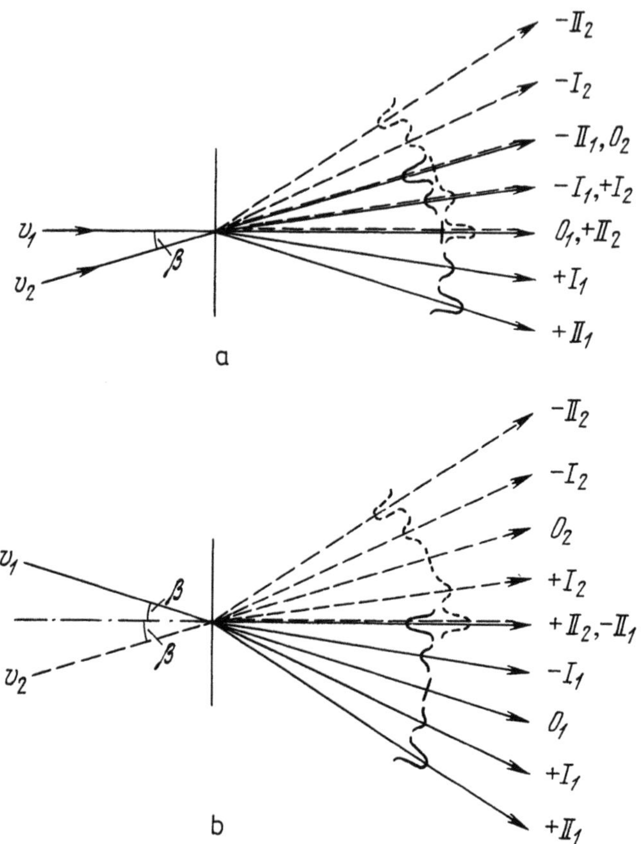

Fig.3.36. Use of higher diffraction orders to increase the sensitivity of holographic interferometry: (a) interference of wave of the 0-th and +2nd order; (b) interference of waves of the ±2nd orders

The interference of these reconstructed waves will result in the appearance of fringes whose displacement is proportional to $m\Delta\varphi$, i.e., m times greater than on a conventional interferogram.

If a hologram is illuminated with two waves symmetrical relative to the normal at the angles $\pm\beta$ (see Fig.3.36b), then reconstructed waves of the positive m-th order for one of these waves and of the negative m-th order for the other will propagate in the direction of the normal. Interference of these waves will produce fringes that correspond to a phase advance of $2m\Delta\varphi$. An example of increase of sensitivity by use of nonlinear effects is shown in Fig.3.37. MATSUMOTO and TAKASHIMA [3.62] achieved a fourteenfold increase of sensitivity of holographic interferometry in this way.

This method of increasing sensitivity amplifies not only the distortions of the wavefront introduced by the object, but also any distortions due to imperfections of the optical elements of the holographic setup. For this reason, holographic setups intended for increase of sensitivity are usually designed on the basis of interferometers. In addition, as we have noted earlier, the higher-order images in setups with diffusing screens have little in common with the object. Hence, a hologram must either be recorded in the plane of the image (an image hologram), or in a setup without a diffusing screen.

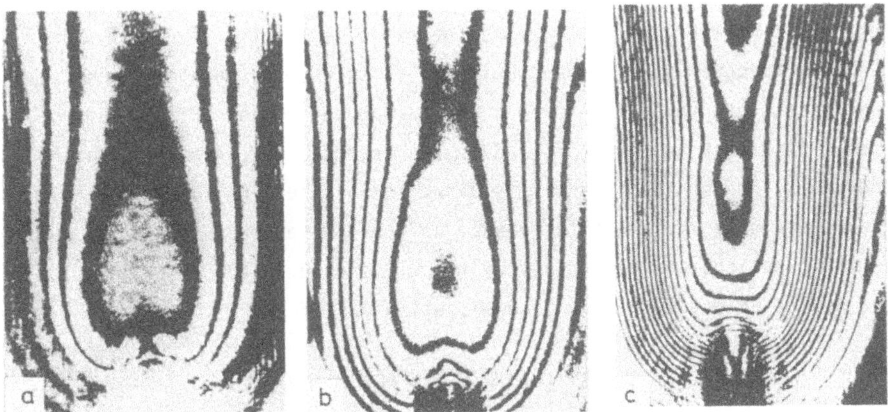

Fig.3.37. Interferograms of an alcohol-lamp flame obtained by use of the same hologram, by interference of waves of the 0-th and 1st orders (a), the ±1st orders (b), and the ±2nd orders (c) [3.61]

Holographic interferograms produced by the double-exposure method are known to be insensitive to the quality of the optical components of the holographic setup. But use of this method in reconstructing an interferogram in higher orders encounters a number of difficulties. The difficulties arise because waves of phases $\ell\varphi_1 + k\varphi_2$, where $\ell + k = m$ and φ_1 and φ_2 are the phases of the waves in the two exposures, also propagate in the direction of the reconstructed m-th order waves that correspond to the two waves that are recorded in two consecutive exposures. The superposition of these waves noticeably distorts the observed interference pattern. VELZEL [3.63] showed possible ways of surmounting these obstacles. He achieved a fourfold increase of sensitivity of measurement of the distribution of density of a gas inside a vessel with nonuniform walls by the method of double-exposure holographic interferometry with reconstruction of the interferograms in higher orders.

3.2.7 Dispersion Holographic Interferometry [3.64-66]

The name dispersion is applied to this method because the displacement of the fringes on an interferogram is determined not by the absolute value of the refractive index of the object [see (3.21)], but by the difference between the refractive indices for two wavelengths, i.e., by the dispersion of the object.

A hologram of the phase object is recorded with the aid of light consisting of two wavelengths that are in the ratio of exactly 2(λ_1, and $\lambda_2 = \lambda_1/2$). The hologram is recorded nonlinearly; therefore, not only waves of positive and negative first orders are formed in reconstruction, but also those of higher orders (Fig.3.38). The second-order wave that corresponds to λ_1 propagates in the same direction as the first-oder wave that corresponds to λ_2. The interference of these waves produces fringes whose configurations are determined by the difference in the refractive indices of the object for light of wavelengths λ_1 and λ_2.

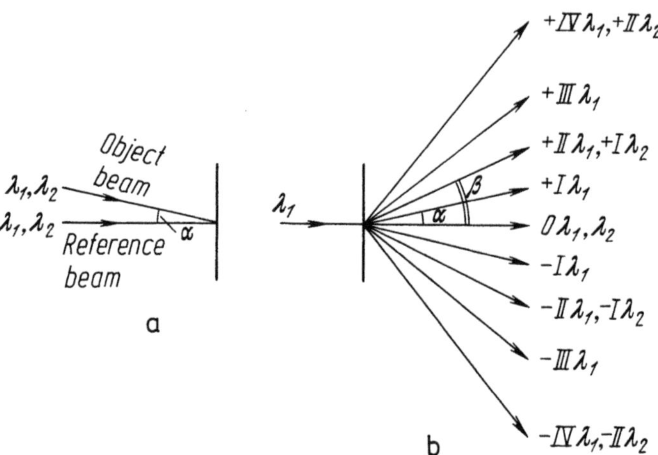

Fig.3.38. Recording a two-wavelength hologram (a) and using it to reconstruct the wavefronts (b)

The phase of the first-order wave for λ_2, according to (3.34), is

$$\varphi_{I,2} = \frac{2\pi x}{\lambda_2} \sin\alpha + \Delta\varphi_2 = \frac{4\pi x}{\lambda_1} \sin\alpha + \Delta\varphi_2 \quad . \tag{3.41}$$

The phase of the second-order wave for λ_1, in accordance with (3.39), is

$$\varphi_{II,1} = \frac{4\pi x}{\lambda_1} \sin\alpha + 2\Delta\varphi_1 \quad . \tag{3.42}$$

It follows from (3.41,42) that both waves propagate at the same angle $\beta = \arcsin 2 \sin\alpha$ and the fringe displacement on an interferogram formed by these waves equals

$$k_D = \frac{\varphi_{II,1} - \varphi_{I,2}}{2\pi} = \frac{2\Delta\varphi_1 - \Delta\varphi_2}{2\pi} \quad . \tag{3.43}$$

If we introduce the values of $\Delta\varphi_1$ and $\Delta\varphi_2$ from (3.30) into this expression and take into account that $\lambda_1 = 2\lambda_2$, we get

$$k_D = \frac{2}{\lambda_1} \int_{z_1}^{z_2} (n_1 - n_2) dz \quad . \tag{3.44}$$

In the absence of dispersion ($n_1 = n_2$), the fringe displacement is zero. Thus, the method is not sensitive to phase objects that have no dispersion of refractive index. The dispersion method is a single-exposure one. Therefore, no rigid stabilization of the holographic setup is needed when it is used.

In [3.56,57], two-wavelength holographic interferograms are also prepared during one exposure, but the reference beams with wavelengths of λ_1 and λ_2 are directed onto the hologram at different angles, in accordance with condition (3.28), which deprives this method of one of the main advantages of holographic interferometry — lack of sensitivity to the quality of the optical components of the holographic setup. In the dispersion method, both the reference and object beams with the wavelengths λ_1 and λ_2 travel along the same path and are distorted to an equal extent by the imperfections of the optical elements. The difference between the distortions of the beams of wavelengths λ_1 and λ_2 by the optical components is associated only with dispersion of the latter, and is completely absent when mirror optical components are used.

The dispersion method involves a number of technical difficulties and difficulties of principle. To obtain highly visible dispersion interferograms, the amplitude of the reconstructed second-order wave for λ_1 must be sufficiently great and close in value to that of the first-order wave for λ_2. The intensities of the higher-order waves depend appreciably on the nature of the nonlinear dependence of the transmittance of the photographic material on the exposure, and also on the conditions of the experiment [3.67,68].

In addition, there are a number of reasons why the fringes in dispersion interferograms are distorted. Particularly, distortions may be due to displacement or to a different scale of the second-order image for λ_1 and the first-order image for λ_2 depending on the chromatic aberration and the dispersion of the elements of the holographic setup. Such distortions can be avoided if setups are used in which only mirror optical components are employed to split the beams into object and reference beams and for broadening them.

Distortion of the second-order images due to nonlinear effects may also result in distortion of the fringes in dispersion interferograms. These distortions are encountered not only in the dispersion method, but also in the methods for increasing the sensitivity of holographic interferometry when high-order waves are used. We have already noted that these distortions are insignificant when image holograms are used.

The use of the dispersion method for the diagnostics of plasma is of special interest. The distribution of the concentration of the electrons in plasma can be determined from the fringe displacements in interferograms. There is no need to take into consideration the influence of the

Fig.3.39. Dispersion interferogram of a laser-induced spark

distribution of heavy particles (atoms and ions) because the refraction of the latter, far from their absorption lines, is practically independent of the wavelength. Figure 3.39 shows an interferogram of the plasma of a laser spark obtained by the dispersion method [3.65].

3.2.8 Method of Resonance Interferometry

Up to now, we have treated methods of changing the sensitivity of holographic interferometry to determination of phase shift. We have not considered in detail the nature of the relationship between the refractivity of a phase object and such of its parameters as temperature, pressure, or concentrations of atoms, ions, and electrons. The resonance method permits us to change the sensitivity of interference measurements of the concentration of atoms and ions, within broad limits. This is achieved by using the dependence of the refractive index on the wavelength in the vicinity of absorption lines. The refraction of atoms and ions near an absorption line having a dispersion contour is described [3.69] by

$$n - 1 = C'\lambda_0^3 Nf \frac{\lambda - \lambda_0}{(\lambda - \lambda_0)^2 + (\Delta\lambda/2)^2} , \quad (3.45)$$

where λ_0 is the wavelength that corresponds to the center of an absorption line, $\Delta\lambda$ is the full-width at half-maximum, λ is the wavelength of the incident radiation, f is the absorption oscillator strength, N is the concentration of the atoms at the absorbing level, and $C' = e^2/4\pi mc^2 = 2.24 \times$ $\times 10^{-14}$ cm.

The changes of refractivity (n - 1) and of absorption coefficient κ near the line are shown in Fig.3.40. When the absorption line is approached, the refractivity of the corresponding atoms increases sharply and may be several orders of magnitude greater than the refractivity of the same atoms far from the line. Therefore, the sensitivity of measurement of concentration of the atoms can be increased appreciably by preparing interferograms with light that has a wavelength close to their absorption line. This method can by employed in both conventional and holographic interferometry, and also in the schlieren and shadow methods of studying phase objects [3.70].

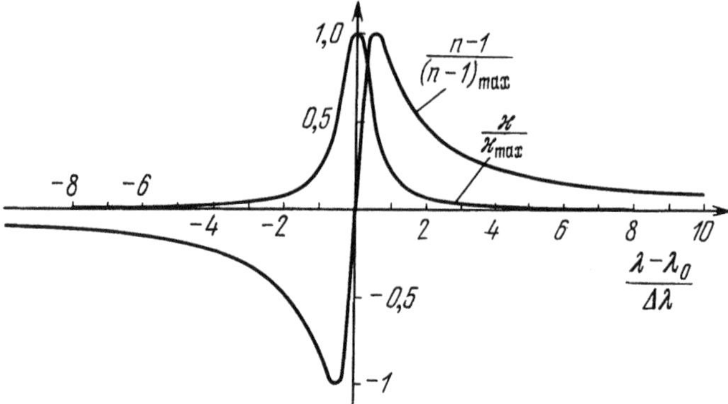

Fig.3.40. Change of the index of refraction and of the absorption coefficient near an absorption line

Preparation of an interference pattern with light of wavelength close to the absorption line of one of the components of the plasma being studied was first proposed in 1961 [3.71]. This method has been used to increase the sensitivity of measurements in both conventional [3.72-76] and holographic [3.76,77] interferometry.

The sensitivity, and the limits of application of the method, for an absorption line whose contour is determined by the combined action of dispersion and Doppler broadening are determined [3.78]. Those calculations were based on the assumption that the incident light is monochromatic. The sensitivity and the limits of application of the method for incident radiation of finite width are calculated in [3.79].

According to (3.45), the displacement of an interference fringe for a homogeneous layer with thickness ℓ is

$$|k| = \frac{(n-1)\ell}{\lambda} \approx C'\lambda_0^2 \ell N f \frac{|\lambda - \lambda_0|}{(\lambda - \lambda_0)^2 + (\Delta\lambda/2)^2} \quad . \tag{3.46}$$

Hence, the minimum detectable concentration of atoms is

$$N_{min} \approx \frac{|k_{min}|}{C'\lambda_0^2 \ell f} \frac{(\lambda - \lambda_0)^2 + (\Delta\lambda/2)^2}{|\lambda - \lambda_0|} \quad , \tag{3.47}$$

where k_{min} is the minimum detectable fringe displacement.

We can vary the limiting sensitivity of the method within broad limits by changing the separation $\lambda - \lambda_0$ between the incident line and the absorption line. The maximum sensitivity is reached when $\lambda - \lambda_0 = \pm\Delta\lambda/2$. Here

$$N_{min} \approx \frac{|k_{min}|\Delta\lambda}{C'\lambda_0^2 \ell f} \quad . \tag{3.48}$$

Assuming that $\Delta\lambda \approx 1$ Å, $k_{min} = 0.1$, $\lambda_0 = 6000$ Å, $f = 1$, and $\ell = 10$ cm, we have $N^{min} \approx 10^{12} cm^{-3}$.

The maximum concentration of atoms that can be measured by the resonance method depends on the width of the probing line. If the width of the incident line is comparable with that of the absorption line, then the fringe displacements that correspond to different parts of the probing line are different. This causes reduction of the visibility of the interference pattern [3.79].

The holographic variant of the method of resonance interferometry was first used in [3.76,77] to study the plasma of an arc that contained potassium vapor and to study a laser-produced plasma on a potassium target. The stimulated Raman scattering in nitrobenzene (λ = 7658 Å) of light from a ruby laser was used as the incident radiation. It is seven

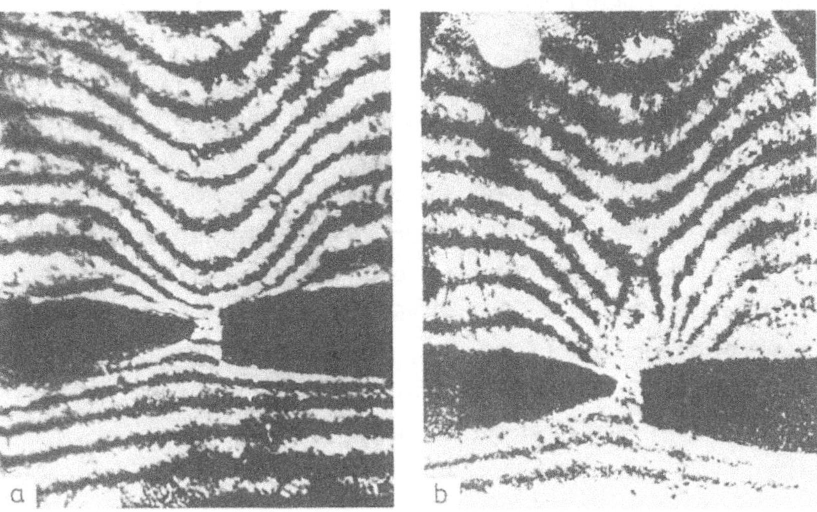

Fig.3.41. Holographic interferograms of plasma of a dc carbon arc with potassium core: (a) λ = 6943 Å; (b) λ = 7658 Å

angstroms from the resonance line of potassium (λ_0 = 7665 Å). Figure 3.41 shows holographic interferograms of an arc, obtained in [3.76].

The use of dye lasers that make it possible to change the wavelength of the incident line continuously, within broad limits, opens up great possibilities for resonance interferometry. It extends the range of objects that can be subjected to resonance interferometry and permits us to approach practically any absorption line in the region from 3500 to 10 000 Å. These lasers are already being used to obtain resonance interferograms in a Mach-Zehnder interferometer [3.71-73].

3.3 Holographic Diagnostics of Plasma

This and the following sections are devoted to applications of the methods of holographic interferometry, set out above, to plasma and gas-dynamic investigations.

The possibility of using holography for diagnostics of plasma was first shown by OSTROVSKAYA and OSTROVSKY [3.80] in 1966. Beginning in 1967, similar methods were introduced for study of plasmas in the USA [3.18,81]. A consideration of the new possibilities of this method and a review of the main results of its application are given in [3.15,82-84]. At this writing, over 100 papers have been published on the holographic diagnostics of plasma carried out both in the Soviet Union and in other countries. They include studies that show the possibilities of various diagnostics procedures, and the results of holographic studies of plasma objects.

In connection with the comparatively low sensitivity of holographic (and of conventional) interferometry, holographic methods are used mainly to study dense plasmas: the plasma produced by the interaction of pulsed laser radiation with gases [3.17,80-95] and solid targets [3.96-99, 77], the plasmas of flash lamps [3.100-102], plasmatrons [3.103-106], electric discharges [3.107-110], exploding wires [3.3,111], z and θ pinches [3.18,81,112-115], and of a neutral current layer [3.116], etc.

3.3.1 Features and Possibilities of Holographic Plasma Diagnostics Methods

A light wave is subjected to both phase and amplitude distortions when it passes through plasma. A hologram makes it possible to record and then reconstruct a light wave that has passed through plasma. The wave can be studied by different optical methods: interference, schlieren, shadow, etc. As in conventional interferometry and in the schlieren and shadow methods, the refractive index of the plasma, or its first and second derivatives, is what is measured. We can determine the coefficient of absorption of the plasma by studying the amplitude distortions it introduces into a wave.

We can obtain information on the parameters of plasma if we know how they are related to the refractive index and the absorption coefficient. In this sense, the holographic method gives nothing new, in principle, in comparison with the conventional optical method. Nevertheless, holographic methods often have significant advantages and provide a number of new possibilities for the diagnostics of plasmas, many of which have been considered earlier:

1) When a wave that has passed through a pulsed plasma has been recorded on a hologram, it can be studied in stationary conditions by several methods.

2) If the plasma was illuminated within a broad solid angle when the hologram was recorded, then a single hologram reconstructs the light waves that have passed through the plasma in different directions, which in principle makes it possible to obtain a spatial distribution of the refractive index in the absence of axial symmetry.

3) The differential nature of holographic interferometry decreases the requirements on quality of optical components of an interferometer. This makes possible the study of plasmas confined in vessels with inhomogeneous walls and interference studies of plasmas of practically unlimited dimensions.

4) Holography opens up broad possibilities for both a general increase of the sensitivity of interference methods and for selective increase of sensitivity for a specific kind of particle (see Sect. 3.2).

5) A hologram records not only the phase distortions introduced by the plasma, but also its effects on amplitude. The absorption coefficient of

the plasma can be determined by study of those effects [3.117]. An advantage of the holographic method over conventional absorption methods is its insensitivity to the intrinsic radiation of the plasma.

3.3.2 Refraction of Plasma

A plasma is a mixture of a large number of various particles: Electrons, atoms, and ions in their ground and excited states; low-temperature plasmas also contain molecules. The contributions of different kinds of particles to the refractivity of plasma can be considered as additive, i.e.,

$$n - 1 = \sum_k c_k N_k \quad , \tag{3.49}$$

where c_k is the refractivity of particles of the k-th species per particle, and N_k is the number of the relevant particles in a unit volume.

The contribution of electrons to the refractivity of a plasma can be found from the expression [3.118]

$$n_e - 1 = -\frac{1}{2}\frac{\omega_p^2}{\omega^2} = -\frac{e^2\lambda^2 N_e}{2\pi mc^2} = -4.49 \times 10^{-14}\lambda^2 N_e \tag{3.50}$$

where $\omega = 2\pi c/\lambda$ is the angular frequency of radiation, $\omega_p = \sqrt{4\pi e^2 N_e/m}$ is the electron plasma frequency, and N_e is the concentration of electrons.

Formula (3.50) holds for the case when the frequency of the light is much greater than that of the collisions between electrons and heavy particles, and also plasma frequency and electron-cyclotron frequency.

The contribution of atoms and ions in the normal and excited states to the refraction of plasma is given [3.119,120] by

$$(n - 1)_\alpha = \frac{2\pi e^2}{m} \sum_{\substack{i,k \\ i \neq k}} \frac{f_{ik} N_i}{\omega_{ik}^2 - \omega^2} \quad . \tag{3.51}$$

Here N_i is the concentration of the atoms or ions at the i-th level, ω_{ik} is the frequency and f_{ik} the absorption oscillator strength that corresponds to the transition between the i-th and k-th levels. The subscript α signifies the stage of ionization.

If a plasma contains atoms and ions in different states of ionization, then to determine the total contribution of all of the heavy particles to refraction, $(n - 1)_\alpha$ must be summed over all possible values of α.

Changing over to wavelengths, we can write formula (3.51) in the form

$$(n - 1)_\alpha = \frac{2\pi e^2}{mc^2} \sum_{\substack{i,k \\ i \neq k}} \frac{f_{ik}\lambda_{ik}^2 N_i}{\lambda^2 - \lambda_{ik}^2} \quad . \tag{3.52}$$

For gases at comparatively low temperatures, neutral atoms in the ground state give the main contribution to refractivity. In addition, the resonance lines of most gases are in the vacuum ultraviolet region. Therefore, the formula (3.52) for the refractivity of gases $(n - 1)_a$ for radiation in the visible region ($\lambda >> \lambda_{1k}$) can be written in the form

$$(n - 1)_a = \left[A + \frac{B}{\lambda^2} \right] \frac{N_a}{N_L} \quad . \tag{3.53}$$

Here N_L is the Loschmidt number, and N_a is the concentration of the atoms in their normal state. The latter expression is known as the Cauchy formula. For the visible region of the spectrum, we usually have $B/\lambda^2 << A$, and the refractivity of the atoms and molecules is practically independent of wavelength.

Formulas (3.51,52) hold when the distance between the absorption lines λ_{ik} and the wavelength of the incident radiation λ is considerably greater than the width of the corresponding absorption line. Near a single absorption line, the refractivity of atoms and ions is described by expression (3.45).

Thus, the refractivity of plasma in the general case is the sum of the refractivities described by formulas such as (3.50, 53, 45). In going from the measured value of the refractive index to the parameters of plasma, we must either have additional information on the plasma that permits us to assess the contribution of various kinds of particles to the total refractivity, or we must measure the refractivity at several wavelengths [3.121,122]. In this case, taking advantage of the different wavelength dependences of the refractivity described by (3.50,53,45), we can separate the contribution to the refractivity introduced by particles of various kinds, and thus determine the concentration of the various particles in the plasma.

Using (3.50,53,45), we can assess the sensitivity of holographic interferometric diagnostics for determination of the concentrations of various particles. In accordance with (3.50), an electron concentration of $N_{e,min} = 5 \times 10^{16} cm^{-3}$ corresponds to a change of optical path by $\lambda/10$ (for $\lambda = 0.5$ μm and a column length of 1 cm). The minimum atom concentration, in accordance with the values of the coefficients in the Cauchy formula (3.53) is about one or two orders of magnitude greater than $N_{e,min}$.

The sensitivity for determining the electron concentration can be appreciably increased by increasing the wavelength of the incident ratiation. Particularly, when transferring from a ruby laser ($\lambda = 0.69$ μm) to a carbon dioxide laser ($\lambda = 10.6$ μm), the sensitivity increases by about 15 times. Notwithstanding difficulties with material for recording holograms, a number of attempts have been made for holographic diagnostics of plasmas in the infrared region of the spectrum [3.123-125].

As we have shown in the preceding section, the sensitivity of the interference method for determination of concentrations of atoms and ions can be appreciably increased by use of resonance methods.

3.3.3 Holographic Investigation of a Laser-Induced Spark. Cineholography

A laser-induced spark is the plasma formed when powerful laser radiation is focused in gases. The diagnostics of the plasma of a laser-induced spark is a very intricate task. A high-pressure region is formed as a result of the release of considerable energy in a small volume during a time of the order of 10^{-8}s. The hydrodynamical dispersion of the plasma results in rapid changes of its parameters in time and space. Therefore, special methods of diagnostics that have high temporal and spatial resolution have to be developed to study laser-induced sparks. Methods of holographic interferometry often meet these requirements. A laser-induced spark was the first object studied by these methods [3.17].

Figure 3.42 shows a cineholographic setup with which a number of holograms of a laser-induced spark can be recorded during one of its flashes. A laser-induced spark was produced with the aid of a monopulsed ruby laser 1 ($E \approx 0.75$ J, $\tau \approx 30$ ns) Q switched by use of a rotating prism. The laser spark was formed at the focal point of the lens 2($f = 30$ mm). A part of

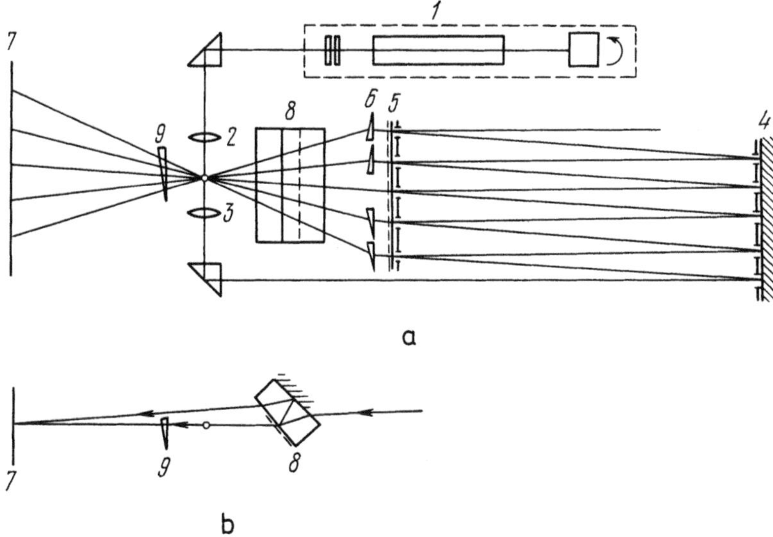

Fig.3.42. Setup for the cineholographic investigation of a laser-induced spark: (a) top view; (b) side view facing the beam-splitting wedge

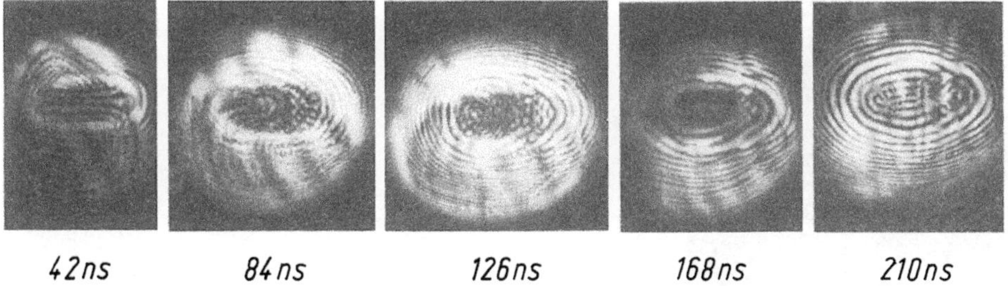

Fig.3.43. Five-frame single-ray cinehologram of a lser-induced spark

the laser pulse that passed through the focal volume before a spark appeared ($E \approx 0.2$ J, $\tau \approx 15$ Ns) was collimated by the lens 3 and directed to the optical delay line formed by the two parallel mirrors 4 and 5 spaced at a distance of $\ell \approx 6$ m.

The light beams that leave semitransparent mirror 5 were delayed with respect to each other by $2\ell/c \approx 40$ ns. The wedges 6 directed these beams onto the laser-induced spark; those beams exposed five single-ray (Gabor) holograms of the spark, corresponding to the consecutive stages of its development, on the film 7 (Fig.3.43). A beam-splitting wedge 8 was introduced into the setup to form double-ray holograms. The wedge divided each of the time-delayed beams into an object beam that passed through the laser-induced spark, and a reference beam (see Fig.3.42b). The object and reference beams were made to coincide in the plane of the photographic film 7. To obtain double-exposure interferograms with fringes of finite width, the wedge 9 was introduced into the object beams during one of the exposures.

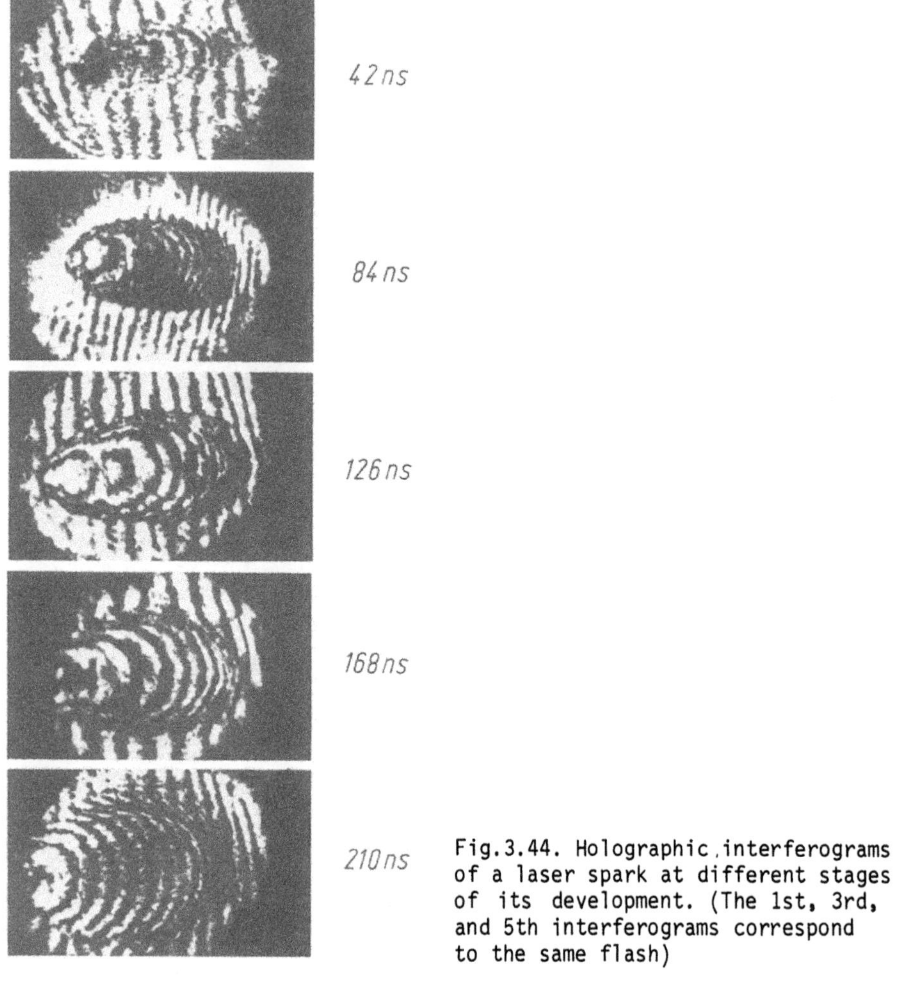

Fig.3.44. Holographic interferograms of a laser spark at different stages of its development. (The 1st, 3rd, and 5th interferograms correspond to the same flash)

Figure 3.44 shows holographic interferograms of a laser-induced spark, obtained in this setup. The fringe displacements on the interferograms of a laser-induced spark are due on the one hand to the appearance of electrons and on the other to the considerable reduction of the density of the gas in the central regions which is associated with the formation of a shock wave. The interferograms obtained in [3.88] do not make it possible to separate the action of these causes; the electron concentration was calculated on the assumption of complete displacement of the gas from the volume of the spark into the shock wave.

Later, radiation of two wavelengths - of a ruby laser and its second harmonic - was used to separate the contributions of the heavy particles and the electrons to the refractivity of the laser-induced plasma [3.89-92].

3.3.4 Two-Wavelength Holographic Interferometry of a Laser-Induced Spark

The setup used is shown in Fig.3.45. A laser-induced spark was formed in a chamber filled with various gases. A potassium dihydrogen phosphate (KDP) crystal was used to produce the second harmonic. Holograms that correspond to different moments after the beginning of a spark were obtained by changing the distance between mirror 5 and the chamber.

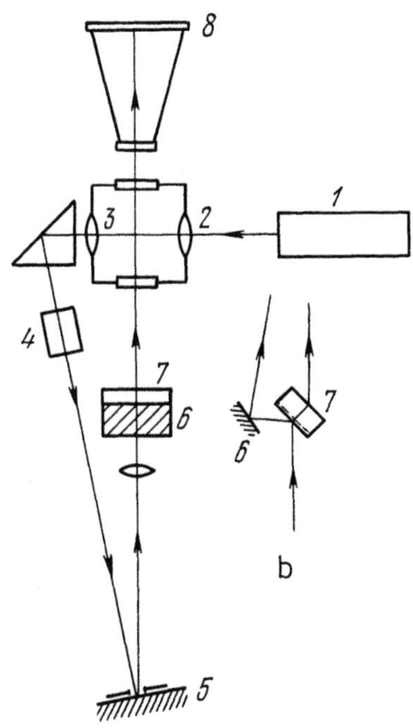

Fig.3.45. Arrangement for studying a laser-induced spark by the method of two-wavelength holographic interferometry (a); side view of the beam splitter (b); 1 - ruby laser; 2 - lens that focuses the radiation into a gas-filled chamber; 3 - collimating lens; 4 - KDP crystal; 5 - mirror of optical delay line; 6, 7 - mirrors that divide the light beam into object and reference beams; 8 - hologram

Interferograms of a laser-induced spark in helium, reconstructed by use of one two-wavelength double-exposure hologram are shown in Fig.3.46. The spatial distribution of the electrons and the gas density in a laser-induced spark were studied as a result of the joint processing of the "blue" (λ = 3472 Å) and the "red" (λ = 6943 Å) interferograms.

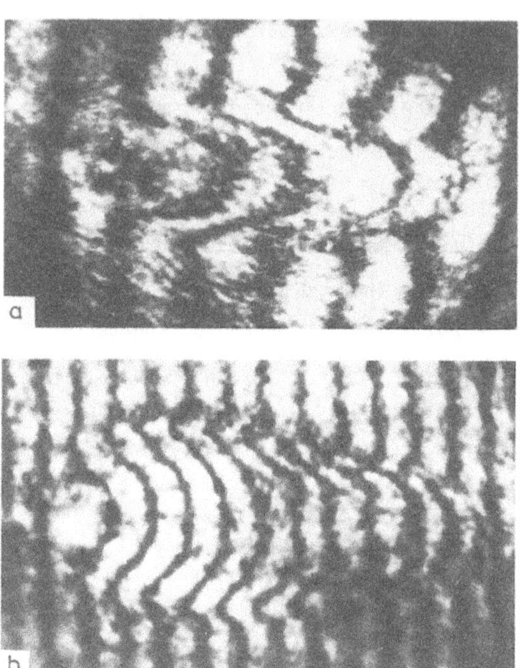

Fig.3.46. Interferograms of a laser-induced spark in helium at a pressure of 6 atm: (a) λ = 6943 Å; (b) λ = 3472 Å

The displacements of the interference fringes along the axis of a laser-induced spark for λ_1 and λ_2, in accordance with (3.50,53) can be written in the form

$$k_1 = k_{a1} - 4.49 \times 10^{-14}\lambda_1 \int_0^{2R} N_e dx$$

$$k_2 = k_{a2} - 4.49 \times 10^{-14}\lambda_2 \int_0^{2R} N_e dx \quad , \tag{3.54}$$

where

$$k_{a1} = \frac{(A + B/\lambda_1^2)}{N_L\lambda_1} \int_0^{2R} (N_a - N_0)dx$$

and

$$k_{a2} = \frac{(A + B/\lambda_2^2)}{N_L\lambda_2} \int_0^{2R} (N_a - N_0)dx$$

are the fringe displacements due to redistribution of the initial gas, mainly associated with the formation of a shock wave; $(N_a - N_0)$ is the corresponding change of the concentration of the atoms, and R is the radius of the spark.

Ignoring the dispersion of the refractive index of the gas, i.e., considering that $A >> B/\lambda^2$, and taking into account that $\lambda_2 = \lambda_1/2$, we have $k_{a2} = 2k_{a1}$. Hence, solution of the system (3.54) yields

$$\begin{aligned} \int_0^{2R} N_e dx &= 2.23 \times 10^{13} \frac{k_2 - 2k_1}{3\lambda_2} \\ \int_0^{2R} (N_a - N_0)dx &= \frac{2N_L\lambda_2}{3A} (2k_2 - k_1) \quad . \end{aligned} \tag{3.55}$$

Figure 3.47 gives the distributions $\bar{N}_e = \frac{1}{2R}\int_0^{2R} N_e dx$ along the longitudinal axis of a spark in air for different times; Fig.3.48 shows the radial distribution of the electron N_e and relative atom N_a/N_0 concentrations obtained by processing the interferograms of Fig.3.46 according to Abel.

3.3.5 Study of a Laser-Induced Plasma on a Solid Target

A number of authors have studied, using methods of holographic interferometry, the plasma produced when laser radiation is focused on the surface of a solid target [3.96-99,77]. SIGEL [3.96,97] obtained a plasma when he focused radiation on thin films of solid hydrogen. He used a setup

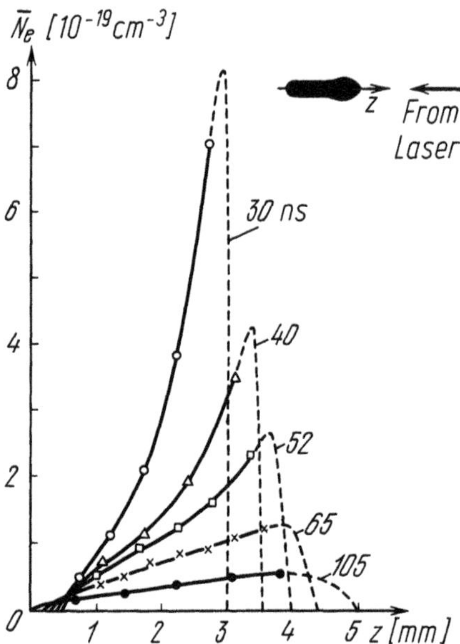

Fig.3.47. Longitudinal distribution of the average electron concentration over a cross section in the plasma of a laser-induced spark in air at different moments

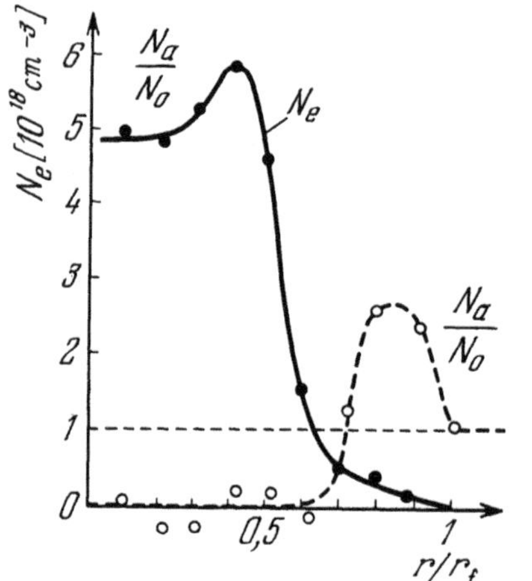

Fig.3.48. Radial distribution of density of electrons and atoms in a laser-induced spark in helium 80 ns after the breakdown. N_0 is the initial concentration of the helium atoms

with a diffusing screen that simultaneously separated the laser beam into object and reference beams similar to those described earlier [3.126]. The interferograms were used to calculate the total number of electrons in the plasma for different stages of its existence.

BELLAND et al. [3.99] studied the plasma produced on the surface of an aluminium target when picosecond pulses generated by a neodymium laser were focused on it. The radiation of the second harmonic of the same laser (λ = 0.53 μm) was used to record the holograms. These authors achieved a record temporal resolution for holographic interferomentry, namely, 7×10^{-12}s.

DREIDEN et al. [3.77] prepared holographic interferograms of a laser plasma on the surface of a potassium target using radiation with three wavelengths – the fundamental frequency and the second harmonic of a ruby laser (λ = 6943 Å and 3472 Å) and the stimulated Raman scattering SRS of the radiation of a ruby laser in nitrobenzene (λ = 7658 Å).

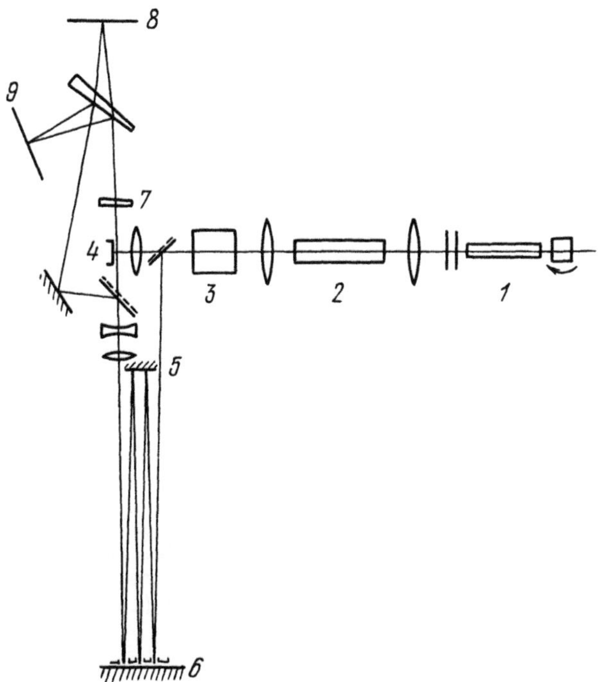

Fig.3.49. Arrangement for three-wavelength investigation of a laser-induced plasma on a potassium target: 1 - ruby laser; 2 - stimulated Raman scattering cell with nitrobenzene; 3 - KDP crystal; 4 - potassium target; 5-6 - optical delay line; 7 - wedge for obtaining fringes of finite width; 8, 9 - holograms

The experimental setup used is shown in Fig.3.49. The laser plasma was obtained in air. The incident three-wavelength pulse passed through an optical delay line. This not only provided its required displacement in time, but also improved the spatial coherence of the radiation, which was especially significant for SRS having a comparatively low degree of spatial coherence [3.127]. Two holograms were recorded simultaneously — a two-wavelength hologram in the light of the fundamental frequency and the second harmonic of the ruby laser, and a second one in the light of the SRS.

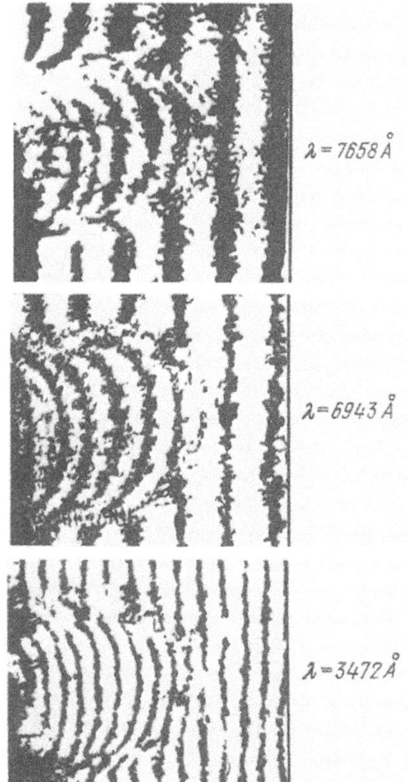

Fig.3.50. Interferograms of a laser-induced plasma on a potassium target in air at atmospheric pressure, recorded during one exposure (the delay time was 115 ns)

Figure 3.50 shows holographic interferograms recorded simultaneously in radiation of different wavelengths. The fringe displacements are due mainly to the simultaneous action of three factors:

a) the appearance of electrons in the plasma of the flare;

b) the appearance of potassium atoms escaping from the target surface;

c) the displacement of air into the shockwave. Owing to the different dependence of the refraction of electrons, atoms of potassium and molecules of air on the wavelength described by (3.50,45,53), respectively, DREIDEN et al. succeeded in separating the contributions of these three components of the total refractivity of the plasma and obtained the spatial distributions of the electrons, potassium atoms and of the density of the air at different stages of plasma development.

3.3.6 Holographic Interferometry of Flash Lamps

As we have already pointed out, the methods of holographic interferometry make it possible to exclude phase inhomogeneities introduced by the optical components of the setup and the walls of the vessel that contains the volume being studied.

In one of the first works on holographic interferometry – that of BROOKS et al. [3.128], an interferogram of the flows of gas in the bulb of a conventional incandescent lamp was obtained by use of the double-exposure method in the light of a gas laser.

Flash lamps used for pumping of laser are cylindrical bulbs of thick quartz with a great number of striae. Before the appearance of holographic methods, the interference investigation of such lamps was impossible. Insertion of windows of interferometric quality may appreciably change the parameters of the plasma being studied. For this reason, study of the plasma of such lamps by methods of holographic interferometry is of special interest.

The procedure for recording holograms in this case is necessitated by the considerable refraction of the light beam by the walls of the quartz bulb. As a result, the superimposition of the mode structure of the object and reference beams is impossible; this, in turn, leads to more strict requirements for the mode composition of the radiation used for holographing. ASCHEULOV et al. [3.100] used a monopulsed laser for the holographic investigation of flash lamps. They selected the mode by introducing a diaphragm into the resonator. The energy of the laser pulse was diminished by such an amount that incoherent illumination of the hologram by the intrinsic radiation of the flash lamp appreciably affected the brightness and quality of the reconstructed interferograms.

Later, higher-quality holographic interferograms of flash lamps were obtained by various investigators [3.101,102]. They used radiation of a single-mode ruby laser that had passed twice through an optical amplifier, for recording the holograms. Figure 3.51 shows holographic interferograms of a flash lamp ИФП-2000 [3.102]. When interpreting such interferograms, the influence of shear, strain, and discharge-tube-wall heating on the interference pattern must be taken into account. These aspects are discussed in [3.129,130].

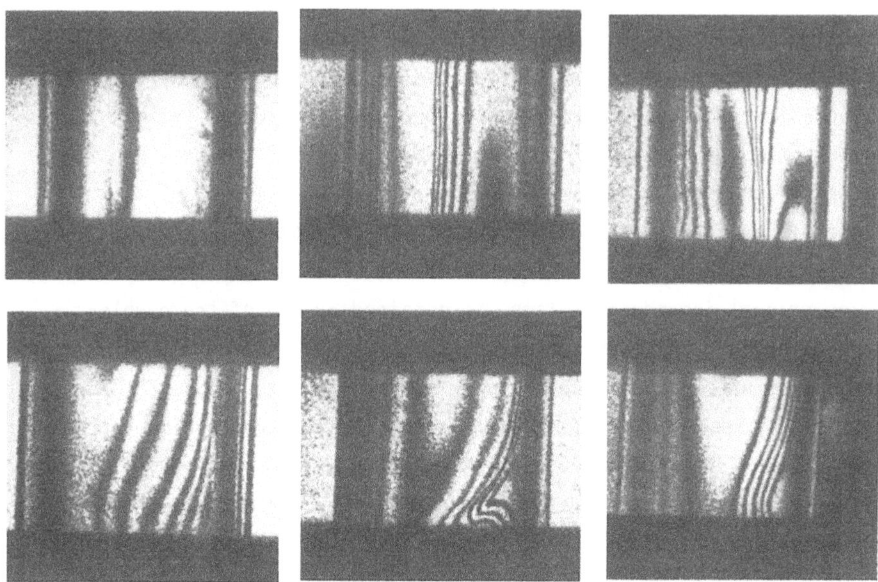

Fig.3.51. Holographic interferograms of a lamp ИФП-2000 corresponding to different moments

3.3.7 Holographic Interferometric Investigation of Plasma Jets

BURMAKOV and OSTROVSKAYA [3.103] used the two-wavelength method of holographic interferometry to study a jet of low-temperature plasma obtained in a dc plasmatron that had 3 kW power with vortex-gas stabilization of the arc. Nitrogen was used as the working substance. The plasmatron arc current was 25 A, and its voltage 120 V.

The double-exposure method was used to record the interference holograms. A monopulsed ruby laser and its second harmonic were the sources of radiation. When the radiation of a helium-neon laser was passed through the two-wavelength holograms, the interferograms of the plasma jet that corresponded to the two wavelengths were reconstructed separately in space.

Such interferograms showed that the electron density in the plasma jet does not exceed 10^{17}cm^{-3}; the shifts of the interference fringes are due mainly to a considerable reduction of the atom density in the zone occupied by the plasma.

A similar procedure was used to study the plasma jet of a pulsed accelerator of the erosion type [3.104]. Two-wavelength interferograms were used to investigate the spatial distribution and the time change of concentration of the electrons, which reached $10^{18}cm^{-3}$.

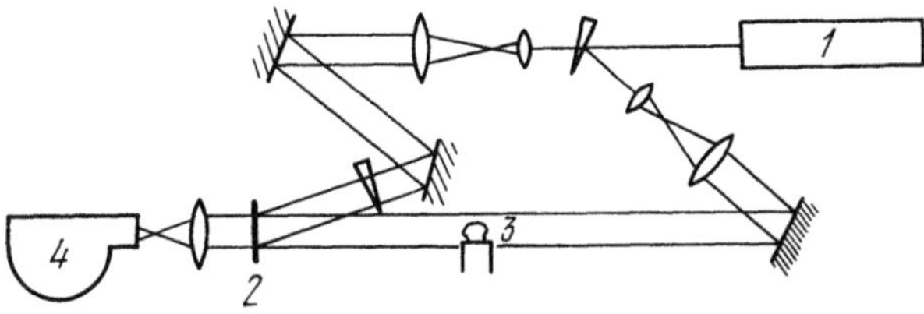

Fig.3.52. Optical diagram of arrangement for real-time cineholographic investigation of pulsed plasma streams

Real-time holographic interferometry was used by BURMAKOV et al. [3.105, 106] to study plasma jets. An optical diagram of the setup is shown in Fig.3.52. A ruby laser (1) that operated in free-generation conditions was the source of radiation. The hologram (2) was exposed in the absence of plasma and was developed in place by use of a special tray (such as that shown in Fig.2.30). The interferograms of the plasma jet (3) were photographed with a super-speed camera CΦP (4). The camera was synchronized with the laser flash and the high-voltage pulse that initiated the plasma. The exposure of each frame was determined by the duration of the corresponding peak and did not exceed 1 μs.

Figure 3.53 shows four interferograms of a pulsed plasma jet, corresponding to different stages of its development [3.106]. Figure 3.54 shows double-exposure holographic interferograms of the plasma jet of a plasmatron for different conditions of its outflow [3.105].

3.3.8 Investigation of a Spark Breakdown

The investigation of a spark breakdown [3.108] is of interest mainly because of the use of real-time interferometry with an electron-optical image converter. The setup is shown schematically in Fig.3.55. After the first exposure (without plasma), the hologram was developed, bleached,

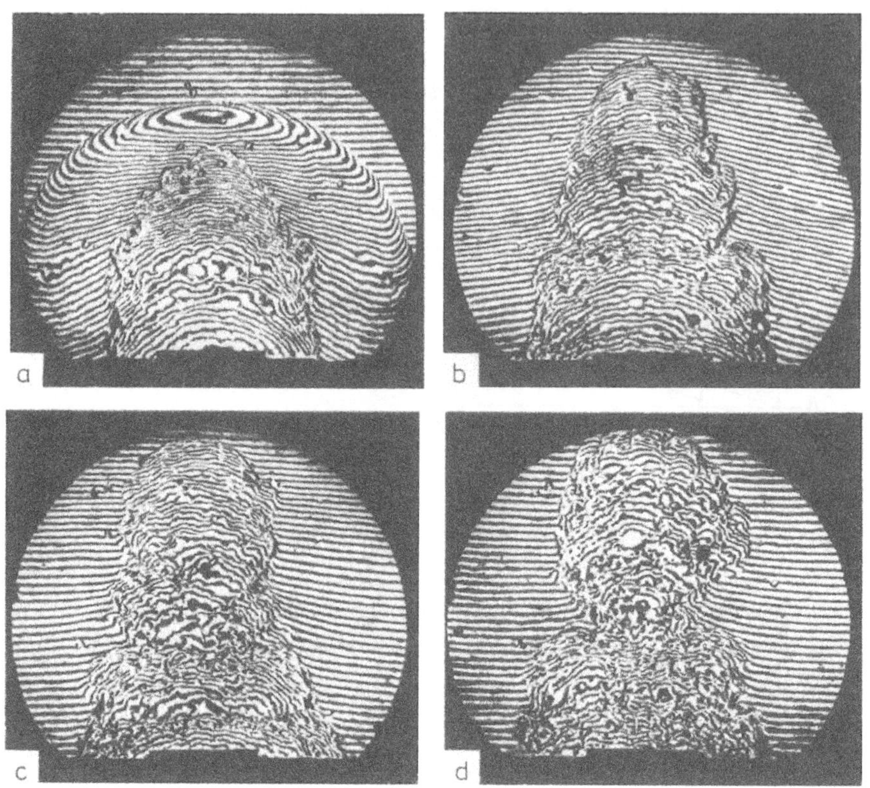

Fig.3.53. Holographic Interferograms of a pulsed plasma stream 70 μs (a), 180 μs (b), 270 μs (c), and 390 μs (d) after beginning of pulse

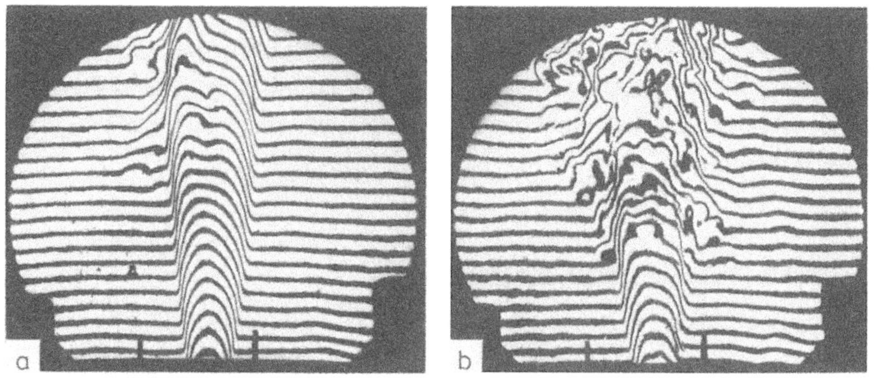

Fig.3.54. Holographic interferograms of a plasma stream for different outflow conditions: (a) laminar flow; (b) transition conditions

and replaced exactly where it was during the exposure. A ruby laser that generated a pulse of about 50 μs was used for reconstruction. If, during wavefront reconstruction, the object beam from the same laser passed through a plasma, then a dynamic interference pattern appeared in the plane of the virtual image. It could be photographed at desired moments from the screens of two electron-optical image converters.

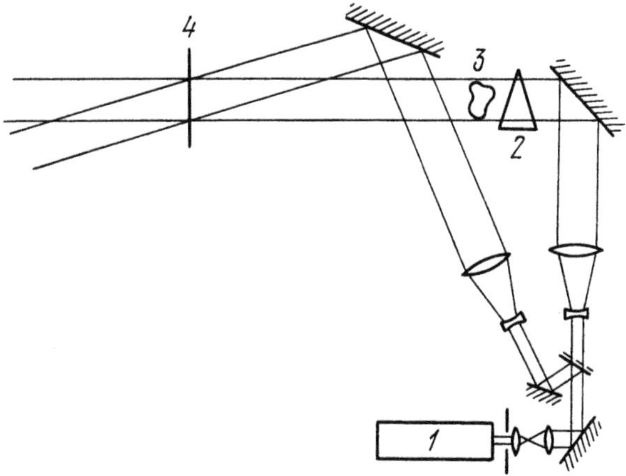

Fig.3.55. Arrangement for the investigation of spark breakdown, by real-time holographic interferometry: 1 - laser; 2 - wedge; 3 - object; 4 - hologram

3.3.9 Holographic Investigations of θ and z Pinches

Plasma installations of this kind are distinguished by their great dimensions, as much as several meters long and several decimeters cross section. This makes it difficult, and sometimes completely impossible, to employ conventional interferometers.

The first holographic investigations of the plasma of a θ pinch were described in [3.18,81]. The experiments were conducted on a Scilla installation. An optical diagram of the holographic setup is shown in Fig.3.56.

A Q switched ruby laser was used to prepare the holograms. The axial modes were separated by means of the diaphragm D (about 0.4 mm in diameter). The laser beam was widened by the diverging lens L_1 and was divided by the glass plate S into an object and a reference beam. The system of mirrors M_1 and M_2 directed the reference beam straight onto the photographic plate P; the object beam illuminated the diffusing screen R that, through the transparent hollow wedge W, was projected by the lens L_2 to the center of the plasma. The lens L_3 projected a full-size image of the diffusing screen in the plane of the hologram. To obtain interference fringes of finite width, the wedge was filled with air for one exposure and with SF_6 for the other.

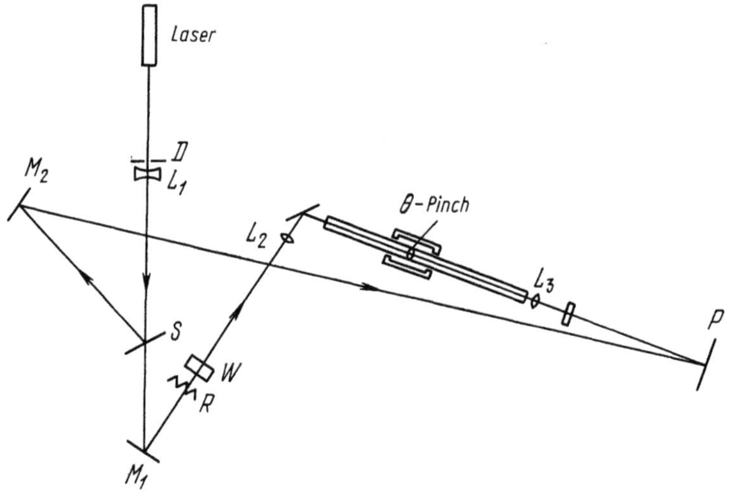

Fig.3.56. Holographic arrangement for investigating the plasma of a θ pinch

High-quality holographic interferograms of a three-meter θ pinch of 10 cm diameter were later obtained by other investigators [3.112,113]. Holograms of the plasma of a θ pinch were obtained in one case without a diffusing screen, by use of a single-mode ruby laser [3.112]. The central plane of the plasma was projected onto the hologram by a special lens that was installed in the object beam, thus forming an image hologram. Figure 3.57 shows one of the reconstructed double-exposure interferograms. To obtain interferograms with fringes of finite width, the mirror that directed the object beam onto the hologram was inclined between the exposures. Different stages of a discharge were studied in its sequential flashes with a change of the delay time between the beginning of the discharge and the pulse that opened the electrooptical shutter of the laser Q switch.

Later, a setup with three ruby lasers was used for more detailed investigation of the plasma of a θ pinch [3.113]. With it, three interferograms at intervals of about 1 microsecond were obtained during one discharge pulse.

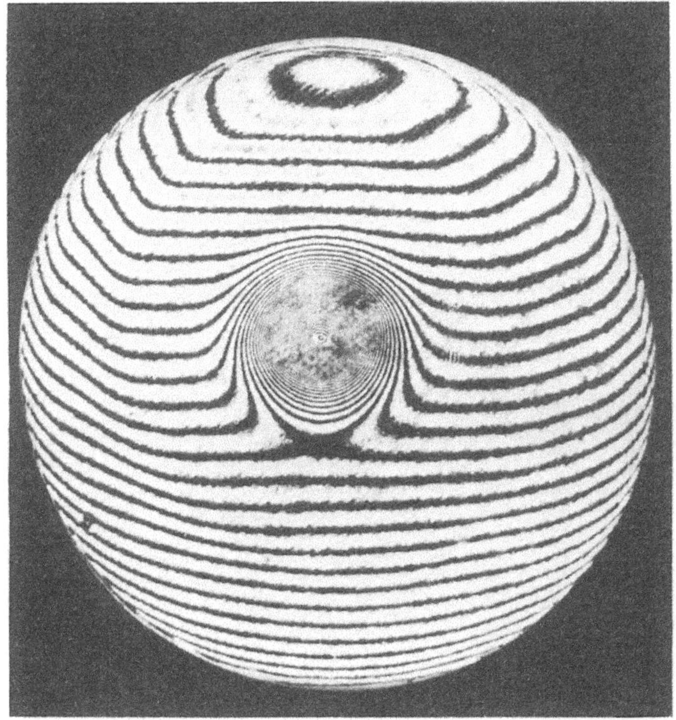

Fig.3.57. Holographic interferogram of a θ pinch [3.112]

The setup is shown schematically in Fig.3.58. The object beams formed by all three lasers were made to coincide in space by use of half-silvered mirrors and were directed into the discharge chamber. The transparent grating diffracted about five per cent of the incident light into each of the three diffraction orders which illuminated three holograms. The reference beam from each of the lasers illuminated only one hologram and interfered only with the part of the object beam that fell on the hologram that had been produced by the same laser. The remaining parts of the object beam which came from the other two lasers played the part of incoherent illumination and had practically no influence on the quality of the reconstructed interferograms.

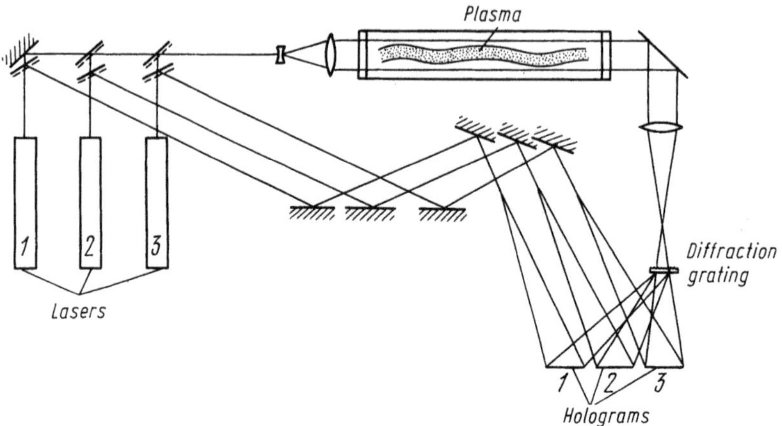

Fig.3.58. Arrangement with three lasers for cineholographic investigation of a θ pinch

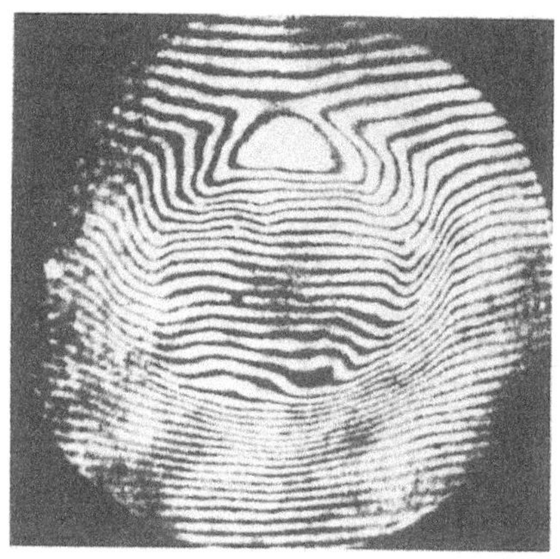

Fig.3.59. Holographic interferogram of the plasma of a θ pinch [3.114]

The results of experiments on holographic diagnostics of plasmas in z and θ pinches 10 cm in diameter are given in [3.114]. One of the interferograms obtained is shown in Fig.3.59. In reconstruction, an interferometer was used with which the width and orientation of the fringes on the interferogram could be adjusted [3.7]. Its action is based on the spatial separation of the reconstructed wavefronts that correspond to the two exposures (see Fig.1.36). To obtain cineholograms of θ and z pinches, an acoustooptical shutter with a moving-wave slit was proposed [3.131]. It permitted different sections of the active element to be switched in consecutively for generation (see Fig.2.5).

3.3.10 Investigation of Plasma in the Vicinity of a Neutral-Current Layer

Neutral-current layers separate magnetic fields of opposite direction and appear in magnetized plasma.

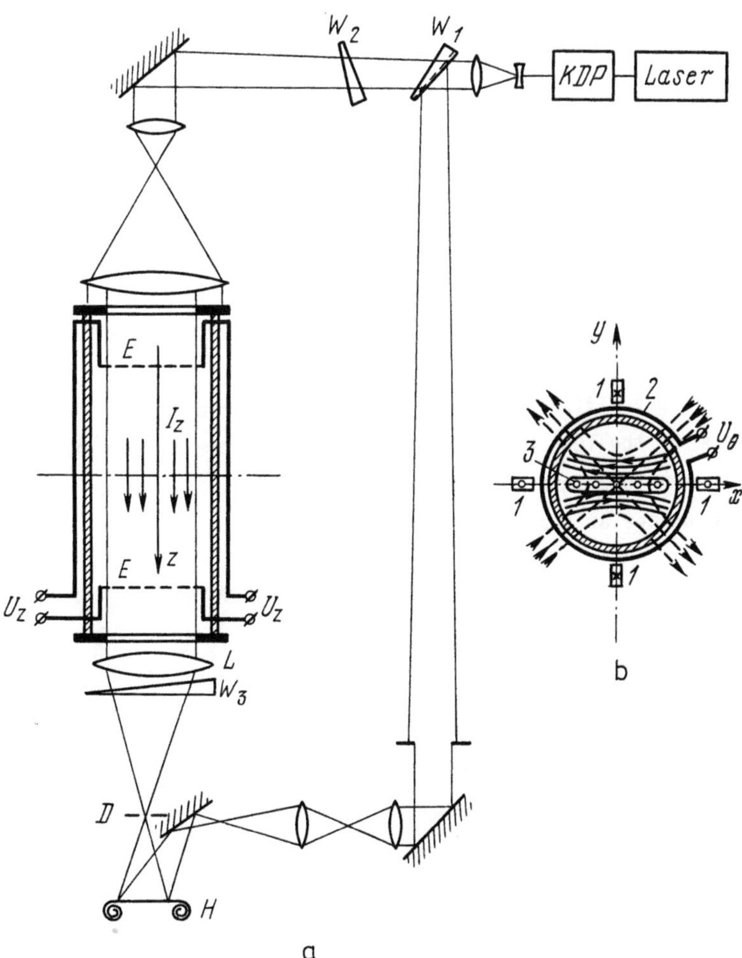

Fig.3.60. Installation for investigating plasma in the vicinity of a neutral current layer: (a) top view; (b) section of the discharge chamber; W_1 - beam-splitting wedge; W_2 - compensating wedge; H - hologram; 1 - current-carrying conductors that produce a quadrupole magnetic field (the dotted lines show the field lines of the quadrupole field); 2 - turns of the θ discharge; 3 - current layer (the direction of the current is shown). The solid lines are the magnetic-field lines in the vicinity of the layer

Holographic interferometric investigations of the plasma in the vicinity of a neutral-current layer, performed on TC-3 installation, are described in [3.116]. The installation produces a high-current direct discharge in a quadrupole magnetic field.

The installation is shown schematically in Fig.3.60a. A multimode ruby laser of type OГM-20 ($E \approx 0.4$ J, $\tau \approx 20$ ns) synchronized with the plasma installation was used as the light source. The setup made it possible to obtain holograms in the light of the fundamental frequency and the second harmonic of the ruby laser. The object beam, 10 cm in diameter, passed along the axis of the discharge chamber. The mode structure of the object and reference beams was superimposed in the plane of the hologram.

A feature of the installation was the presence of grid electrodes E in the path of the object beam. They formed a longitudinal electric ("fast") field that produced a plane current layer whose section is shown in Fig.3.60b.

To eliminate the light diffracted by the grids, a diaphragm D was placed in the focus of the lens L. It spatially filtered the radiation. Wedge W_3 was provided to obtain interferograms in fringes of finite width. It turned about the axis z between exposures.

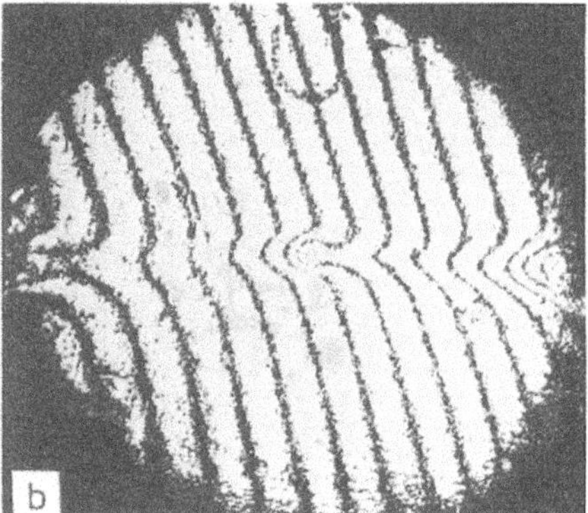

Fig.3.61. Holographic interferograms of the plasma in the vicinity of the neutral current layer [3.116]

Figure 3.61 shows, as examples, interferograms of the plasma obtained at a moment close to the maximum of the fast field and 0.5 μm after the maximum. The interferograms were processed on the assumption of homogeneity of the plasma layer in the direction of the axis of the transmitted radiation (axis z). The length of the layer was 40 cm.

The spatial distributions of the electron concentration in the current layer obtained by processing the interferograms (see Fig.3.61) are plotted against x and y in Fig.3.62.

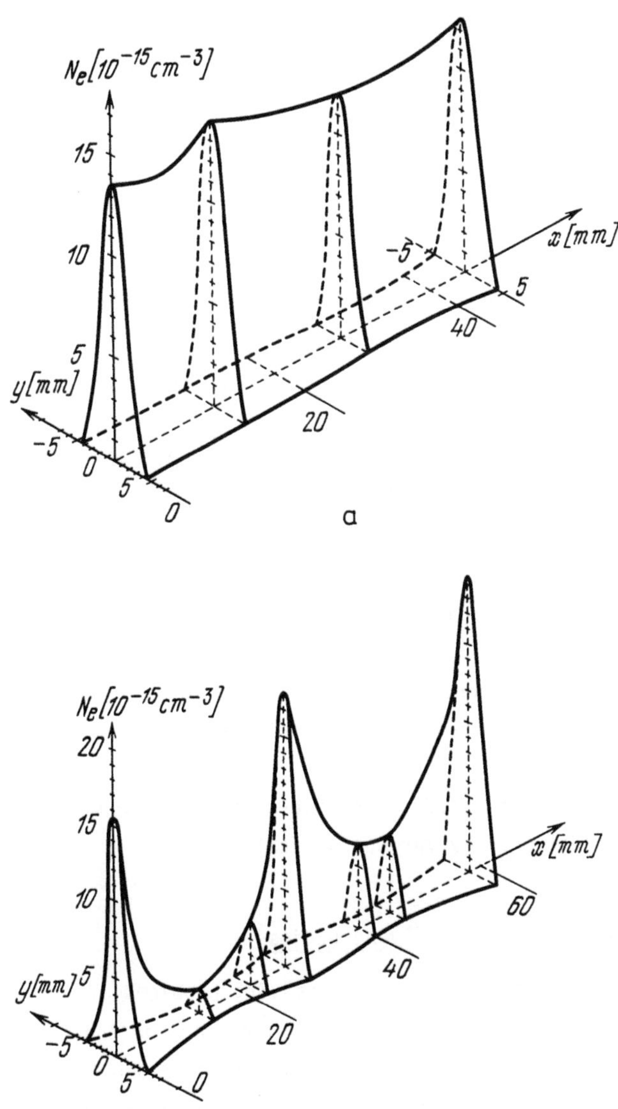

Fig.3.62. Distribution of electron concentration in the current layer, obtained by processing the interferograms of Fig.3.61

3.4 Use of Holographic Interferometry in Gas-Dynamic Investigations

Here we shall treat the use of holographic interferometry for the investigation of gas streams, shock waves, streams around flying bodies, etc. Although some of these phenomena are attended by considerable heating of the gas and accompanying ionization and exitation, nevertheless the contribution of the electrons, ions, and excited atoms to the total refraction may be disregarded in the majority of cases, and account need be taken of only local changes of the gas density. In this connection, we can limit ourselves, as a rule, to obtaining holographic interferograms with one wavelength in such investigations.

With the exception of this simplifying circumstance, the methods of investigating gas-dynamic phenomena do not differ appreciably from those used to study plasmas, described in the preceding section. The following advantages of holographic interferometry in comparison with the conventional procedure in such investigations are:

(a) the possibility of investigations in setups with large optical cross sections (in large shock and wind tunnels, aeroballistic ranges, etc.);

(b) the possibility of studying nonaxisymmetrical distributions of density of a gas by transmission of light in many directions through gas-dynamic objects of intricate shape;

(c) the possibility of increasing the sensitivity of interference measurements so that gas-dynamic systems can be investigated at low gas densities;

(d) the possibility of *a posteriori* study of single-pulsed gas-dynamic processes under stationary conditions, by various optical methods.

3.4.1 Investigation of Flow Around Freely Flying Bodies

The effectiveness of holographic interferometry for solving problems of aeroballistics was demonstrated in one of the first publications on the subject [3.132]. Figure 3.63, taken from that publication shows a doubly-exposed holographic interferogram of the shock wave that accompanies supersonic flight of a bullet. The latter flies in front of a homogeneously illuminated diffuser. The first exposure records a hologram of the diffuser before the flight of the bullet, and the second during its flight. Thus, the light scattered by the stationary diffuser is recorded during both exposures instead of the light scattered by the flying bullet. Accordingly, the requirements for the temporal resolution of the setup are determined not by the speed of the bullet (the light scattered by it is not recorded by the hologram), but by the rate of change of the advance of the object-wave phase as a result of the change of the refractive index in the region that adjoins the bullet. The displacement of the bullet during exposure (which was several scores of wavelengths for the case shown in Fig. 3.63) causes only the corresponding blurring of the shadow from the bullet. This, as a rule, is of no significance. Also of no significance is the shift of the interference pattern due to motion of the bullet during the exposure – it does not exceed a tenth of a fringe.

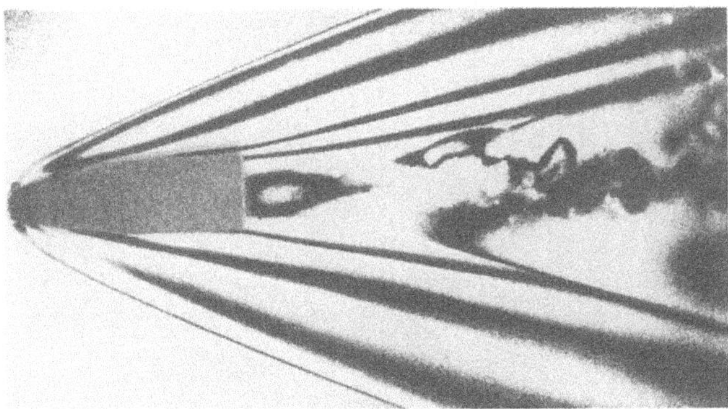

Fig.3.63. Holographic interferogram of the shock wave produced by a flying bullet [3.132]

A number of publications have been devoted to study of streams on an aeroballistic range [3.8,9,51,133,134]. In one case, a holographic setup was used that was assembled on the basis of a Mach-Zehnder interferometer with a beam diameter of 150 mm [3.133] (Fig.3.64). The densities of the gas near spherical models that flew with speeds of about 700 m/s in air at reduced pressures were studied. The laser flash was synchronized with the moment when the model passed through the section of the object beam, as follows. Two light beams from helium-neon lasers 6 and 7 intersected the trajectory of the model. When the first of them was interrupted, the pumping lamps of the ruby laser were ignited; the second controlled the electrooptical Q switch. Nonlinear effects were used to amplify the phase shifts. Figure 3.65 shows one of the series of interferograms obtained when different diffraction orders were used.

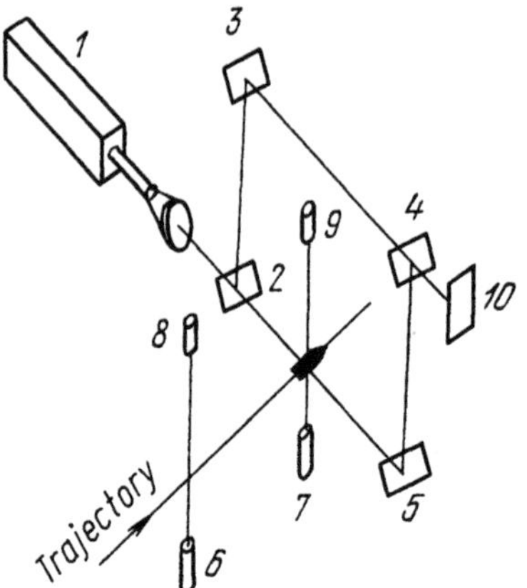

Fig.3.64. Experimental setup for obtaining holographic interferograms of gas streams on an aeroballistic trajectory: 1 - ruby laser; 2-5 - mirrors of Mach-Zehnder interferometer; 6, 7 - helium-neon lasers; 8, 9 - photomultipliers; 10 - hologram

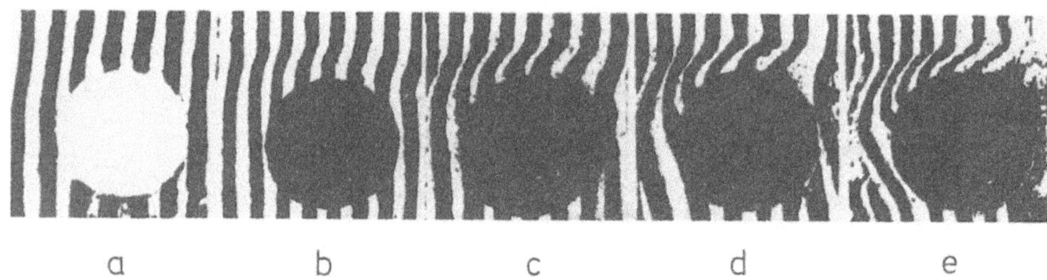

Fig.3.65. Interferograms of the field of flow around a sphere, obtained by use of various diffraction orders: (a) 0-th and +1st orders; (b) ±1st orders; (c) ±2nd orders; (d) ±3rd orders; (e) ±4th orders [3.134]

A holographic interferometer was also assembled on the basis of a standard shadow instrument ИАБ-351 [3.8,9]. An optical diagram of the setup is shown in Fig.3.66. The object beam was spread to a diameter of 230 mm. The nonexpanded reference beam (2.5 mm in diameter) passed through the same optical system. Image holograms were recorded at the output of the instrument, so that the image could be reconstructed with light from a mercury or incandescent lamp. The image was reconstructed by use of the same instrument provided with an attachment that made possible the production of either an interferogram or a schlieren photograph (see Fig.3.4).

In the deciphering of interferograms obtained by use of monochromatic radiation, difficulties appear that are associated with a sharp jump of the density near the front of the shock wave. This jump results in an uncertainty of an integer in the fringe-number change over the entire pattern. In conventional interferometry, an achromatic fringe is used to determine the number of a fringe on both sides of the jump. For this purpose, the interferogram is photographed with a source of white light. In holographic interferometry, the same information can be obtained with a monochromatic source by using some information concerning the object [3.135,136]. The quantitative results of the density distribution of the air stream that surrounds the models being studied, given in [3.8,9] have been obtained by use of this procedure for the calculations.

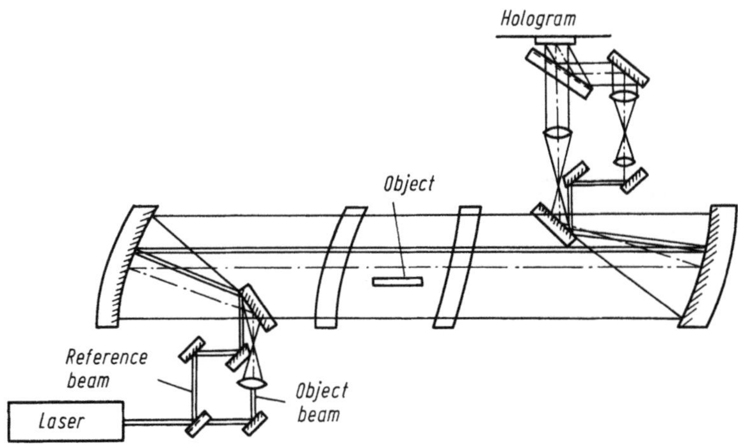

Fig.3.66 Diagram of a holographic interferometer based on the shadow instrument ИАБ-451 [3.9]

Multipath methods are also used (see Fig.3.29) to increase the sensitivity of holographic interferometry in the aeroballistic investigations of low-density gas streams [3.51].

3.4.2 Investigations in Wind Tunnels

We cite [3.137] as an example of investigations in which the method of holographic interferometry was used to study the flow of air around bodies in a wind tunnel. A feature of these investigations is that they were conducted in real conditions in a large tunnel. The flow of a subsonic stream around a 2.4 m wing profile was studied. The laser and the remaining elements of the holographic setup were arranged above the wind tunnel (Fig.3.67). The object beam was expanded by the lens L and was directed onto a diffuse 57 × 57 cm aluminium reflector on the floor of the tunnel. The scattered light was recorded on a hologram. The reference beam, after passing the delay line which reduced the path difference to a minimum, also impinged on the hologram. Thus, the object beam passed twice through the air stream being studied. The light passed through this volume, within the limits of the solid angle subtended by the diffuse reflector. Although such a system presents appreciable difficulties in interpreting interferograms, its merit is that the diffuse reflector is the only large optical component.

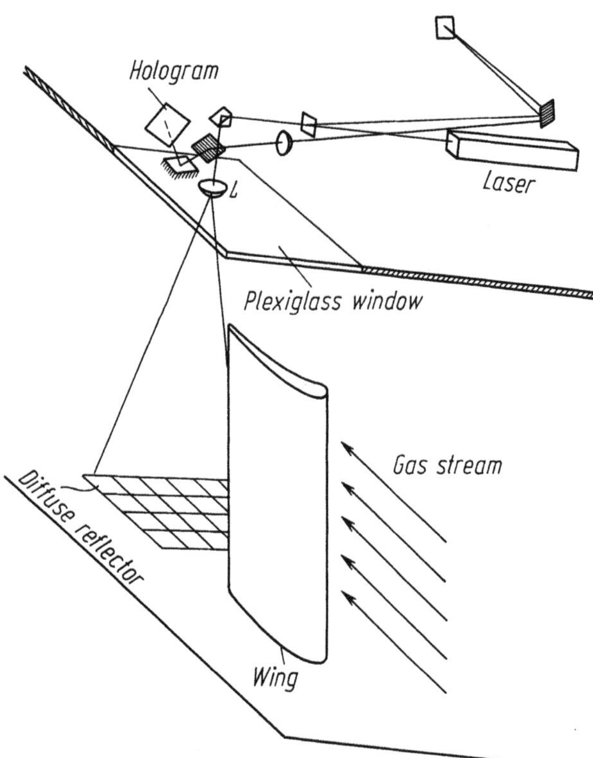

Fig.3.67. Diagram of arrangement for holographic investigation of gas streams in a wind tunnel

In the conditions of the aerodynamic experiment, the fringe displacements of the doubly exposed interferograms were due not only to local changes of the density of the air stream, but also to uncontrolled shifts of the diffuse reflector during the time between the two exposures. This also introduced difficulties into interpretation of the interferograms. It was nevertheless found possible to construct a field of the densities in the stream averaged for the lines of observation.

Various applications of holographic interferometry for studying gas streams have also been considered in [3.138,139].

4. Investigation of Displacements and Relief

4.1 The Process of Interference-Pattern Formation in Holography

Because an interference-fringe pattern is the main source of information on the changes that occur in an object whose hologram is recorded, attention must be given to the nature of the formation and the physical properties of an interference pattern.

A frequent view is that hobographic interferometry is a combination of holographic techniques and the formation and interpretation of interference patterns known from classical optics [4.1]. The reason why this view appeared is quite obvious. It is due to the tendency to simplify the method of holographic interferometry as far as possible and to convert it into a convenient and simple method of engineering in a very short time. In addition to what has been set out in Chap. 3, we shall consider some arguments that show that this viewpoint, which certainly simplifies the interpretation, is restricted and may often lead to erroneous results.

1) Unlike classical interferometers, in which wavefronts are compared whose shapes differ only slightly from simple analytical surfaces (plane, spherical, cylindrical), holographic interferometry deals with light fields of much more complicated structures. The correlation properties of these fields, which determine the properties of an interferogram, depend radically on the properties of the scattering surface. Thus, the nature of an interferogram is due not only to the behavior of an object, but also to its light-scattering properties. This exceedingly important conclusion will be discussed in detail in the present chapter.

2) The processes that occur when a hologram is recorded, and in the reconstruction of the image are transformations [1.35] of the information recorded on the hologram. These processes are equivalent to a complicated amplitude-frequency filter that acts on the optical signal that describes the field of the object. The nonlinear response of the recording medium (Chap. 2) and the dependence of this response on the microinterference-

structure frequencies have the result that the field of the reconstructed wave may differ greatly from the original object, especially in minute details. The resulting distortions of the object-field characteristics may affect the properties of the interferogram.

3) In holographic interferometry, especially in vibrometry (Chap. 5), the recording and subsequent reconstruction from a hologram of not two, but in principle of an infinitely great number of wavefronts with various weight coefficients are possible. This also affects the properties of an interference pattern.

Therefore, one of the prerequisites for successful practical application of the methods of holographic interferometry and the proper interpretation of the resulting interferograms is the solution of physical notions concerning the mechanism of light scattering on the surface of real objects, and concerning the properties of the resulting scattered field that influence the formation or disappearance of the interference-fringe pattern in holography.

If we limit ourselves to a scalar approximation, i.e., consider only the component of the scattered light field that, for example, is polarized in the same way as the laser radiation that illuminates the object (see Chaps. 1, 2), then at the arbitrary point of space P (Fig.4.1) the light field a(P) is related to the field on the surface of the object $a(\vec{\xi})$ by the known Kirchhoff-Fresnel integral [4.2]

$$a(P) = -\frac{i}{\lambda}\int \frac{a(\vec{\xi})\exp[ikR(\vec{\xi},P)]\cos\theta(\vec{\xi},P)}{R(\vec{\xi},P)}\,d\vec{\xi} \quad . \tag{4.1}$$

Here $a(\vec{\xi})$ is the field at the arbitrary point $\vec{\xi}$ with the coordinates (ξ, η) on the surface being studied, R is the distance from this point to the point P for which the field a(P) is being computed, and θ is the angle between a normal to the surface and the vector $\vec{R}$.

The analytical relationship (4.1) confirms the unambiguous correspondence between the field on the surface and that in the surrounding space. This follows quite naturally from the unambiguity of the solutions of Maxwell's equations when the characteristics of the field $a(\vec{\xi})$ are given as the boundary conditions.

What are the properties of the field $a(\vec{\xi})$? They depend on the characteristics of the field $a_0(\vec{\xi})$ that illuminates the surface being studied

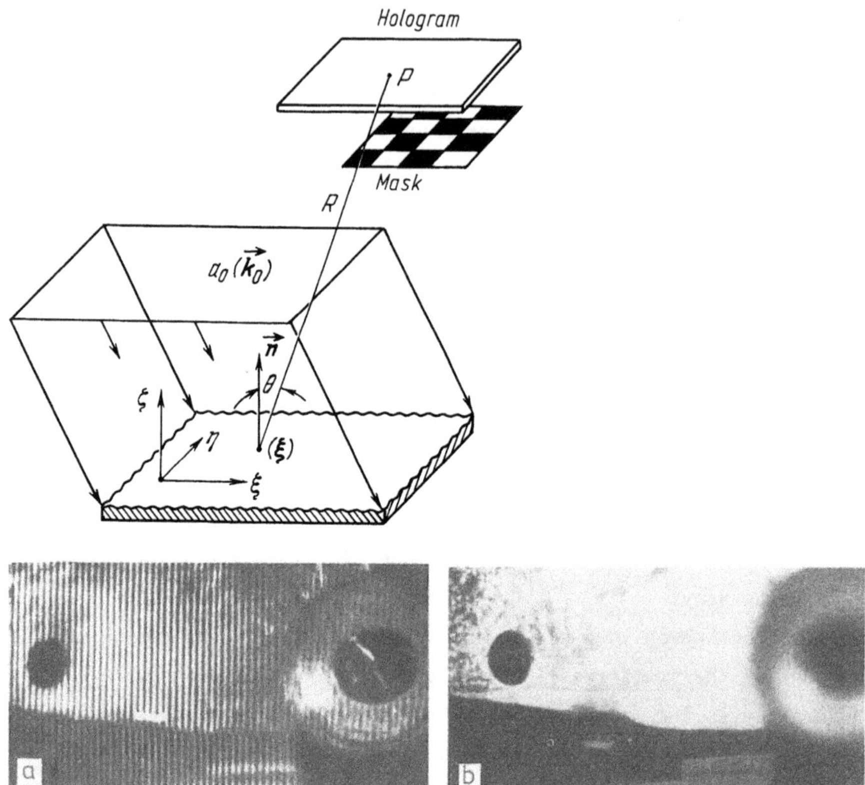

Fig.4.1. Calculation of field of light wave a(P) scattered by surface $\zeta(\vec{\xi})$ at an arbitrary point P of a hologram. Also shown (at the bottom) are images of the object obtained by the double-exposure method, from different parts of the hologram (a) not covered by a mask; (b) exposed through a binary mask that was moved between the exposures [4.3]

[4.3]. We shall consider, for simplicity, that the surface is illuminated along a normal with a plane quasimonochromatic light wave, i.e., that $a_0(\vec{\xi}) = \text{const}(\vec{\xi})$.

The characteristics of the light-scattering surface are a second basic factor that affects the properties of the field $a(\vec{\xi})$. We have already noted that the most important advance realized by holographic interferometry is that it has become possible to study objects whose surfaces have not only arbitrary shapes (not necessarily plane, spherical, etc.), but also arbitrary microstructure. The surface of a real object (if we consider its dimensions to exceed the length of the light wave that illuminates it) is, as a rule, an exceedingly complicated function of the coordinates. For example, a metal surface, after various kinds of processing (grinding, chemical pickling, electrolysis, painting) will be described by an intricate function $\zeta(\vec{\xi})$. The function $\zeta(\vec{\xi})$ is not regular, and we can rarely succeed in expressing it analytically. Hence, without going into the details of the process of light reflection [4.4], we can write a general functional relationship between the coefficient of reflection V at the given point $(\vec{\xi})$ and the shape of the surface $\zeta(\vec{\xi})$,

$$V(\vec{\xi}) = \hat{V}\zeta(\vec{\xi}) \quad , \tag{4.2}$$

where $\hat{V}$ is a functional operator that allows us to calculate the coefficient V at each point [4.4,5].

To what extent can we apply the results to other surfaces? As a rule, two absolutely identical (in the microscopic sense) surfaces never exist, even if they were obtained as a result of identical operations. Therefore, only properties of surfaces and scattered fields can be taken into account that are characteristic of large complexes, or ensembles, that have identical statistical properties. The statistical properties of a surface that are basic to the measurable characteristics of a scattered field are:

a) The <u>mean value</u> of the surface function $\zeta(\vec{\xi})$, i.e., its mean height above the plane $\zeta = 0$,

$$\overline{\zeta(\vec{\xi})} = \int_{-\infty}^{\infty} \zeta W_1(\zeta,\vec{\xi})d\zeta \quad . \tag{4.3}$$

Equation (4.3) includes the very important characteristic of a statistic function $W_1(\zeta,\vec{\xi})$ — the conditional density of the probability that the height of the surface at the point $\vec{\xi}$ is ζ.

b) The <u>dispersion of the heights</u> σ^2 of the surface,

$$\sigma^2 = \overline{\zeta^2(\vec{\xi})} = \int_{-\infty}^{\infty} \zeta^2 W_1(\zeta,\vec{\xi})d\zeta \quad , \tag{4.4}$$

which characterizes the scatter of the heights relative to the zero level.

c) The correlation function Φ of the ζ values at two separated points $\vec{\xi}_1$ and $\vec{\xi}_2$,

$$\Phi = \sigma^{-2} \int\int_{-\infty}^{\infty} \zeta_1\zeta_2 W_2(\zeta_1, \vec{\xi}_1, \zeta_2, \vec{\xi}_2) d\zeta_1 d\zeta_2 \quad . \tag{4.5}$$

d) The correlation radius

$$\ell = \int_{-\infty}^{\infty} \Phi(\vec{\xi}) d\vec{\xi} \quad , \tag{4.6}$$

i.e., the radius of a circle within which the deviations of the surface from the mean plane are correlated.

The calculation of most statistical characteristics associated with averaging over an ensemble can be imagined as applied to the measurement of a scattered field as follows. We imagine measuring the amplitude a and intensity I of the scattered field at all of the points of interest. We imagine performing a series of such measurements on many different light-scattering surfaces that have identical statistical characteristics, which are called the "ensemble". The data thus obtained at each point are averaged over, i.e., for all members of, the ensemble. The resulting function contains only systematic variations; the random variations among members of the ensemble result in speckle structure of the field [4.6] vanish, owing to their being averaged.

The statistical characteristics of the field scattered by the surface of an object are of interest not only because they can be used, in principle, to reconstruct the structure of the surface, but also because the properties (visibility and localization) of the fringes observed on a holographic interferogram also depend on those characteristics.

Assume that two fields a_I and a_{II} that correspond, for instance, to the initial and final states of the surface of an object (the first and second exposures) are reconstructed from a doubly exposed hologram and simultaneously observed. The intensity of the total field, when account is taken of the averaging over the ensemble of surfaces of one class required to eliminate the speckle structure [4.7], will be [4.8]

$$I(P) = \langle a_I a_I^* \rangle + \langle a_{II} a_{II}^* \rangle + \langle a_I a_{II}^* \rangle + \langle a_I^* a_{II} \rangle =$$

$$= I_I(P) + I_{II}(P) + \frac{2|\mu|\cos\delta(P)}{(I_I I_{II})^{1/2}} \quad , \tag{4.7}$$

where I_I and I_{II} are the intensities of the field that correspond to the first and second exposures, observed at the point P when the fields a_I and a_{II} are reconstructed independently. The third term of (4.7) determines the form of the interference pattern. The visibility of the pattern is the higher, the greater the coefficient of the harmonic factor that describes the periodic change of the intensity from the point P to adjacent points.

Equation (4.7) contains the degree of mutual coherence μ of the fields a_I and a_{II}. It is related in a simple way to their mutual-correlation function (Chap. 1),

$$\mu = |\mu|\cos\delta = \frac{\langle a_I a_{II}^* \rangle}{(I_I I_{II})^{1/2}} \quad . \tag{4.8}$$

Equations (4.7,8) show that the correlation properties of the light waves that are recorded on a hologram during the first and second exposures, and observed simultaneously by reconstruction, determine the visibility of the interference pattern. If we assume that the spacing of the interference fringes determined by the spatial dependence of the phase term $\delta(P)$ is not great (therefore I_I, I_{II}, and μ change only slightly over a distance equal to several fringe periods), then the visibility of the interference pattern p [4.2] is, in accordance with (1.42) and (4.7),

$$p = \frac{I_{max} - I_{min}}{I_{max} + I_{min}} = \frac{2(I_I I_{II})^{1/2}}{I_I + I_{II}} |\mu| \quad , \tag{4.9}$$

which, in the customary case when $I_I = I_{II}$, gives the simple relationship

$$p = |\mu| \quad . \tag{4.10}$$

The procedure of averaging over an ensemble made it possible to exclude the influence of the noise due to a speckle structure on the characteristics of the light field. This noise will, of course, be observed in every actual experiment; its details will depend on the conditions of the experiment. The basic characteristics of an interferogram, particularly the fringe visibility and spacing, will not change from experiment however.

A simple experimental confirmation of the close relationship between the correlation properties of a field and the visibility of the fringes can be obtained as follows (see Fig.4.1). Let us place before part of a hologram a binary periodic mask that consists of regularly alternating transparent and opaque squares of equal area (a checkerboard). During the first exposure, the mask exposes definite parts of the hologram. Before the second exposure, the mask is moved half the period of square alternation and exposes only the previously unexposed parts [4.8].

When the field of such a hologram is reconstructed, we see the object and all its details, but there is no interference pattern. However, through the part of the hologram that was not covered by the mask, interference fringes are seen [see Fig.4.1a (at the bottom left)]. The fields reconstructed from two different parts of the hologram exposed separately are not correlated because this phenomenon is similar to an attempt to synthesize the same function by use of different parts of its spatial spectrum. Consequently, in accordance with (4.8), the visibility of the interference pattern vanishes.

Not only the shape and number of interference fringes are of interest to an experimenter, but also the region of their spatial localization. A clearly observed region of spatial localization of fringes, i.e., of maximum visibility, is not characteristic of classical interferometers with coherent sources. In them, the fringe pattern is observed in the entire region where the two light beams overlap in space. Proper physical interpretation of fringe localization in holographic interferometry provides deeper understanding of the specific nature of this field of optical investigations. It transforms this effect from a harmful hindering one into a source of additional information on the properties of an object. Thus, the light-scattering properties of a surface, which depend on its microstructure, determine the nature of the scattered field and, consequently, the volume and shape of the fringe-localization region. We shall attempt to trace this relationship, which may facilitate the development of correct notions of the effect in which we are interested.

To date, there is no single and sufficiently simple physicomathematical approach to the description of the processes of light scattering on a statistically uneven surface, even in a scalar approximation [4.7]. We can establish a relationship between the characteristics of a scattered field and a surface for two extreme cases that, in their pure form, are very rare in practice (Fig.4.2).

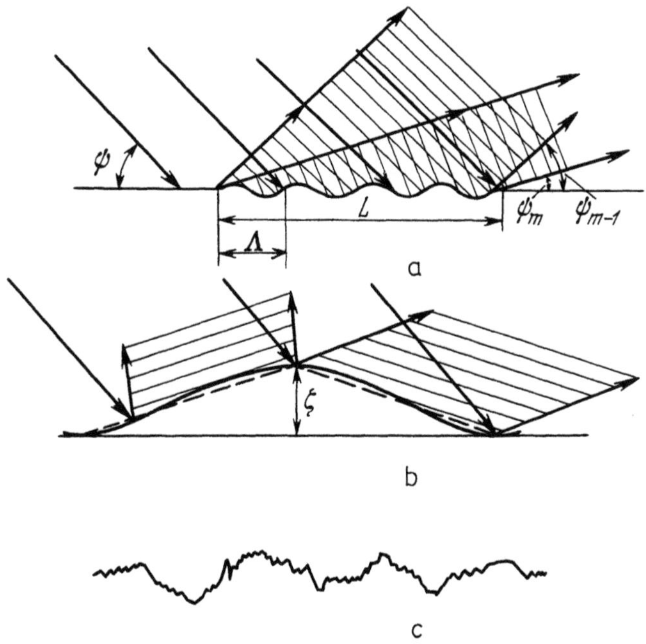

Fig.4.2. Calculation of scattered-light field by different methods (a) perturbation theory; (b) Kirchhoff's method; (c) microprofilogram of a real rough surface

The first case relates to surfaces that have small dispersions of the heights ($\sigma \ll \lambda$) and inhomogeneities with sufficiently gentle slopes ($\sigma \ll \ell$). Such surfaces are not very frequent in practice, but, for example, a light-scattering layer of sprayed-on magnesium oxide has properties close to those of surfaces of this class. The method of perturbation theory (Fig.4.2a), considered already by RAYLEIGH [4.9], is generally employed in the analysis of the scattering properties of such surfaces and the correlation properties of the scattered field.

A surface with slight irregularities can be represented in the form of a statistical superposition of spatial harmonics of different spatial periods Λ_i and amplitudes ζ_i that comply with Rayleigh's condition ($2k\zeta_i \sin\psi << \pi/2$) for grazing angles ψ of the illuminating light wave. Resonance scattering occurs for each of those spatial harmonics, over the length of the illuminated section L: the wave that falls at the grazing angle ψ is diffracted only in discrete directions whose grazing angles ψ_m are given by

$$\cos\psi_m = \cos\psi + m\frac{\lambda}{\Lambda} \quad . \tag{4.11}$$

No diffraction spectra are seen for a real surface in scattered light. Owing to statistical interaction of various spatial harmonics, we must consider the integral decomposition of the field at each point along plane waves,

$$a(P) = \iint\limits_{-\infty}^{\infty} \Omega(\vec{k},\vec{R})N(\vec{k})e^{i\vec{k}\vec{R}}d\vec{k} \quad . \tag{4.12}$$

Here $N(\vec{k})$ is the angular spectrum of the plane waves, which is related to the surface function $\zeta(\vec{\xi})$ by the Fourier transform

$$N(\vec{k} - \vec{k}_0) = \frac{1}{(2\pi)^2}\int\limits_{-\infty}^{\infty} e^{i(\vec{k}-\vec{k}_0)\vec{\xi}}\zeta(\vec{\xi})d\vec{\xi} \quad ,$$

(where $\vec{k}_0$ is the wave vector of the illuminating wave). Equation (4.12) also contains the aperture function $\Omega(\vec{k},\vec{R})$, which describes the spatial filtering of the scattered waves by the finite apertures of the hologram and the observation system.

For surfaces with average irregularity sizes much greater than a wavelength, the perturbation method cannot be used; in such cases, the characteristics of the scattered field must be calculated by use of Kirchhoff's method. In that method (see Fig.4.2b), the effects of large inhomogeneities ($\sigma >> \lambda$) on the incident light are considered to be similar to those of a set of reflecting surfaces arranged tangent to the inhomogeneities (see the dashed lines in Fig.4.2b) and having the corresponding dimensions. Scattering is thus considered to be the sum of the actions of separate surfaces that reflect in accordance with the laws of geometric optics. The dimensions of each such reflector are small in comparison with the illuminated region of the surface, but considerably exceed the wavelength.

Experimental investigations have shown that although most real surfaces are multiple-parametric, i.e., are characterized by the simultaneous presence of micro- and macroroughnesses (see Fig.4.2c), the large-scale inhomogeneities make the main contribution to the characteristics of the scattered field. Let us consider, in view of the foregoing, how interference fringes are localized in space. We shall take, as an example, a comparatively simple kind of displacement of the object between exposures that describes, however, many kinds of vibrational displacements (Fig.4.3); it is a normal (piston) displacement over the distance g_0 and subsequent rotation of the object as a whole through the angle β.

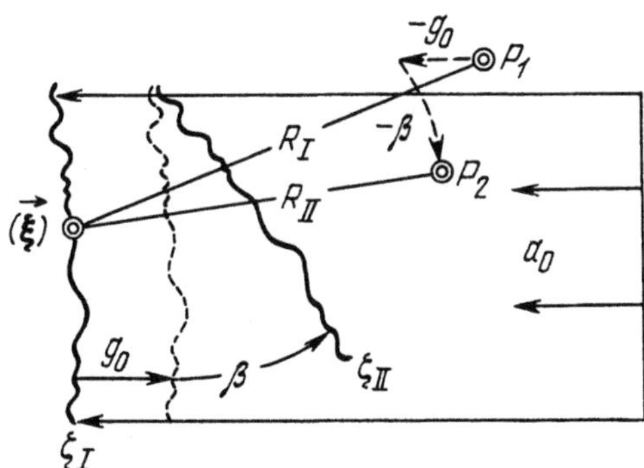

Fig.4.3. Calculation of the correlation characteristics of fields scattered by two displaced identical surfaces ζ_I and ζ_{II}

Let us calculate the field $a(P_1)$ at the point P_1. We introduce the two-dimensional function $\zeta(\vec{\xi})$ to describe the surface. The field on the surface during the first exposure is

$$a_I(\vec{\xi}) = a_0\hat{V}_1(\vec{\xi}_1)\zeta_1(\vec{\xi}) \quad , \tag{4.13a}$$

where the subscript I denotes the first exposure, the coefficient of reflection V_I at the point $(\vec{\xi}_I)$ has been written according to (4.2), and a_0 is the amplitude of the illuminating wave. Similarly, the field during the second exposure is

$$a_{II}(\vec{\xi}) = a_0\hat{V}_{II}(\vec{\xi}_{II})\zeta_{II}(\vec{\xi}_{II}) \quad . \tag{4.13b}$$

The field a_{II} differs from a_I by the constant phase $\Delta\varphi = 2\pi g_0/\lambda = kg_0$ as a result of displacement of the surface toward the illuminating wave and by a more complicated quantity as a result of the change of the local angles of illumination and observation as a result of rotation. In view of the small rotation angle $\beta << \pi$ and the slow dependence of the operator function $\hat{V}$ on the angles of incidence [4.5], we can rewrite (4.13b) in the form

$$a_{II}(\vec{\xi}) \approx a_0 \exp(-ikg_0) \exp(-ik\beta\xi)\hat{V}_{II}(\vec{\xi}_{II})\zeta_I(\vec{\xi}_I) \quad . \tag{4.13c}$$

We assumed, in deriving (4.13c), that the displaced position ζ_{II} of the surface ζ is related to its initial position ζ_I by the expression

$$\zeta_{II} \approx \zeta_I + g_0 + \beta\xi \quad ,$$

i.e., the surface does not change its shape and moves only as a rigid body. Hence,

$$\hat{V}_{II}\zeta_{II} \approx \exp(-ikg_0)\exp(-ik\beta\xi)\hat{V}_{II}\zeta_I \quad ,$$

which was taken into consideration in obtaining (4.13c).

Let us now consider the field $a_{II}(P_1)$ produced by the displaced surface at the point P_1. We see that it differs from the field $a_I(P_1)$ for two reasons: first, because of the difference between the field $a_I(\vec{\xi})$ and $a_{II}(\vec{\xi})$ just considered, and, second, because of the displacement of the point P_1 relative to the surface $\zeta(\vec{\xi})$. The magnitude of this displacement depends on the distance $\vec{R}$ to the point P_1, and also on the surface-displacement parameters $\vec{g}_0$ and β. Taking (4.1,8,13) into account, we can write an expression for the visibility of the interference fringes at the point P_1,

$$p \sim \int\limits_{(s)} \int\limits_{(s)} \frac{\exp\{ik[R_I(\vec{\xi},P_1) - R_{II}(\vec{\xi},P_1)] - ik(g_0 - \beta\xi)\}}{R_I(\vec{\xi},P_1)R_{II}(\vec{\xi},P_1)} \times$$

$$\times \cos\theta_I(\vec{\xi},P_1)\cos\theta_{II}(\vec{\xi},P_1)\hat{V}_I\hat{V}^*_{II}\Phi(\vec{\xi}_I,\vec{\xi}_{II})d\vec{\xi}_I d\vec{\xi}_{II} \quad . \tag{4.14}$$

In writing (4.14), we used (4.5) for the correlation function of a surface. Let us bear in mind that, in comparison with the fast-varying functions Φ and $\exp(ikR)$, the functions $\hat{V}$, $\cos\theta(\vec{\xi},P_1)$, and $\exp[-ik(g_0 - \beta\xi)]$ are slowly varying functions of the arguments $\vec{\xi}_I$ and $\vec{\xi}_{II}$. We thus see that an approximate value of the integral (4.14) is determined by the fast-varying functions, whereas the slowly varying ones may be replaced by constants and put outside the integral (the stationary phase method) [4.10]. Physically, this conclusion signifies that the main decorrelation of the fields a_I and a_{II} results not from the change of the field $\delta a(\vec{\xi}) = a_I(\vec{\xi}) - a_{II}(\vec{\xi})$ on the surface of the object, but from the change of location of the point P_1 relative to the first and second locations of the object. For this reason, formula (4.14) can be approximately rewritten [4.5] as

$$p \sim \int\limits_{(s)} \int\limits_{(s)} \exp\{ik[R_I(\vec{\xi},P_1) - R_I'(\vec{\xi},P_2)]\}\Phi(\Delta\vec{\xi})d(\Delta\vec{\xi})d\vec{\xi} \quad . \tag{4.15}$$

Thus, the visibility p of the fringes at the point P_1 is determined by the mutual correlation of the fields produced by the undisplaced surface $\zeta_I(\vec{\xi})$ at this point P_1 and at a point P_2. The latter is arranged relative to the undisplaced surface ζ_I in the same way as the point P_1 is relative to the displaced surface ζ_{II} (see Fig.4.3).

We have arrived at the important conclusion that an interference pattern carrying information concerning the change of location of an object is produced by displacement of the source of illumination relative to the object, in accordance with (4.13a,c) whereas destruction of the pattern of interference fringes is associated with the relative displacement of the point of observation and the resultant decorrelation of the fields $a_I(P_1)$ and $a_{II}(P_1)$.

How can we calculate the mutual correlation characteristics of the fields $a_I(P_1)$ and $a_I(P_2)$ so as to use formula (4.15)? We can conclude, by analizing the calculations presented in [4.5], that the correlation characteristics of fields calculated from perturbation theory and by Kirchhoff's approximation are similar if we consider that the large inhomogeneities have gentle slopes and that the illumination is close to normal.

To assess the degree of mutual correlation of the field a_I at the point P_1 and P_2 (with a small angular and spatial separation of these points), we must draw mirror-image rays from these points to the illuminating source S (Fig.4.4). The mutual correlation of the fields $a_I(P_1)$ and $a_I(P_2)$ is the same as the correlation surface function (4.5) whose argument equals the distance between the points Q_1 and Q_2 of mirror reflection.

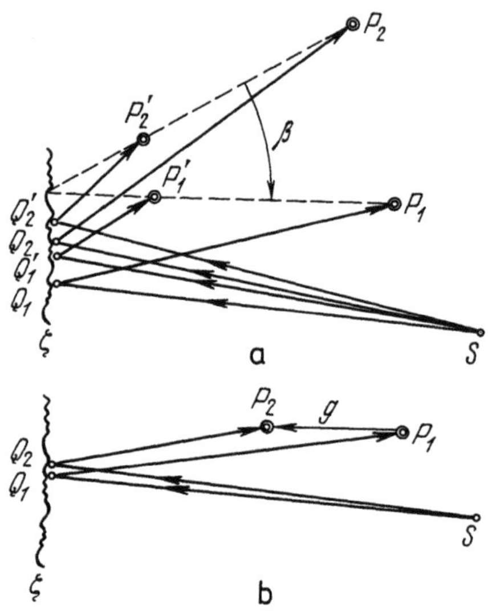

Fig.4.4. Determining the degree of mutual correlation of fields at the point P_1 for different surface displacements: (a) rotation through the angle β; (b) piston displacement by the vector $\vec{g}$

This conclusion gives us a direct indication of the region of localization (i.e., of greatest visibility) of the interference fringes. The latter are observed only in the region of space for whose points the distance Q_1Q_2 between the corresponding points of mirror reflection becomes smaller than (or equal to) the radius of correlation of the irregularities on the surface being studied,

$$|Q_1Q_2| \leqslant \ell \quad .$$

Consequently, for pure rotation, there is a tendence for fringe localization near the surface (Fig.4.4a). In piston displacement, the fringes seem to be farther from the surface when the direction of observation is closer to the direction of illumination (Fig.4.4b).

Here it is essential to introduce a more precise definition of the correlation radius ℓ. It is quite plausible that the position and dimensions of the fringe-localization region depend critically on the value of ℓ. For example, for mirror surfaces, for which $\ell \sim L$ (L is the dimension of the illuminated part of the surface), the fringes are localized in a considerably greater region than for diffuse surfaces. This is confirmed by the data of Fig.4.5, which show interferograms of a surface that was subjected to an angular displacement between exposures [4.11]. The image of the surface 10 cm behind the hologram was photographed by use of an optical system that has a depth of field of about 1 cm. The right-hand part of the surface was polished. It can be seen that the interference fringes are localized on the surface if the latter is diffusely reflecting, and exist in the entire region between the observer and the surface if the latter is polished.

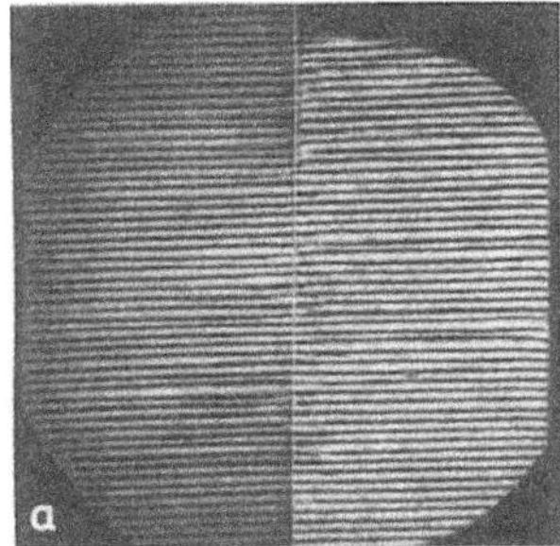

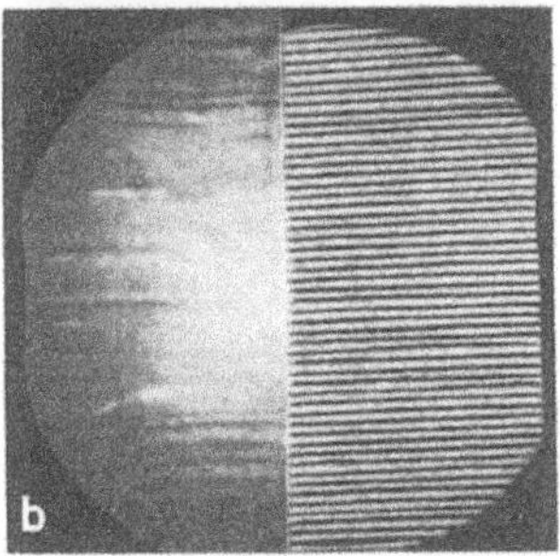

Fig.4.5. Localization of interference fringes by holograms of specularly (right) and diffusely (left) reflecting surfaces [4.11]: (a) focusing on object plane; (b) focusing outside of object plane

This is the case, however, only when the polished surface is observed in the direction of mirror reflection. In other directions, the correlation radii associated with scattering by imperfections of the mirror surface (usually of very small size) may be smaller than for a conventional diffusely reflecting surface.

The assumptions concerning the kind of displacement made in deriving (4.15), are not obligatory. It is not difficult to see that for small displacements,

$R_I, R'_I \gg |\vec{g}|$,

similar rules for constructing a second point P_2 near the point of space P_1 of interest to us can be derived. The position of the point P_2 depends on the kind of displacement of the object and the distance to the point P_1 from the surface. The relationship between the characteristics of a scattering surface and the values of the correlation volume of the scattered field is also, as we have already mentioned, well known for a very large range of inhomogeneities of the surface $\zeta(\vec{\xi})$.

Therefore, the foregoing approach is most general, which permits us to determine the region of localization of the interference fringes for objects with different degrees of surface finish and which undergo different displacements between exposures. In addition to a knowledge of the general laws that govern the formation of a localization region, however, it is often necessary to work out simpler, although less general, criteria that will permit us to describe this effect clearly.

The method proposed by ALEKSANDROV and BONCH-BRUEVICH [4.12], called the "method of corresponding points", has found greatest favor. This approach was developed in works of Viénot and described later in [4.13]. A more detailed substantiation of this method is given by WALLES [4.14]. According to the ideas of this group of investigators, the laws of localization of an interference pattern can be presented on the basis of the following reasoning (Fig.4.6a). Regular interference, i.e., interference that

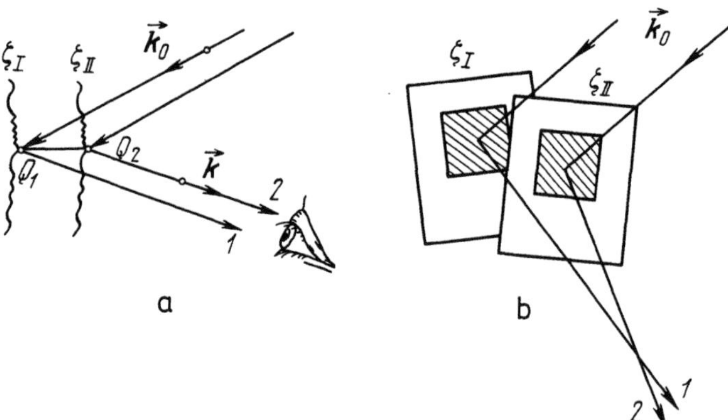

Fig.4.6. Explanation of effect of interference-fringe localization: (a) by the method of corresponding points; (b) on the basis of the perturbation theory

produces a clear fringe pattern, is possible when two surfaces ζ_I and ζ_{II} are observed simultaneously through a hologram only by light rays that travel to the observer from identical points Q_I and Q_{II} on these surfaces. It is obvious that the vector $\overrightarrow{Q_IQ_{II}}$ in Fig.4.6 describes the displacement ($\vec{g}$) between exposures. The phase shift between the rays 1 and 2 that travel to the observer from these points depends on the changes of phase of the illuminating wave that arrives at the points Q_I and Q_{II} from a laser, and of the phase of the scattered wave that travels to the observer,

$$\Delta\varphi = (\vec{k}_0 - \vec{k})\vec{g} \quad . \tag{4.16}$$

The existence of a fringe-localization region leads to an effect similar to parallax: by slightly changing our point of observation, the fringe pattern is displaced relative to the surface being studied. Hence, the localization region can be obtained for each specific kind of surface displacement by equating to zero two derivatives (with respect to orthogonal coordinates in the plane of the surface) of the phase increment $\Delta\varphi$ written in the form of (4.16). Thus, according to the method of corresponding points, the localization region can be determined as the surface or volume in which the increments of the phases of light rays that travel from corresponding points of two surfaces do not depend on small changes of the observer's position.

In a real experiment, the method of corresponding points gives unambiguous results only when small displacements of perfect diffusely reflecting surfaces are studied (i.e., surfaces that produce no bright spots by specular reflection, [4.11]. The correlation method described earlier is preferable when a broader range of objects is studied.

To supplement our ideas of the different ways of interpreting the fringe localization effect, we must mention the work of TSURUTA et al. [4.15] who described this phenomenon in terms of the so-called principle of selective scattering that results from the perturbation theory and formula (4.12). In this approach, the surface element responsible for formation of a wavefront that travels in a given direction is considered as diffraction grating. The orientation of its lines and its grating constant are such that, according to an expression of the type of (4.11), the illuminating wave is diffracted exactly in the given direction (Fig.4.6b). What will happen to that grating when that element of surface is displaced between exposures? If the surface is rigid, then the orientation and position of the grating will change. If the surface is deformed at the given place, then the grating constant will also change. The light diffracted by the displaced or deformed grating may intersect the light from the undisplaced grating in a certain (real or imaginary) region of space, which according to [4.15] is the region of interference-fringe localization.

A notable merit of the method proposed in [4.15] is that it can be used to describe the influence not only of displacements of an object, but also of certain deformations (for example stretching) on fringe localization. This approach is convenient when the experimenter is skillful and has a well-developed spatial imagination. Strictly speaking, however, when applied to diffusely reflecting surfaces (in the approximation of the perturbation theory) it is valid only in the zone of "spectrum separation" which ranges from

$$R \gg k\ell L \sin^2\psi \quad , \tag{4.17a}$$

where R is measured from the surface being studied [4.7], to the region of Fraunhofer diffraction by the illuminated part L of the surface

$$R \gg kL^2 \quad . \tag{4.17b}$$

We remind the reader that ℓ is the correlation radius of roughnesses, and $k = 2\pi/\lambda$.

Generally, in holography, the geometrical relationships between the dimensions of the illuminated section L and its distance R to the point P where the field is studied do not satisfy the weaker requirement (4.17a).

The foregoing analysis shows that analysis of the correlation properties of the fields a_I and a_{II} that are reconstructed from a hologram is the most general and unambiguous approach for understanding the processes of interference-fringe formation and localization [4.16].

4.2 Methods of Interpreting Holographic Interferograms when Displacements are Studied

The expansion of the possibilities of interferometry associated with the development of holography is attended by a growing complication of the methods used to interpret the interferograms, i.e., of the transition from the stage of recording an interference pattern to that of obtaining complete information on the changes that occur in an object during the time between two exposures (for example – surface displacements).

Complex deformations of an object's surface can be interpreted as combined effects of a number of simpler movements:

a) movements of the object as a whole (parallel displacement of the surface in directions either in its own plane or outside of it; rotation of the surface about axes in the plane of the surface and orthogonal to that plane);

b) deformations of separate parts of the surface of the object (compression and tension, shear, bending, twisting).

The main difficulty in interpreting holographic interferograms is that quite different displacements of an object may result in the production of identical interference patterns. Inspection of Fig.4.7 shows that identical systems of fringes are obtained when a surface rotates about an axis in its plane and when it rotates about an axis normal to its plane [4.12].

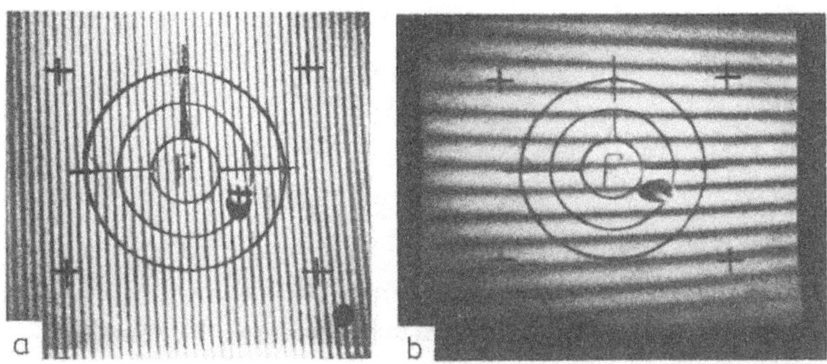

Fig.4.7. Nature of the interference fringes formed by different kinds of displacement of a surface [4.12]: (a) rotation about an axis in the object plane; (b) rotation about an axis normal to the object

Experimental skill and also study of the localization features of an interference pattern permit us partly to eliminate this difficulty. This is not adequate, however, for quantitative interpretation of the interferograms, and a significant amount of experimental work and calculation has to be done. Let us consider the path of the light rays in the general case when a holographic interferogram of a part of a surface near the point Q_1 (Fig.4.8) is recorded. The light source is at the point O, which is chosen as the origin of coordinates, and the observer is at the point P. The element of surface being studied is displaced between exposures to the point Q_2, i.e., $\overrightarrow{Q_1Q_2} = \vec{g}$. Let $\vec{k}_1$ and $\vec{k}_3$, $\vec{k}_2$ and $\vec{k}_4$ be the wave vectors that describe the direction of the illuminating and scattering light rays, respectively, and relative to the points Q_1 and Q_2.

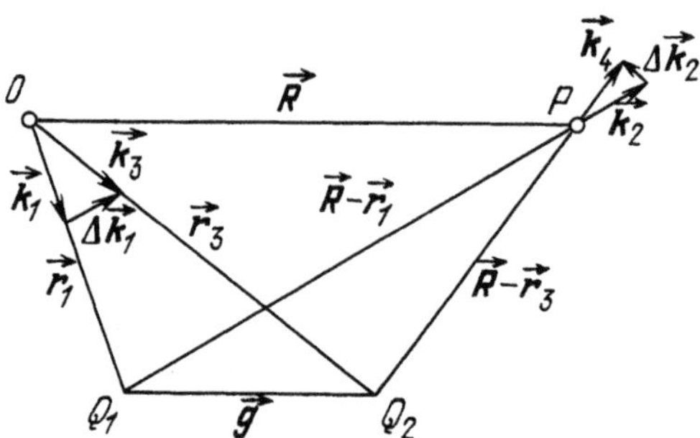

Fig.4.8. Paths of light rays from corresponding points Q_1 and Q_2 of an object

Calculation of the phases of the rays that interfere at the point P(OQ_1P and OQ_2P) gives

$$\left.\begin{aligned} \varphi_1 &= \vec{k}_1\vec{r}_1 + \vec{k}_2(\vec{R} - \vec{r}_1) + \varphi_0 \quad , \\ \varphi_2 &= \vec{k}_3\vec{r}_3 + \vec{k}_4(\vec{R} - \vec{r}_3) + \varphi_0 \quad . \end{aligned}\right\} \quad (4.18)$$

The quantity φ_0 in (4.18) signifies the constant phase shift of a light wave of wavelength λ when it is reflected from the material of which the surface being studied consists. Expressing the wave vectors in accordance with Fig.4.8, i.e.,

$$\vec{k}_3 = \vec{k}_1 + \Delta\vec{k}_1 \quad \text{and} \quad \vec{k}_4 = \vec{k}_2 + \Delta\vec{k}_2 \quad ,$$

we get an expression for the phase difference between two rays,

$$\varphi = \varphi_2 - \varphi_1 = -(\vec{k}_1 - \vec{k}_2)(\vec{r}_1 - \vec{r}_3) - \Delta\vec{k}_1\vec{r}_3 + \Delta\vec{k}_2(\vec{R} - \vec{r}_3) \quad . \quad (4.19a)$$

The condition

$$|\vec{g}| = |\vec{r}_3 - \vec{r}_1| << |\vec{r}_1| \approx |\vec{r}_3| \quad , \tag{4.20}$$

which is usually satisfied experimentally, permits us to consider that $\Delta\vec{k}_1$ and $\Delta\vec{k}_2$ are orthogonal to the vectors $\vec{r}_3$ and $(\vec{R} - \vec{r}_3)$, respectively; therefore we can get [4.12]

$$\varphi = -(\vec{k}_1 - \vec{k}_2)(\vec{r}_1 - \vec{r}_3) = (\vec{k}_1 - \vec{k}_2)\vec{g} \quad . \tag{4.19b}$$

Expression (4.19b) must be analyzed for two basic cases:

a) the two-dimensional case — all of the vectors in this expression are in one plane, which, without harm to generalization, can be considered the plane of Fig.4.8;

b) the three-dimensional case — the directions of the vectors are arbitrary; for example, the displacement vector $\vec{g}$ is not in the plane that passes through the vectors $\vec{k}_1$ and $\vec{k}_2$. Such a plane always exists, however, in accordance with the definition of these vectors. Figure 4.9a shows the directions of the vectors that interest us, and the angles between them for the two-dimensional case [4.17]. We can arrange to have this case in practice when we know beforehand in what plane the displacement vector lies and correspondingly choose the directions of illumination and observation.

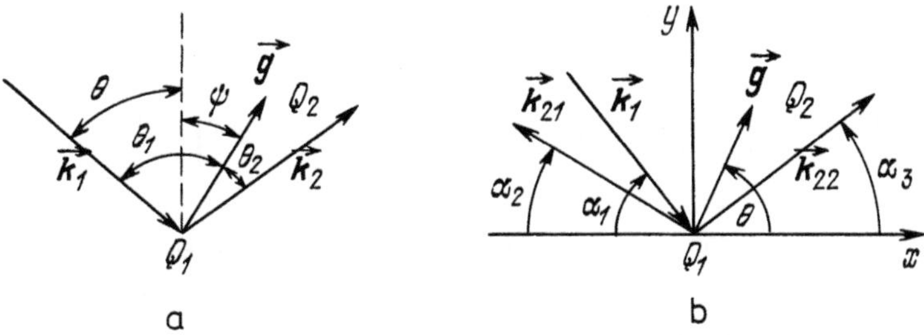

Fig.4.9. Designation of wave vectors and angles of illumination and observation for an experiment in which the plane containing the displacement vector $\vec{g}$ is known beforehand: (a) magnitude of the vector $\vec{g}$ is not known; (b) with an indefinite direction of the displacement vector $\vec{g}$ (within the limits of the plane of the drawing)

Transforming (4.19b) in view of $|\vec{k}_1| = |\vec{k}_2| = 2\pi/\lambda$ and $\cos|\pi - \theta_1| =$ $= -\cos\theta_1$, we get

$$\varphi = \frac{2\pi}{\lambda} g(\cos\theta_1 + \cos\theta_2) \quad . \tag{4.21}$$

Introducing the angles θ and Ψ according to Fig.4.9a, i.e., $\theta = (\theta_1 + \theta_2)/2$ and $\Psi = (\theta_1 - \theta_2)/2$, we can transform (4.21) into a form known from the theory of radio-signal phase modulation,

$$\varphi = \frac{2\pi}{\lambda} g \cos\theta \cos\Psi \quad . \tag{4.22}$$

Equation (4.21) can be used quite successfully when we know the direction of the displacement vector $\vec{g}$ but do not know its value. This is often encountered in the vibrometry of objects in which the vector of the vibrations is normal to the surface.

If the direction of the vector $\vec{g}$ (with account taken of the assumption that permits us to reduce the case to a plane problem) is unknown, however, then (4.21) is not sufficient for determining the two quantities $|\vec{g}|$ and Ψ (or θ_2). We then have to choose two directions of observation (see Fig.4.9b) that correspond to the wave vectors $\vec{k}_{21}$ and $\vec{k}_{22}$. Introducing new designations of the angles, we get a system of two equations for the phase shifts observed at these two angles (α_2 and α_3),

$$\left.\begin{aligned} \varphi_{12} &= -\frac{4\pi}{\lambda} g \cos\frac{1}{2}(\alpha_2 - \alpha_1)\cos\left[\theta + \frac{1}{2}(\alpha_2 + \alpha_1)\right] \quad , \\ \varphi_{13} &= -\frac{4\pi}{\lambda} g \sin\frac{1}{2}(\alpha_3 + \alpha_1)\sin\left[\theta - \frac{1}{2}(\alpha_3 - \alpha_1)\right] \quad . \end{aligned}\right\} \tag{4.23}$$

Now, to find φ_{12} and φ_{13} from the experimental data and to use them to determine the unknown quantities g and θ, we have to count the number of interference fringes that separate the given point Q_1 on the surface of the object observed at the angle α_2 or α_3 from a point that is not displaced, i.e., determine the number of the interference fringe N_{12} or N_{13}

$$N_{12} = \frac{\varphi_{12}}{2\pi} \quad \text{and} \quad N_{13} = \frac{\varphi_{13}}{2\pi} \quad . \tag{4.24}$$

Usually the zero fringe in an experiment is defined as the one that does not change its position when the point of observation is displaced: the fringe pattern runs in accordance with (4.23,24), while the zero fringe remains stationary.

Equations (4.23) are simplified somewhat if we introduce some symmetry into the illumination and observation schemes [4.18]. We can consider the following cases (see Fig.4.9b).

1) $\alpha_1 = \alpha_3$. One of the directions of observation corresponds to that of specular reflection. Here the components of the vector $\vec{g}$ are determined that are parallel and perpendicular to the surface from which the angles in Fig.4.9b are measured (i.e., g_x and g_y),

$$g_x = \left[\frac{2\pi}{\lambda}(\cos\alpha_2 + \cos\alpha_1)\right]^{-1}\left[\frac{1}{2}\varphi_{13}\left(1 + \frac{\sin\alpha_2}{\sin\alpha_1}\right) - \varphi_{12}\right] ,$$

$$g_y = \left[\frac{2\pi}{\lambda}(\sin\alpha_2 + \sin\alpha_1)\right]^{-1}\left[\frac{1}{2}\varphi_{13}\left(1 + \frac{\sin\alpha_2}{\sin\alpha_1}\right)\right] = \frac{\lambda\varphi_{13}}{4\pi\sin\alpha_1} . \quad (4.25)$$

The values of g_x and g_y obtained by use of (4.25) allow us to find the required quantities g and θ,

$$\theta = \arctan\frac{g_y}{g_x} \quad \text{and} \quad g = \sqrt{g_x^2 + g_y^2} . \quad (4.26)$$

2) $\alpha_2 = \alpha_3$. Both directions of observation are mirror images of each other. In this case, we determine the components of the displacement vector $\vec{g}$ directed along the sum and the difference of the vectors $\vec{k}_{21}$ and $\vec{k}_{22}$, i.e., perpendicular and parallel to the surface,

$$g_x = \lambda\frac{(\varphi_{13} - \varphi_{12})}{4\pi\cos\alpha_2} ,$$

$$g_y = \frac{\lambda\varphi_{13}[1 + (\cos\alpha_1/\cos\alpha_2)] + \varphi_{12}[1 - (\cos\alpha_1/\cos\alpha_2)]}{4\pi(\sin\alpha_2 + \sin\alpha_1)} . \quad (4.27)$$

The most-general three-dimensional case, when we have no a priori information concerning the displacement vector $\vec{g}$ is illustrated in Fig.4.10a.

All that remains from our previous assumptions is that which relates to the small value of the displacement vector (4.20).

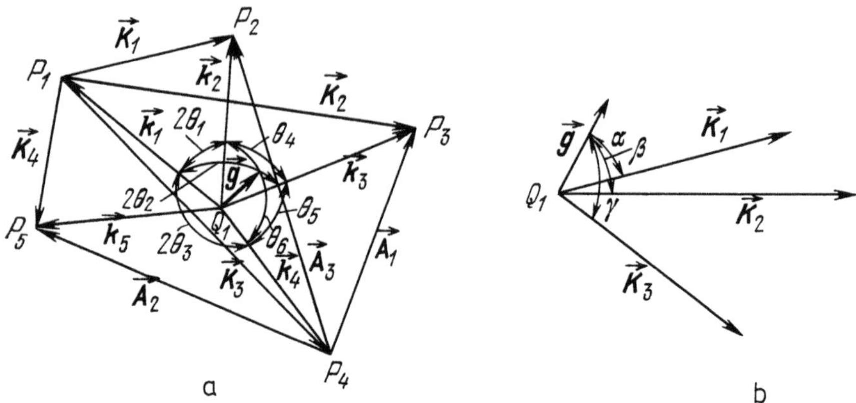

Fig.4.10. Designations of the wave vectors and angles for interpreting interferograms in the general case: (a) the points of the object are illuminated in the direction $\vec{k}_1$, observation is in the direction $\vec{k}_2$, $\vec{k}_3$, $\vec{k}_4$, $\vec{k}_5$; (b) determination of the angles α, β, and γ

Let us consider two variants of the three-dimensional case.

1) If "zero" fringes can be detected on the surface, i.e., fringes that remain stationary when the observer moves, the multiple-hologram method should be used to interpret the interference pattern [4.17].

2) If no "zero" fringes are detected, the method of ALEKSANDROV and BRONCH-BRUEVICH [4.12] is more suitable.

To understand the part that belongs to each of these methods, let us imagine that we have a hologram in the form of a hemisphere with a radius much greater than the expected displacements. The hemisphere is centered around the point Q_1 in which we are interested, and a doubly exposed hologram is recorded on it. In the reconstruction stage, we see the image of the object and an interferogram. If we translate our point of observation and look at the point Q_1 through different parts of the hologram, then the interference fringes move. By counting, for example, how many dark fringes pass through the line "point Q_1 to observer", we get a measure of the phase shift $\varphi = 2\pi N$, where N is the number of fringes.

If two segments are now taken out of the hemisphere, we can no longer count continuously the fringes by passing over from one hologram to another. We can determine the number of each fringe, however, if the "zero" fringe is visible. Performing this calculation for each segment, we find the required number N by simple subtraction of one number from the other (the remainder may be a fraction). This remainder will equal the number that we would get by continuously changing the direction of observation corresponding to the transfer of the observation point from one hologram to another.

Thus, the technique of the multiple-hologram method depends on finding the "zero" fringe, whereas that of the single-hologram method depends on counting the number of fringes when the point of observation changes within the limits of one hologram.

Let us return to the general model in Fig.4.10a. Here $\vec{k}_1$ is the wave vector of the illuminating wave, $\vec{k}_2, \ldots, \vec{k}_5$ are the wave vectors that are associated with the different viewing directions. If we can count the number of fringes that pass through the given point for different directions of observation, then the basic equation (4.19b) can be transformed into a whole set of equations:

$$\left.\begin{aligned} 2\pi N_1 &= \varphi_1 = (\vec{k}_2 - \vec{k}_1)\vec{g} = \vec{K}_1\vec{g} \quad , \\ 2\pi N_2 &= \varphi_2 = (\vec{k}_3 - \vec{k}_1)\vec{g} = \vec{K}_2\vec{g} \quad , \\ 2\pi N_3 &= \varphi_3 = (\vec{k}_4 - \vec{k}_1)\vec{g} = \vec{K}_3\vec{g} \quad , \\ 2\pi N_4 &= \varphi_4 = (\vec{k}_5 - \vec{k}_1)\vec{g} = \vec{K}_4\vec{g} \quad . \end{aligned}\right\} \qquad (4.28)$$

The interpretation of (4.28) will vary, depending on the procedure used.

4.2.1 Procedure for Multiple-Hologram Investigation

In this case we know the direction of illumination $\vec{k}_1$, use three holograms the normals to which are directed toward the point Q_1 being studied, and form the wave vectors $\vec{k}_2$, $\vec{k}_3$, and $\vec{k}_4$. After constructing the difference vectors $\vec{K}_1$, $\vec{K}_2$, and $\vec{K}_3$, as shown in Fig.4.10a and transforming the first three equations of (4.28) to the form

$$\left.\begin{aligned} \varphi_1 &= \frac{4\pi}{\lambda} \sin\theta_1 \cdot g \cos\alpha \quad , \\ \varphi_2 &= \frac{4\pi}{\lambda} \sin\theta_2 \cdot g \cos\beta \quad , \\ \varphi_3 &= \frac{4\pi}{\lambda} \sin\theta_3 \cdot g \cos\gamma \quad , \end{aligned}\right\} \tag{4.29}$$

where the angles α, β, and γ are determined from Fig.4.10b, we can determine the projections of the vector $\vec{g}$ onto the directions $\vec{K}_1$, $\vec{K}_2$, and $\vec{K}_3$ according to the values of φ_1, φ_2, and φ_3 calculated from the interferograms and the angles θ_1, θ_2, and θ_3 found from the geometry of the scheme. This permits us to find the value and direction of the vector $\vec{g}$.

4.2.2 Procedure for Single-Hologram Investigation

Here we must have four detectors that can count the fringes when they are placed within the limit of the hologram aperture and view the given point from different directions $\vec{k}_2, \ldots, \vec{k}_5$.

Let N_{ij} be the number of fringes counted by a detector when it passes from the direction $\vec{k}_i$ to the direction $\vec{k}_j (i, j = 2,\ldots,5)$,

$$\Delta\varphi_{ij} = 2\pi N_{ij} \quad . \tag{4.30}$$

On the other hand, subtracting one equation of (4.28) from another, we get

$$\begin{aligned} \Delta\varphi_{34} &= \varphi_3 - \varphi_4 = (\vec{K}_3 - \vec{K}_4)\vec{g} \quad , \\ \Delta\varphi_{32} &= \varphi_3 - \varphi_2 = (\vec{K}_3 - \vec{K}_2)\vec{g} \quad , \\ \Delta\varphi_{13} &= \varphi_1 - \varphi_3 = (\vec{K}_1 - \vec{K}_3)\vec{g} \quad . \end{aligned} \tag{4.31}$$

Equations (4.31), like (4.29), determine the three projections of the vector $\vec{g}$ on the directions $\vec{A}_1$, $\vec{A}_2$, and $\vec{A}_3$,

$$\vec{A}_1 = \vec{K}_3 - \vec{K}_2, \; \vec{A}_2 = \vec{K}_3 - \vec{K}_4, \quad \text{and} \quad \vec{A}_3 = \vec{K}_1 - \vec{K}_3 \quad , \tag{4.32}$$

that can be constructed on the basis of the spatial arrangement of four detectors in a holographic setup (the points P_2, P_3, P_4, and P_5 in Fig. 4.10a).

Equations (4.28) and the multiple-hologram method that they describe require a knowledge of the direction of illumination and the position of the "zero" fringe. The single-hologram method and the equations (4.31) that describe it do not require this a priori information. On the face of it, this makes the single-hologram method considerably more practicable and convenient. This is indeed true if the solid angle subtended by the hologram permits the direction of observation to be so displaced that it is possible to view the passing of at least one interference fringe through each point of interest on the surface. The effort to subtend the greatest possible solid angle with a hologram by bringing it closer to the object must not violate requirement (4.20).

Calculations described in [4.18] show convincingly that for a 20 × 25 cm hologram 60 cm from the object, the angle between the difference vectors $\vec{A}_1$, $\vec{A}_2$, or $\vec{A}_3$ is 168 degrees, i.e., these vectors are almost coplanar and are in the plane of the hologram. Hence, the single-hologram method makes it possible to determine exactly the components of the displacement vector parallel to the plane of the hologram. The normal components are determined with much less accuracy.

The multiple-hologram method is considerably more convenient for finding the normal components, especially when we can easily identify the "zero" line (for example, from physical considerations, from the conditions of fastening of the object).

A geometrical interpretation of the interference patterns observed as a result of recording doubly exposed holograms that is interesting and very helpful for understanding the phenomena is given by ABRAMSON in a series of investigations [4.19-24]. It is important to note that these investigations also resulted in the creation of a device for the practical evaluation of interferograms [4.23] and extension of the procedure proposed by the author to a great number of other problems of optics and radio physics in which the interference of signals is used [4.24].

According to Abramson's development of the fundamental nature of interference fringes [4.1], any holographic scheme that consists of a source of diverging light flux A, a hologram B, and an object C, determines a family of ellipses (for the three-dimensional case — of ellipsoids) that are the locus of points of equal optical-path differences along the path "source — object — hologram" (Fig.4.11a).

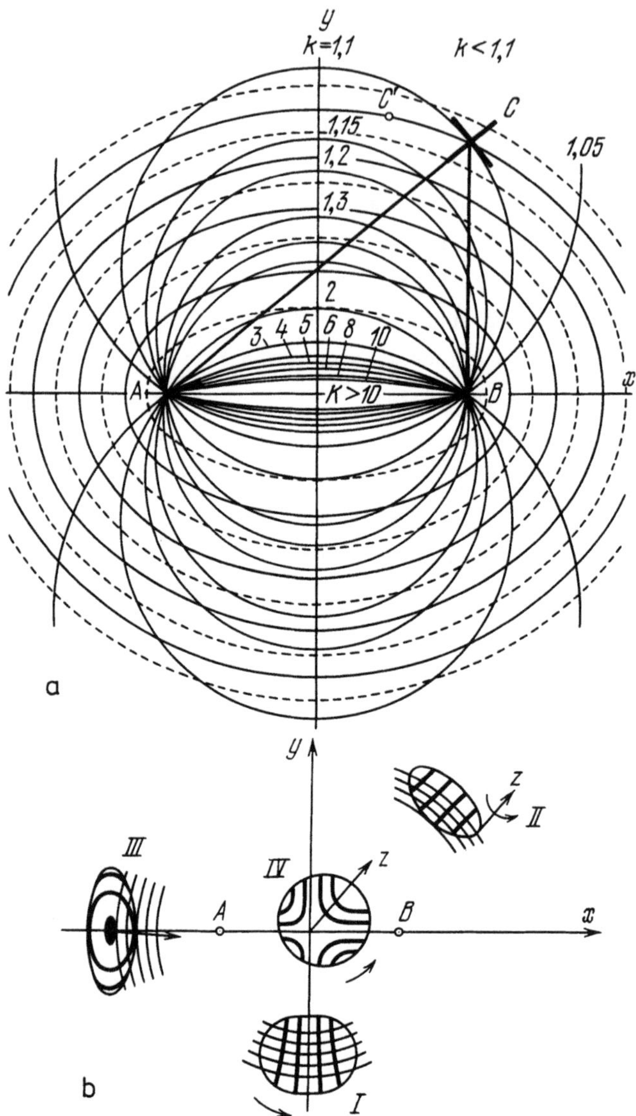

Fig.4.11. Holographic diagram of ABRAMSON [4.19]: (a) general view of diagram: A - source; B - hologram; C - object; (b) view of interferograms for different positions and kinds of displacement of objects: I - the plane of the object is parallel to the plane xy, the axis of rotation is parallel to the z axis; II - the plane of the object is normal to the plane xy, the axis of rotation is parallel to the plane of the object and the z axis; III - the plane of the object is normal to the plane xy, the displacement vector is parallel to the x axis" IV - same as for case I, but the directions of illumination and observation are close to grazing

By the hologram B and object C, we understand only small regions of them that can be represented by the points B and C, and by the source A the point in the diagram from which the illuminating beam expands (for instance, the location of a pinhole diaphragm).

Let us ascribe to two neighboring ellipses the property that from one of them to the other the optical-path length ACB differs by λ. It is now a simple matter to explain the appearance of the interference fringes by basing our explanation on the moiré effect [4.25]. Although the content of this topic is closer to Sect.4.1, we prefer to treat it here so as to present Abramson's ideas [4.19-24] consistently. Abramson reasons that when the beam ACB interferes with the reference beam at the point B, a microstructure, which in the first approximation can be considered similar to a local diffraction grating, appears on the hologram and is recorded on its light-sensitive material. We shall call this microstructure the primary interference fringes. If between the exposures, the object C is moved along the ellipse on which it was originally, the phase difference between the reference beam and the beam AC'B does not change, and the primary fringes were not displaced on the hologram. In the reconstruction, the point C of the object is seen with brightness twice what it would have in a singly exposed hologram. What is taken to be the "zero" line of a holographic interferogram, or the secondary interference pattern, is seen.

If, between exposures, the object C is displaced to a point between two adjacent ellipses, then the maxima of the primary fringes that correspond to the first exposure will coincide with the minima of the fringes that correspond to the second exposure. As a whole, a uniform gray will be recorded on the hologram B, and in the reconstruction, the part C of the object will be invisible (a dark fringe of the secondary interference pattern).

Every intersection of an ellipse by the point C will correspond to a change of the optical-path length of ACB by λ and, consequently, to one interference fringe in the secondary pattern. What distance does the object C have to travel to pass over from one ellipse to another? This distance obviously depends on the position of the point C on the ellipses. In the vicinity of the x axis (the conditions of normal illumination and observation), the sensitivity of the setup is greatest because the required displacement is $\lambda/2$. In the vicinity of the y axis, the sensitivity depends on the number of the ellipse – the more it is extended, the lower is the sensitivity (the conditions of grazing illumination). The lines of constant distance between the ellipses are arcs of circles passing through the points A and B (see Fig.4.11a). These arcs are called k lines, where k corresponds to the distance between the ellipsoids expressed in units of $\lambda/2$.

Thus, knowing the position of the object (Fig.4.11a, point C), we can compute its displacement in a direction normal to an ellipse as

$$g = \frac{nk\lambda}{2} \quad . \tag{4.33}$$

For the case shown in Fig.4.11a, the point C is in the region between the arcs $k = 1.1$ and $k = 1.15$, therefore the displacement $g \approx 1.12\lambda$ corresponds to one interference fringe.

Although Abramson's interpretation of interference-pattern formation is excessively simplified, and the primary interference structure can rarely be even approximately identified with a diffraction grating [4.6,26], the diagram shown in Fig.4.11a undoubtedly had a practical use. That diagram allows us to choose the proper configuration of a holographic setup, and also to predict the nature of the fringes for a number of simple displacements (Fig.4.11b). If an object whose surface coincides with the plane xy of the drawing rotates about an axis parallel to the z axis, then those of its points that move along the ellipses do not change the phase of the reflected light that arrives at the point B. Hence, the lines of maximum sensitivity and, consequently, the interference fringes are directed normal to the family of ellipses (object I). An object whose plane and axis of rotation are perpendicular to the xy plane (object II) will also be covered with a family of fringes that are almost parallel to one another and are similar in appearance to the ones obtained in the preceding case. These conditions are confirmed by the data of Fig.4.7. An object III that moves parallel to itself along the x axis will be covered with concentric interference fringes. When an object is rotated in the same way as was object I, but with grazing illumination and observation, the fringes are hyperbolas (object IV).

The possibilities of the methods of holographic interferometry are usually verified experimentally. For this purpose, the diagrams of the displacements obtained by evaluating an interferogram (or several interferograms) are compared with the results of independent measurements or calculations.

Figure 4.12 shows diagrams of the displacements of different points of a cantilever beam [4.16] and [4.27] (for small and big loads). Attention is drawn to the excellent correspondence of theory and experiment. However, the case of a cantilever beam is one of the simplest ones in holography: the fringes are localized on the surface of the object, the zero

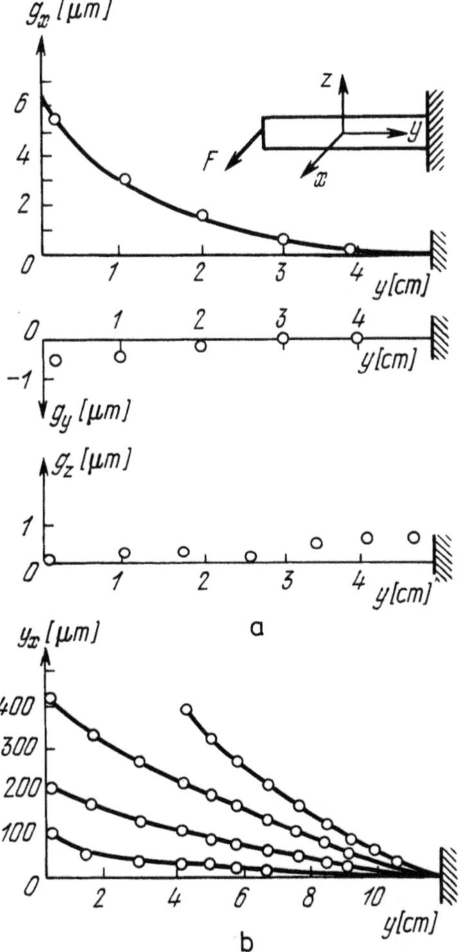

Fig.4.12. Relationship between theoretical and practical results in study of a loaded cantilever beam: (a) the three components of displacement along the axes x, y, and z (points) and the theoretical curve; (b) experimentally measured components of the displacement in the direction of loading (points) and the corresponding theoretical curves for different loads

fringe is known from the conditions of constraint, and the direction of the displacement vector is also known, namely, the x axis (see Fig.4.12a),

$$g_x >> g_y, g_z \quad .$$

As a result of improvement of the quality of holographic experiments, primarily by increase of the brightness and visibility of interference patterns, and decrease of the noise level [4.28], the characteristics of the reconstructed hologram can be used more completely. In other words, we are able to extract information not only from shapes and the numbers of interference fringes, but also from their visibility and localization, as we assumed in Sect.4.1. Let us consider how we can use these characteristics [4.29] and thus diminish the number of holograms or points needed to measure the vector $\vec{g}$ in accordance with (4.28,31).

Let us change Fig.4.8, introducing into it the surface Σ being studied and the region of localization of the interference pattern R, and also the coordinate axes (x,y) and (x',y') constructed in planes parallel to the surface being studied and originating from the point Q and the point of observation P (Fig.4.13).

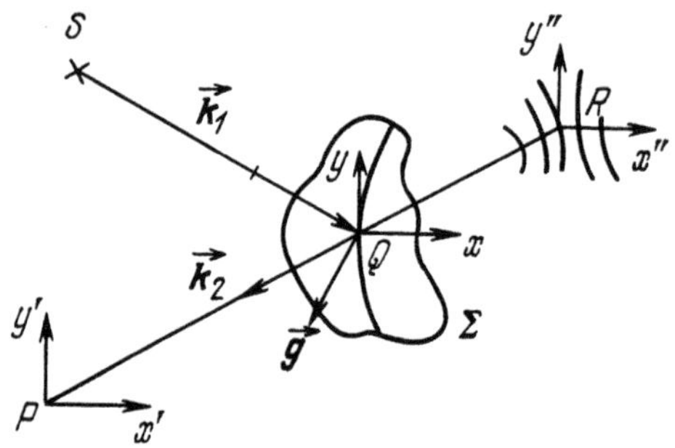

Fig.4.13. Determination of the displacements by measurement of the visibility of the interference fringes [4.29]

Let us move the observation point, moving away from the axis PQ, but remaining in the plane x'y'. The phase difference φ between exposures that result from displacements of the point Q expressed by (4.19b), can be expanded in the form of Taylor series,

$$\varphi(x'y') = \varphi(0,0) + \alpha x' + \beta y' \quad , \tag{4.34}$$

where the coefficients α and β are determined on the basis of Fig.4.13,

$$\left.\begin{aligned}\alpha &= \frac{t-r}{t}\left.\frac{\partial\varphi(0,0)}{\partial x}\right|_Q + \left.\frac{\partial\varphi(0,0)}{\partial x'}\right|_P \quad , \\ \beta &= \frac{t-r}{t}\left.\frac{\partial\varphi(0,0)}{\partial y}\right|_Q + \left.\frac{\partial\varphi(0,0)}{\partial y'}\right|_P \quad .\end{aligned}\right\} \tag{4.35}$$

Here t and r are the distances from the observer to the region of fringe localization (PR) and to the object (PQ), respectively, and $\varphi(0,0)$ is the phase difference produced in the object beam in accordance with the displacement of the point Q by the amount $\vec{g}$ (4.19b). This phase difference corresponds to observation strictly along the axis PQ.

We can show, on the basis of Fresnel's diffraction integral (4.1), that for small movements of the observer (x',y') the relationship between the fields a_I and a_{II} that correspond to the first and second exposures, in the plane (x'',y'') can be expressed in the form [4.29],

$$a_{II}(x'',y'') = a_I(x'' + \alpha t\lambda,\ y'' + \beta t\lambda)\exp[i\varphi(0,0)] \quad . \tag{4.36}$$

In accordance with (4.8,10,36), we can relate the visibility of an interferogram p to the complex autocorrelation function Φ of the statistical field $a_I(x'',y'')$, i.e., to the amplitude of the speckle structure at the point P,

$$p = \Phi_{a_I}(\alpha t\lambda, \beta t\lambda) \quad . \tag{4.37}$$

According to the definition, $\Phi(0,0) = 1$, i.e., $\Phi = 1$ when $\alpha = \beta = 0$. The latter condition can be satisfied in accordance with (4.35) only if

$$\left.\frac{\partial\varphi(0,0)}{\partial x}\right|_Q \left.\frac{\partial\varphi(0,0)}{\partial y'}\right|_P = \left.\frac{\partial\varphi(0,0)}{\partial x'}\right|_P \left.\frac{d\varphi(0,0)}{\partial y}\right|_Q \quad . \tag{4.38}$$

Therefore, an interference pattern of maximum visibility is observed at a distance t from the point P measured in the direction of the point R for which (4.38) is satisfied.

The particular case when t = r, i.e., the fringes are localized on the surface Σ being studied, corresponds to parallelism of the vectors $\vec{g}$ and $\vec{k}_2$, i.e., to observation in the direction of the displacement. This follows from an analysis of (4.19b) by use of (4.34,35,36). For some displacements, this conclusion corresponds to practical observations.

If we measure the visibility of the fringes not in the plane of localization, where $p \equiv 1$, but on the surface of the object, then in the general case

$$p(x,y) = \Phi(g_x,g_y) \leqslant 1$$

because the spatial displacement of the scattered fields a_I and a_{II} on the surface Σ exactly follows the displacement of the pertinent points of the scattering surface.

When the angular dimensions of the aperture at the point P are smaller than the scattering indicatrix of the irregularities $\zeta(x,y)$ on the surface Σ, then as shown by VELZEL [4.29], the autocorrelation function Φ is described by the equations of Fraunhofer diffraction on this aperture. For a round aperture, for example, the correlation length is $\ell_{opt} \approx 0.6r\lambda/d$, where d is the diameter of the aperture. Therefore, when observing an interference pattern on a surface through an optical system with a variable aperture, we begin to note the vanishing of this pattern as soon as the tangential component of the displacement becomes greater than the correlation dimension ℓ_{opt} of the aperture[6],

$$g_x,g_y > \ell_{opt} \quad . \tag{4.39}$$

The example given in Fig.4.14 confirms this very important conclusion and agrees with what was said in Sect.4.1. Figure 4.14 shows an interferogram of a surface that was rotated about a vertical axis in its plane. The scattering properties of the surface (the indicatrix opening exceeds 50°) satisfied the aforementioned relationship with the aperture of the observation system. Introducing the numerical values of the experimental data ($r = 20$ cm, $d = 3$ cm, $\lambda = 0.63$ μm) into the expression for ℓ and tak-

[6] This result agrees with the conclusions of Sect.4.1. Indeed, the greater the displacement, the greater is the distance Q_1Q_2 between the points of mirror reflection situated in accordance with the construction of Fig.4.4. By reducing the aperture of observation, we increase the correlation dimension ℓ, and the fringes become noticeable when the condition $|Q_1Q_2| \leqslant \ell$ is obeyed.

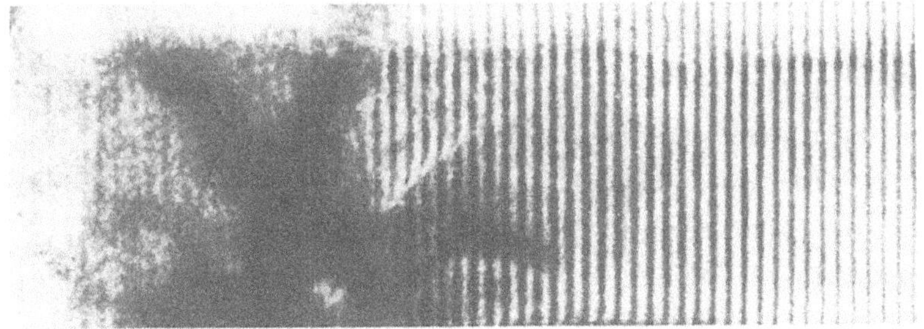

Fig.4.14. Interferogram of a surface rotated about a vertical axis in the plane of the drawing. The reduction of visibility of the interference fringes with increase of the absolute value of the displacement is due to decor relation of the scattered fields and is equal to the calculated value

ing into consideration that both observation and illumination were practically along a normal, we find the value of $\ell_{opt} \approx 2.5\ \mu m$. The tangential component of the displacement g obviously equals

$$g_y = \frac{gy}{r} \quad ,$$

where y is the horizontal coordinate perpendicular to the direction of the fringes. We find from the condition $g_y \approx \ell$ that the vanishing of the interference pattern should be observed at the coordinate y that satisfies the conditions

$$n \frac{\lambda}{2} y \frac{y}{r} \approx 2.5\ \mu m \quad ,$$

where n is the number of interference fringes per centimeter along the horizontal coordinate, and $ny\lambda/2$ is the displacement g of the points of the surface with this coordinate y. Considering that $n \approx 10$ lines/cm, we find that y is about 4.5 cm, which qualitatively corresponds to the results of experiments.

This procedure for evaluating an interferogram can be improved for the purpose of determining the two tangential components of the displacement vector (g_x and g_y), separately. For this purpose, an interferogram must be studied with the aid of a slit aperture that rotates about the axis QP at the point Q and has a variable transverse dimension. By placing the aperture in the position of the maximum visibility of the fringes, we find the direction of the component of g in the xy plane. Next, be changing the size of the aperture and making the fringes disappear, we use the foregoing procedure to find the value of this component of the displacement vector. We must note once more that the described method may be used only if interferograms of highest quality are obtained (with unit visibility), and the noise level in the reconstructed image does not interfere with accurate measurement of the parameters of the interferogram and determination of the appearance and vanishing of the interference fringes in the plane being studied [4.29].

All of the methods of interpreting interferograms described in the foregoing are based on the fact that interference is possible only between two rays that arrive at the point of observation P (see Fig.4.8) from the point Q_1 being studied and its copy Q_2 in the image of the object that corresponds to the second exposure. In essence, this assumption which underlies the hypothesis of corresponding points is valid if the optical system used to observe the reconstructed image does not have resolution sufficient to separate points Q_1 and Q_2, as we have just shown in expression (4.39). Here we are confronted with an interesting fact that is rarely discussed in works on holographic measurement of displacements. It is that, depending on the parameters of the optical system of observation, there are two ways of studying displacements: the first is <u>direct optical resolution</u> of identical elements on the displaced and undisplaced surface and measurement of the corresponding displacements. This method is realized with a high resolution of the optical system. When the resolution becomes poorer, that method passes over into the other – <u>holographic interferometry</u>, to which the present chapter is devoted.

This fact suggests the question: if we separate in some way the corresponding points in the images of the surface being studied and suppress the remaining radiation, under what conditions will it be possible to observe an interferogram of two spherical wavefronts whose centers correspond to the positions of the displaced and undisplaced points on the surface being studied? If this is possible, then one interferogram would provide us with information concerning displacements of any region of interest without need to count fringes in different directions.

This question can be answered as a result of the following experiment [4.30] (Fig.4.15a). A light beam from a laser is directed through a pinhole diaphragm and a microscope objective 1 by the beam splitter 2 and lens 3 onto the surface 5 of the object. The lens 3, apart from collimi-nating the beam, focuses an image of the surface 5 in the plane of the hologram 7. Another lens 4 and the mirror 6 form the reference beam. Hence, here we have one of the variants of a setup for holographic investigation using image hologram techniques [4.31,32].

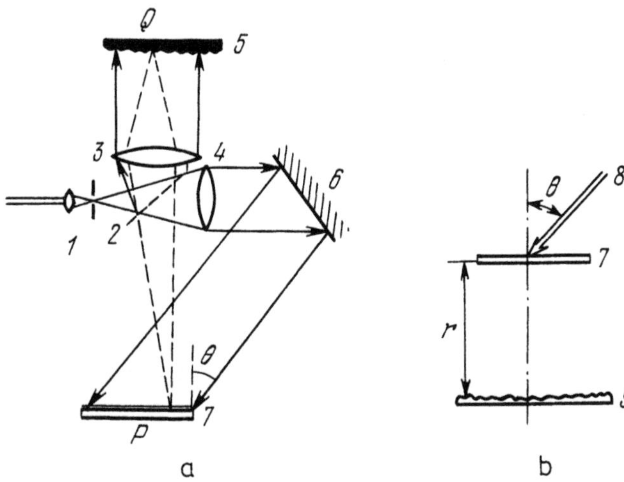

Fig.4.15. Study of displacements by use of interferograms of separate points on a surface [4.30]: (a) arrangement for recording the interferograms; (b) arrangement for reconstruction

Thus, with the aid of the reference wave, the hologram records two wavefronts that correspond to the initial and final positions of the surface 5. Of course, these wavefronts differ from true ones because of the filtering properties of the space between the object and the hologram, the lens and the photographic material [1.35], but we shall disregard this effect in the first approximation. In this approximation, the displacement of the wavefronts identically repeats that of the surface.

Therefore, if in the reconstruction stage we illuminate the processed hologram 7 with the narrow laser beam 8 (see Fig.4.15b), then the images of two elements of the surface that correspond to its displaced and undisplaced positions will be reconstructed in the plane of the hologram. These two elements are equivalent to two extended quasimonochromatic sources displaced in space relative to each other. This displacement equals exactly that of the section Q of the surface 5 we are interested in (the magnification of lens 3 is unity). Consequently, a pattern of interference fringes whose parameters completely describe the displacement of the surface in the region of the point Q will be observed in the ground glass 9 (Fig.4.16).

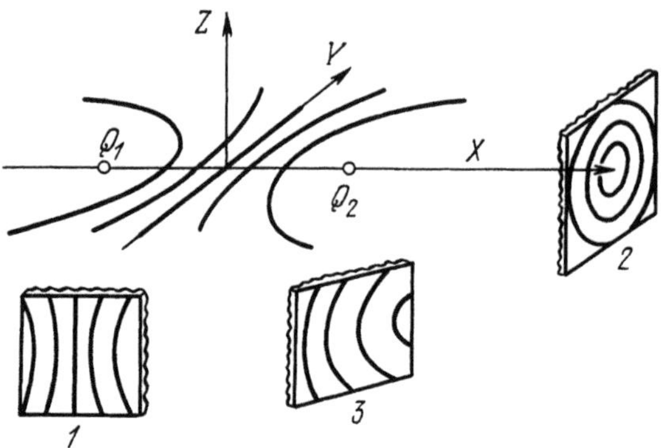

Fig.4.16. Form of interference fringes for different kinds of displacement of a point relative to the observer: 1 - tangential displacement; 2 - normal displacement; 3 - intermediate case

The interference pattern of the two point sources Q_1 and Q_2 at $(0 + \Delta X, 0, 0)$ and $(0 - \Delta X, 0, 0)$ in the system of coordinates shown in Fig.4.16 is known to consist of hyperboloids of revolution whose axis XX is determined by the equation for the dark interference fringes,

$$\frac{X^2}{a^2} - \frac{Y^2 + Z^2}{b^2} = 1 \quad ,$$

where

$$\left.\begin{aligned} 2a &= \left(k + \frac{1}{2}\right)\lambda \\ b &= \sqrt{(\Delta X)^2 - a^2} \\ 2(\Delta X) &= g \end{aligned}\right\} . \qquad (4.40)$$

Thus, if the displacement $Q_1Q_2 = \vec{g}$ is in the plane being studied, then the interference fringes correspond to the position of the ground glass 1 in Fig.4.16 and are in essence Young fringes. If the displacement $\vec{g}$, on the other hand, is normal to the ZY plane, then the fringes are similar to

Newton rings and have the shape shown on the ground glass 2 (see Fig. 4.16). The displacements of the fringes have an intermediate shape for general cases.

The interferograms obtained in this way for various kinds of displacements were used to compute and construct the graphs depicted in Fig.4.17. They show the correspondence between the calculated and the experimentally determined displacement vectors for the rotation of a surface about an axis perpendicular to the surface, and for simultaneous displacement both in a plane and along a normal to it (Fig.4.17a,b, respectively) [4.30].

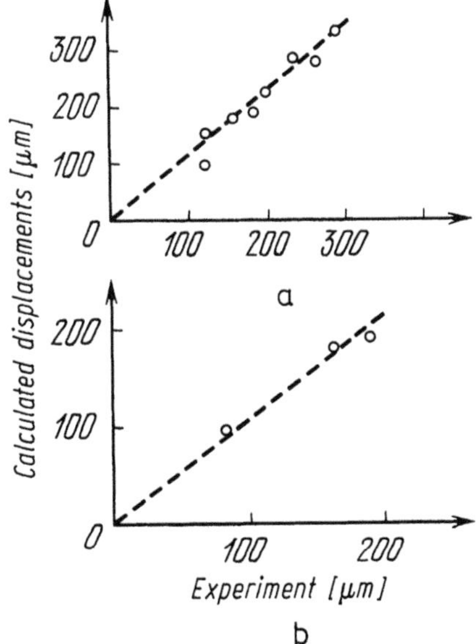

Fig.4.17. Results of experimental verification of the method proposed in [4.30] for different kinds of displacements: (a) rotation of a surface about an axis in its plane; (b) complex motion

These results show the great potential of this method, but they are inadequate for objective comparison of it with the methods set out in the first part of Sect. 4.2. It is clear, at present, that the prospects of this method are due to its suitability for determining comparatively great displacements, and also to its closeness to the methods of socalled speckle-interferometry. Consideration of the latter is beyond the scope of the present book and is the subject of separate study [4.33-35].

4.3 Investigation of Surface Relief

The task of investigating the relief of a surface consists of finding the deviation of the relief from that of a comparison surface. The latter is usually a plane. In this case, the contour lines that characterize the relief of the surface being studied are cross sections of an object with a set of equally spaced planes parallel to the comparison (reference) plane. As we noted in Chap.1, conventional interferometry makes it possible to obtain such contour maps. For this purpose, for example, we can replace one of the reflecting mirrors in a Michelson interferometer (see Fig.1.10) with the surface being studied. The value of one fringe, i.e., the distance between the adjacent planes that intersect the surface being studied, is $\lambda/2$ in this case. Thus, this method has a very high sensitivity and makes it possible to reveal deviations from a plane of the order of tenths and even hundredths of a wavelength.

However, the very great sensitivity of this method prevents use of such interferometry to investigate intricate surfaces that have great depth and gradient of relief, and also surfaces that have microstructures. In the first of these cases, the number of contour lines and their spatial frequencies would be too great, and the fringes would be unresolvable. In the second case, the arrangement and structure of the contour fringes would be determined not only and not so much by the macrorelief of the surface being studied as by its random microstructure.

As we have seen in the present chapter, holographic methods allow us to study surfaces that have a microstructure. What is compared, however, are two displaced or deformed surfaces whose microstructure is identical. The contours obtained characterize the displacements of corresponding points of the surfaces that are holographic replicas of the same deformed surface, and not its relief.

The problem of obtaining contour maps of the macrorelief of surfaces that have microstructure could be solved by obtaining interferograms in light of a wave length that is much greater than the depth of the elements of the microrelief. In this case, the value of one fringe would be correspondingly greater, and the microrelief of the surface would not affect the shapes of the interference fringes. Thus, one way of solving the problem is to use far infrared or even microwave ranges. A serious difficulty here, however, is the absence of materials of the required quality for recording the holograms.

Another way of solving the same problem is to use methods of holography that have reduced sensitivity. We considered some of them in Chap.3. Indeed, we showed that when radiation with two wavelengths λ_1 and λ_2 is used, the sensitivity of the interference method is only $\lambda_2/(\lambda_1 - \lambda_2)$ times that with wavelength λ_1. This is equivalent to using radiation with the wavelength $\lambda_1\lambda_2/(\lambda_1 - \lambda_2)$.

Actually, the same result is obtained when some of the methods of holographic interferometry are used to study relief, namely, the two-wavelength and the immersion methods.

4.3.1 Two-Wavelength Method

This method of obtaining the contours of a surface relief was first proposed by HILDEBRAND and HAINES [4.36]. The method consists of using two nearly equal wavelengths $\lambda_1 = \lambda$ and $\lambda_2 = \lambda + \Delta\lambda$ to obtain a hologram of the surface being studied. As we have already noted, this leads to a reduction in the sensitivity of $\lambda/\Delta\lambda$ times, i.e., the value of one interference fringe (when the surface is illuminated from the side of the hologram) is equal to $\lambda^2/2\Delta\lambda$. Such a two-wavelength hologram can be considered as the superposition of two holograms. In the reconstruction process, each of these holograms gives an image of the surface in the light of the same wavelength of the reconstructing source λ_3.

We have noted [see (1.121,123)] that use in reconstruction of a wavelength that differs from the one with which the hologram was recorded causes a displacement of the image of the object and a change of its scale depending on the value of μ. Because μ differs for holograms prepared with λ_1 and λ_2 (i.e., $\mu_1 = \lambda_3/\lambda_1$ and $\mu_2 = \lambda_3/\lambda_2$), the two reconstructed images differ somewhat in scale and are displaced relative to each other in both transverse and longitudinal directions.

When the waves that correspond to these images interfere, interference contours of the surface relief are produced. The differences the lateral dimensions of the images and also their overall transverse displacements, however, result in transverse displacement of the corresponding points of the microstructure of the reconstructed images. On the one hand, this leads to deformation of the contour fringes, and on the other to displacement of the surface of their localization, relative to the surface of the object. Similar effects that result from the displacements and deformations of object were considered in the preceding sections of this chapter.

We can eliminate the lateral magnification of the image and the undesirable effects associated with it by using plane reference and reconstructing waves.

Indeed, assuming in (1.123) that $z_r = z_c = \infty$ and $m = 1$, we get $M_{lat} = 1$.

Under the same conditions, the equation for the lateral coordinate x_i of the image (1.121) has the form

$$x_i = x_0 + \left(\frac{1}{\mu}\frac{x_c}{z_c} - \frac{x_r}{z_r}\right) z_0 \quad . \tag{4.41}$$

Considering that the reference and the reconstructing beams are in the plane xz, we have

$$x_i = x_0 + \left(\frac{1}{\mu}\alpha_c - \alpha_r\right) z_0 \quad , \tag{4.42}$$

where $\alpha_c = x_c/z_c$ and $\alpha_r = x_r/z_r$ are the angles made by the reconstructing and the reference beams, respectively, with the z axis. [It should be remembered that (1.121) was obtained by assuming that the angles are small.]

To prevent displacement of the reconstructed images corresponding to different wavelengths, relative to each other, i.e., to have $x_{i1} = x_{i2}$, it is sufficient, as follows from (4.42), that

$$\frac{\alpha_c}{\lambda_3} = \frac{\alpha_{r1}}{\lambda_1} = \frac{\alpha_{r2}}{\lambda_2} \quad . \tag{4.43}$$

This corresponds to the previously obtained condition (3.28). The condition (4.43) can be satisfied by forming reference beams of wavelengths λ_1 and λ_2 by use of a diffraction grating, similar to the setup in Fig. 3.33. The condition (4.43) is automatically satisfied without any contrivances when holograms are recorded by use of Denisyuk's setup with opposed beams ($\alpha_c = \alpha_{r1} = \alpha_{r2} = 0$). The hologram, in this case, has no spectral dispersion, and there is no displacement of the reconstructed images.

Thus, by satisfying condition (4.43), and also by recording a hologram and reconstructing the image in parallel beams, we eliminate undesirable lateral displacements of the corresponding points on the reconstructed images of the surface being studied.

Let us consider the phases of the waves that emerge from these points. The difference between these phases will determine the nature and arrangement of the interference pattern obtained. Let us consider a setup (Fig. 4.18) in which the object being studied is illuminated by a plane light wave containing radiation of two wavelengths from the side of the hologram with the aid of the half-silvered mirror M. We shall consider a point of the object B with the coordinates x_0 and z_0. We shall measure the phase from the plane A which, like the hologram, is at the distance z_0 from the point B.

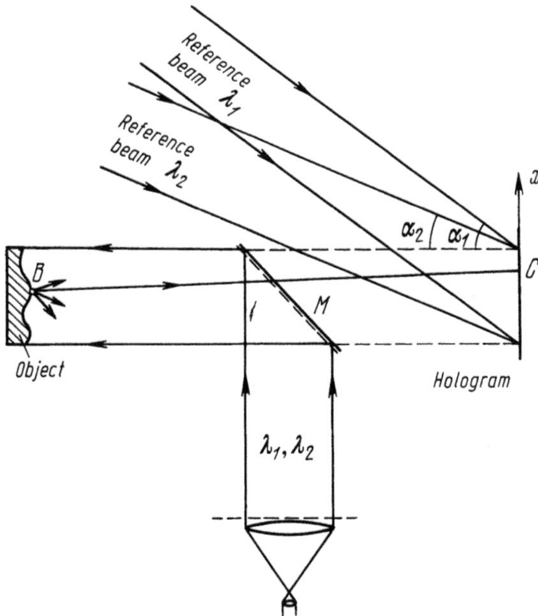

Fig.4.18. Calculation of phase difference in the two-wavelength method of studying relief

The phases of the illuminating wave for the two wavelengths at the point B will be

$$\varphi_{o1} = \frac{2\pi z_0}{\lambda_1} \quad \text{and} \quad \varphi_{o2} = \frac{2\pi z_0}{\lambda_2} \quad . \tag{4.44}$$

The spherical waves that arrive from the point B at the hologram have the phases

$$\varphi_1 = \frac{2\pi z_o}{\lambda_1} + \frac{2\pi r}{\lambda_1} \quad \text{and} \quad \varphi_2 = \frac{2\pi z_o}{\lambda_2} + \frac{2\pi r}{\lambda_2} \, , \tag{4.45}$$

where r is the distance from the point B to an arbitrary point C on the hologram with the coordinate x measured from the center of the hologram,

$$r = \sqrt{z_o^2 + (x - x_o)^2} \approx z_o \left[1 + \frac{1}{2}\left(\frac{x - x_o}{z_o}\right)^2 \right] \quad . \tag{4.46}$$

Using the approximation for small angles, we have $r \approx z_o$, and in this case

$$\varphi_1 = \frac{4\pi z_o}{\lambda_1} \quad \text{and} \quad \varphi_2 = \frac{4\pi z_o}{\lambda_2} \quad . \tag{4.47}$$

Let us direct two reference waves into the hologram at the angles α_1 and α_2. The phases of these waves are

$$\varphi_{r1} = \frac{2\pi x\alpha_1}{\lambda_1} \quad \text{and} \quad \varphi_{r2} = \frac{2\pi x\alpha_2}{\lambda_2} \quad . \tag{4.48}$$

The hologram will record the phases of the elementary spherical waves (4.47) relative to the corresponding reference waves (4.48)

$$\left.\begin{aligned} \varphi_1 - \varphi_{r1} &= \frac{4\pi z_o}{\lambda_1} - \frac{2\pi x\alpha_1}{\lambda_1} = \frac{2\pi}{\lambda_1}(2z_o - x\alpha_1) \, , \\ \varphi_2 - \varphi_{r2} &= \frac{4\pi z_o}{\lambda_2} - \frac{2\pi x\alpha_2}{\lambda_2} = \frac{2\pi}{\lambda_2}(2z_o - x\alpha_2) \, . \end{aligned}\right\} \tag{4.49}$$

By illuminating the hologram with a plane wave directed at the angle α_3 with the wavelength λ_3, we reconstruct the initial distributions of the phase of the waves that emerge from separate points of the object relative to the reconstructing wave, i.e., the phases of the reconstructed elementary waves in the plane of the hologram will be

$$\left.\begin{aligned}\varphi_{1c} &= \varphi_1 - \varphi_{r1} + \frac{2\pi x\alpha_3}{\lambda_3} = \frac{2\pi}{\lambda_1}(2z_o - x\alpha_1) + \frac{2\pi}{\lambda_3}x\alpha_3 \\ \varphi_{2c} &= \varphi_2 - \varphi_{r2} + \frac{2\pi x\alpha_3}{\lambda_3} = \frac{2\pi}{\lambda_2}(2z_o - x\alpha_2) + \frac{2\pi}{\lambda_3}x\alpha_3 \ .\end{aligned}\right\} \tag{4.50}$$

The phase difference of these waves in the plane of the hologram is

$$\Delta\varphi = \varphi_{1c} - \varphi_{2c} = 4\pi z_o\left(\frac{1}{\lambda_1} - \frac{1}{\lambda_2}\right) - 2\pi x\left(\frac{\alpha_1}{\lambda_1} - \frac{\alpha_2}{\lambda_2}\right) \quad . \tag{4.51}$$

When condition (4.43) is satisfied, the second addend in the right-hand side of (4.51) is zero, and we have

$$\Delta\varphi = \frac{4\pi z_o \Delta\lambda}{\lambda_1\lambda_2} \ , \tag{4.52}$$

i.e., the phase difference is proportional to the distance from the hologram to the points of the surface.

This phase difference corresponds to waves that fall on the hologram from the point B of the object. The other points of the object with other coordinates z_o and x_o also send waves to the same point of the hologram. As a result of superposition of a great number of pairs of elementary waves with phase differences that change in accordance with the relief of the surface z_o, we shall not see a regular interference pattern in the plane of the hologram. The waves that emerge from different points of the object will be separated most on the surface of the object itself, which is exactly a surface of localization of the interference pattern. The phase difference, in the approximation of small angles, will be determined by (4.52) for any plane, including that of localization.

A change of the phase difference of 2π corresponds to the distance between the maxima of neighboring interference fringes, i.e.,

$$\Delta z_o = \frac{\lambda_1\lambda_2}{2\Delta\lambda} \approx \frac{\lambda^2}{2\Delta\lambda} \ , \tag{4.53}$$

which determines the value of one contour fringe.

A more accurate treatment that does not use the small-angle approximation and collimated illumination and reference beams shows [4.36] that the contour lines are intersections the surface being studied with a family of ellipsoids of revolution whose common major axis connects the center of the hologram and the source that illuminates the object. If we make the illuminating source coincide with the center of the hologram, then the ellipsoids degenerate into spheres. The sections tend to planes with increasing distance from the object to the hologram and the illuminating source.

Of very great importance in the practical application of the method is the choice of the source of radiation. The latter must produce wavelengths that correspond to the desired fringe value. For this purpose, different pairs of lines of argon and krypton lasers are used [4.36-38], and also dye lasers [4.39] and pulsed ruby lasers [4.40]. In the latter case, when a pile or a Fabry-Perot interferometer is used as the output mirror, a doublet with the wavelength difference $\Delta\lambda = \lambda^2/2d$ is formed at the laser output (where d is the distance between the reflecting surfaces). Hence, the value of a fringe is

$$\Delta z_0 = \frac{\lambda^2}{2\Delta\lambda} = d \quad , \tag{4.54}$$

i.e., it equals the thickness of the étalon.

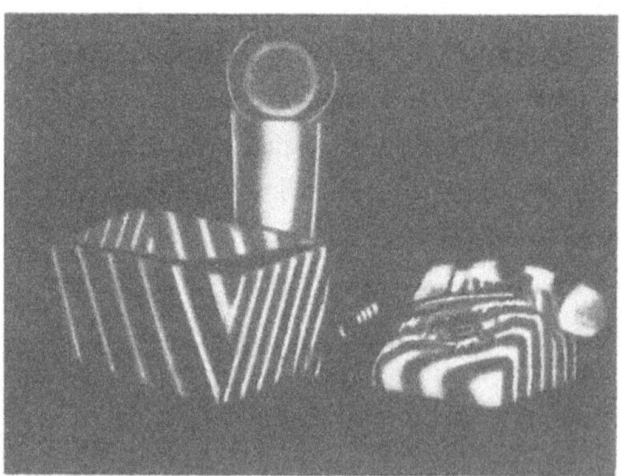

Fig.4.19. Holographic contours of a relief produced by the two-wavelength method ($\Delta\lambda$ = 1/8 Å, Δh = 23 mm) [4.40]

Figure 4.19 shows the contours of a surface relief obtained in this way [4.40] by use of a pulsed ruby laser.

4.3.2 Immersion Method

The double-exposure immersion method for obtaining the contours of surfaces proposed in [4.41] consists of placing the object being studied in an immersion cell (Fig.4.20) with a plane window. Before the first exposure, the cell is filled with a transparent liquid or gas of refractive index n_1, and before the second exposure it is filled with a medium of refractive index n_2. As we shall show below, this is equivalent to using an effective wavelength of about $\lambda/(n_1 - n_2)$, which makes it possible to change the value of a fringe within broad limits.

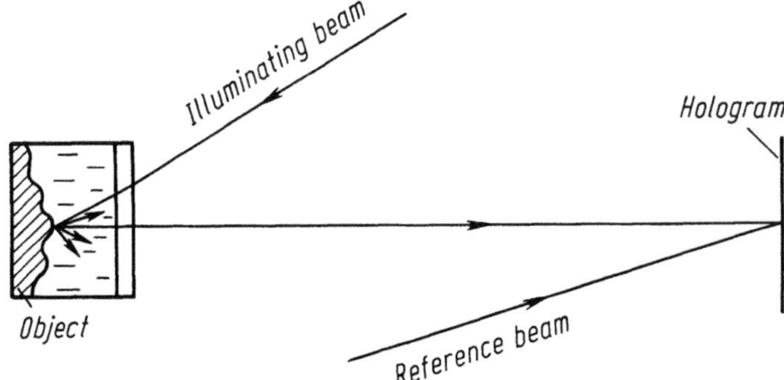

Fig.4.20. Immersion method of obtaining contour lines of a relief

If the plane of the cell window is normal to the direction of propagation of the object wave, then when the refractive index is altered, the angle of incidence of the object beam onto the hologram does not practically change. Because the wavelength and the angle of incidence of the reference beam also do not change, there is no lateral displacement or change of scale of the reconstructed images. This results in localization of the fringes on the surface of the object.

Let us calculate the path difference for two rays that arrive at the same point B of the surface and propagate from it in a direction normal to the immersion-cell window (Fig.4.21). Inspection of the figure shows that

$$\Delta \ell = (CB + h)n_1 - (AB + h)n_2 - DE \quad . \tag{4.55}$$

Taking into account that $CB = h/\cos\beta_1$, $AB = h/\cos\beta_2$, $DE = h(\tan\beta_1 - \tan\beta_2)\sin\alpha$, and also that $\sin\alpha = n_1 \sin\beta_1 = n_2 \sin\beta_2$, we get, after simple transformations,

$$\Delta\ell = [(1 + \cos\beta_1)n_1 - (1 + \cos\beta_2)n_2]h \quad . \tag{4.56}$$

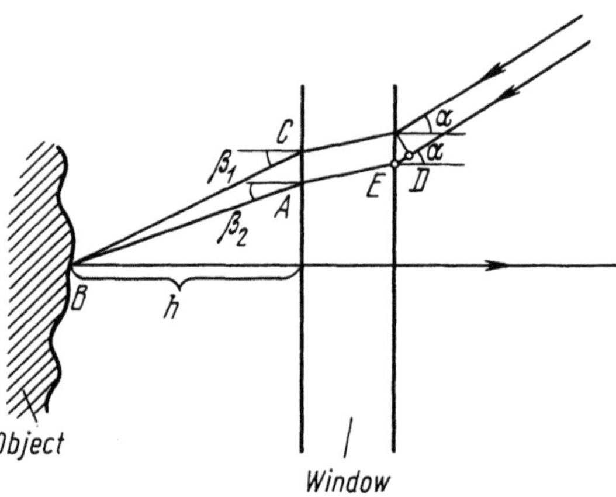

Fig.4.21. Calculation of path difference in the immersion method

The value of one fringe corresponds to a change of $\Delta\ell$ by one wavelength, i.e.,

$$\Delta h = \frac{\lambda}{(1 + \cos\beta_1)n_1 - (1 + \cos\beta_2)n_2} \quad . \tag{4.57}$$

When the object is illuminated from the side of the hologram ($\cos\beta_1 \approx \cos\beta_2 \approx 1$) by use of a setup similar to that shown in Fig.4.18, (4.57) becomes

$$\Delta h = \frac{\lambda}{2(n_1 - n_2)} \quad . \tag{4.58}$$

Gaseous freon was used for immersion in [4.41]. When its pressure is changed by 1 atm, Δh is 0.3 mm. Mixtures of water ($n = 1.333$) and ethylene glycol ($n = 1.427$), and also of water and alcohol ($n = 1.361$) make it possible to change the value of Δh within broad limits. Figure 4.22 shows a contour interferogram of a die for manufacturing turbine blades prepared in this way ($\Delta h = 2$ mm).

For the application of the immersion method, see also [4.37,42,43].

Fig.4.22. Contour interferogram of a die obtained by the immersion method (mixture of ethylene glycol and water, Δh = 2 mm) [4.43]

4.3.3 Double-Source Method

Apart from the two-wavelength method, the method of two illuminating sources shown schematically in Fig.4.23 has also been proposed [4.36]. Here A and B are the positions of the sources that illuminate the object during two consecutive exposures, or the positions of two coherent sources that illuminate the object simultaneously in the single-exposure variant of the method. In the latter case, the interpretation of this method is most evident and simple. Indeed (see Sect.1.1), a system of nodes and antinodes that have the shapes of hyperboloids of revolution is formed in the space that surrounds two coherent point sources. An object introduced into this system will be covered with contour fringes that are intersections of the object with these hyperboloids. In this variant of the method, it is not necessary to record a hologram; the contour fringes can be observed visually, or directly photographed.

In the holographic variant of this method, a hologram is exposed twice; the illuminating source is shifted from position A to B between the exposures. Because the coordinates of the image do not depend on the position of the source, the two reconstructed images completely coincide in space. Only the phases of the scattered waves differ, owing to the shift of position of the source. It is obvious that the positions of the fringes in the first and the second variants of the method are absolutely the same.

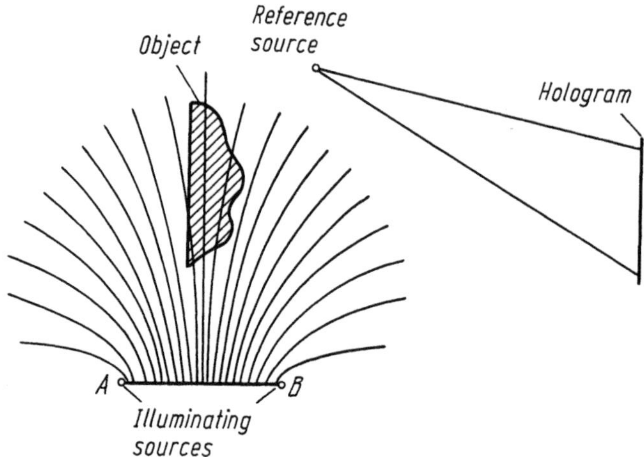

Fig.4.23. Obtaining contour maps by use of two illuminating sources

If the positions A and B of the source are sufficiently far from the surface being studied or, even better, if they are removed to infinity by use of collimators, then the intersecting surfaces, with sufficient accuracy, are equidistant planes that are separated from one another by spaces $\Delta h = \lambda/(2 \sin\alpha/2)$, where α is the angular distance between the sources.

The lines obtained are contour lines of the surface relief if the planes are oriented at right angles to the line of observation. A difficulty encountered with this arrangement arises because portions of the surface of the object are in a shadow, i.e., cannot be illuminated with such an arrangement of the sources.

A solution of this difficulty is to illuminate the object from many sides. An optical arrangement can be used for this purpose in which the light beam, after reflection from a rotatable mirror, is split and illuminates the object from opposite directions [4.44].

Another solution of the problem is to illuminate the object at an angle oblique to the line of observation. This improves the homogeneity of illumination and prevents the formation of shadows. The contour lines obtained are not lines of equal relief depth, but, as shown in [4.45], can serve as the basis for calculating such contour lines.

4.4 Flaw Detection by Holographic Interferometry

Important information can sometimes be obtained about objects studied by methods of holographic interferometry without resorting to interpretation of an interference pattern, using the methods described in Sect.4.2. Development of these methods is at a stage in which it is too early to consider automation of the processing of interferograms and, consequently, the design of industrial holographic installations that would make it possible to obtain all of the essential information about the field of displacements (vibrations) of an object in a short time.

Nevertheless, there is a broad class of applied problems for whose practical solution the interference patterns do not have to be interpreted quantitatively. In these problems, the information needed to make a decision is contained in the shapes and locations of the interference fringes over an object, in its "interference portrait". Features of the arrangement of interference fringes on an object directly provide information that indicates various defects in the internal structure of industrial specimens that serve various purposes such as:

1) Regions of reduced strength in the shells and outer casings of pneumatic and hydraulic devices and instruments (membrane pressure transmitters, rubber and plastic chambers and balls, life-rafts, etc.).

2) Defects in the connection of layers and plies in multilayer structures (printed-circuit cards, motor-vehicle and aircraft tires, sections of aircraft wings, multicomponent devices such as electromechanical transducers in which the quality of the joints between the piezoceramics and metal, metal and rubber, etc., determine their mechanical and acoustic properties, and so on).

3) Structural defects in the vital components of various prime movers (cracks and fatigue zones in turbine blades, high-pressure vessels, etc.).

4) The quality of manufacture of elements used in electromechanical devices intended for service in oscillating conditions (loudspeakers, acoustic receivers, piezotransformers, electromechanical delay lines).

Experiments show that to solve each concrete problem the investigator must choose, first of all, the optimal way of loading the object to detect most effectively the expected defects. The optimal method of loading is such that characteristic deformations of the object will be produced when an anomaly of the interference pattern is formed in the vicinity of the defect. This anomaly makes it possible to locate the defect and sometimes to assess how serious it is. The anomaly must be as clear as possible and noticeable either when the interference pattern produced by the defective object is directly observed, or when it is compared with the pattern of a reference specimen. The comparison of the "interference portraits" of the defective and reference specimens must be facilitated as far as possible by providing such loading conditions that the normal patterns have the simplest possible appearance (for example parallel or concentric fringes).

The following ways of loading objects are possible:

1) thermal (by heating or cooling particular portions of an object or the entire object);

2) mechanical (by compression, bending, twisting, or a combination of them);

3) vibration (the holographic method of studying vibrations is treated in Chap.5).

Figures 4.24,25 show examples of holographic interferograms. These examples show that the region where a defect or flaw appears is usually characterized by a zone of interference fringes in which the spatial frequency of the fringes considerably exceeds that of the adjacent (nondefective) regions. The interferogram of a tire [4.46] was obtained with the double-exposure method (Fig.4.24). The pressure in the tube was changed insignificantly between the two exposures. It is clear that in the parts of the tire with internal defects (for example, cracks, torn cord, etc.) the deformations of the tire surface are considerably greater than in adjacent regions that have normal strength. Therefore, the appearance of interference zones unerringly reveals defects of the tire.

Figure 4.25 shows a time-average interferogram of the surface of a multilayer electromechanical transducer when it was vibrationally excited. As in the preceding case, the regions of the surface over faults of glueing of the layers are mechanically "disconnected" from the substrate (ceramics) and, consequently, have greater amplitudes of oscillation. In this way, defects of glueing and internal cavities up to 0.1 mm thick can be made visible in the form of interference zones. Examination of the defective specimen (cutting in the region of the interference zone) shows that such cavities and defects actually always exist in the region of the interference zone, whereas removal of the layers between the defect and the normal pattern enables us to ascertain that, there, the substrate oscillates normally (see Fig.4.25b, right-hand side).

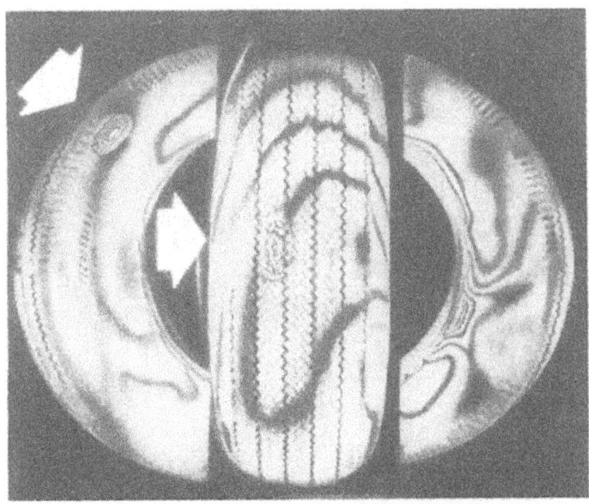

Fig.4.24. Interferogram of a tire [4.46]. The interference zones correspond to concealed defects of the tire (indicated by arrows)

Holographic investigation of loudspeaker diaphragms makes it possible to obtain amplitude and phase patterns of oscillation distribution over the surface of these acoustic devices (Fig.4.26) and, in a number of cases, to ascertain the anomalies that correspond to defects of manufacture. Articles of this kind, however, lend themselves very poorly to investigation because they operate under essentially nonlinear conditions; no success has been achieved in establishing direct correlation between the quality of a loudspeaker and the appearance of an interferogram (for example, its degree of symmetry).

The examples in Fig.4.27 show that, for simpler articles, vibrational excitation makes it possible to answer the question about the presence of

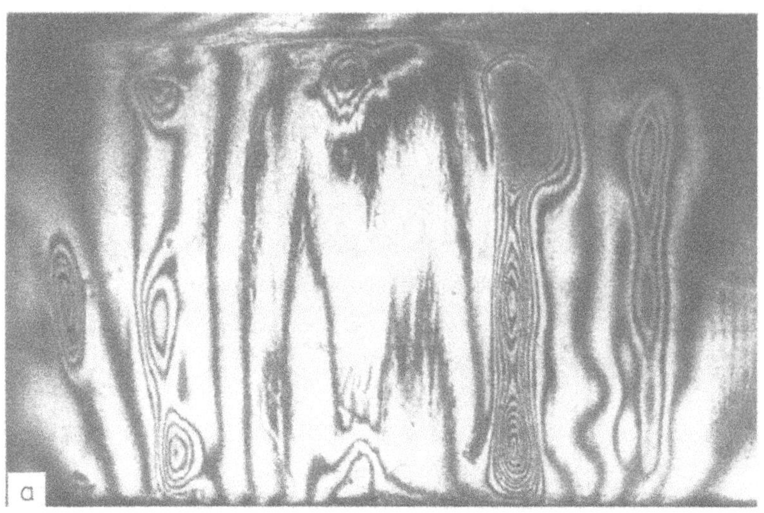

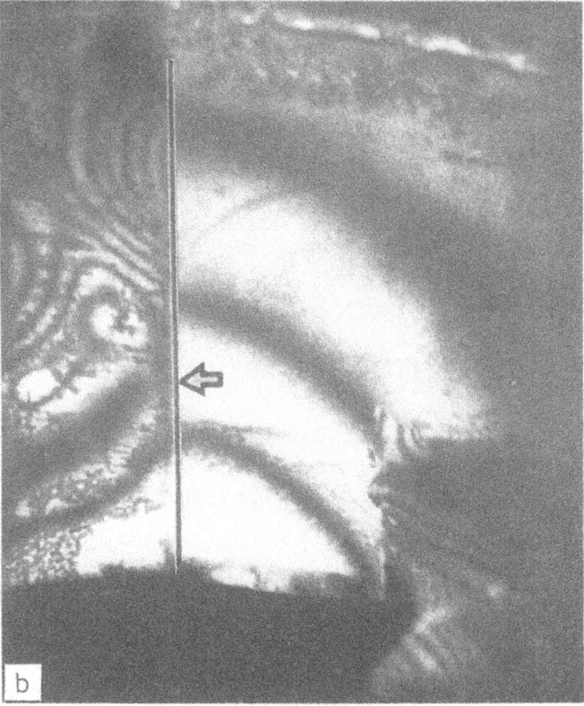

Fig.4.25. Interferogram of a multilayer cermet structure before (a) and after (b) partial removal of the layer between the surface and the defect

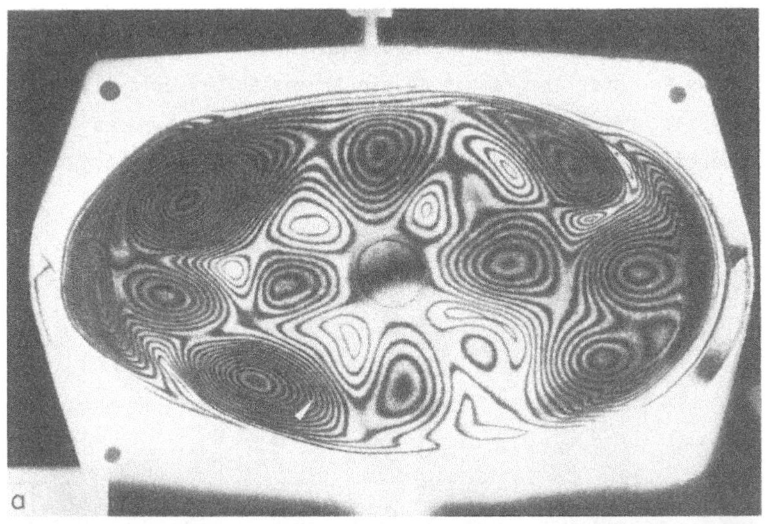

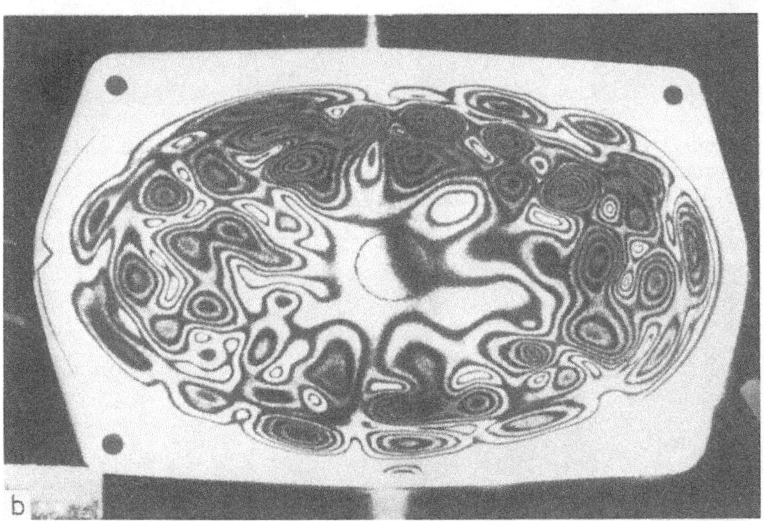

Fig.4.26. Interferograms of a loudspeaker at different frequencies: (a) 1 kHz; (b) 2.9 kHz

defects (thin spots) in the walls of hollow turbine blades. Such defects result in shifts of the resonance frequency of an article. The real-time method is used in this case to determine the resonance frequency and the shape of the oscillations. After this, a hologram of the blade is recorded, at the frequency found by the time-average method (see Chap.5). The interferograms in Fig.4.27 correspond to the third resonance frequency. For a fit specimen (see Fig.4.27a) it was 12.16 kHz, whereas for defective ones it varied from 12.04 to 10.31 kHz (see Fig.4.27b-e) [4.47].

Figure 4.28 shows an interferogram of an electromechanical delay line. The quality of manufacture of the element indicated by the arrow does not ensure its operation at the fundamental oscillating frequency [4.48].

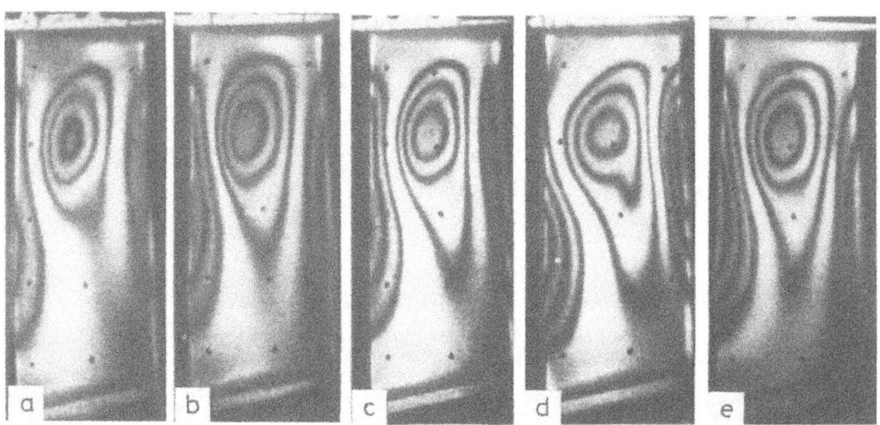

Fig.4.27. Interferograms of five hollow-turbine blades, obtained at the third resonance frequency, which were (a) 12.16 kHz; (b) 11.52 kHz; (c) 10.72 kHz; (d) 10.31 kHz; (e) 10.5 kHz [4.47]

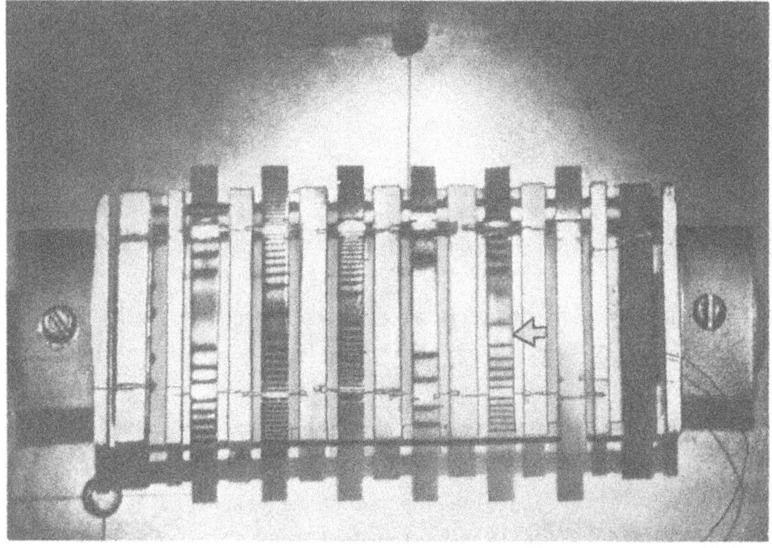

Fig.4.28. Interferogram of an electromechanical delay line. The arrow indicates the element whose oscillations do not correspond to the fundamental resonance

It is most convenient to determine defects in the form of cracks by means of mechanical loading and the double-exposure method. The method of studying cracks is based on the fact that under load the field of deformation of the surface of an object loses its continuity in the vicinity of a crack. Cracks result in disruption or sharp bends of the interference fringes if the load is selected so that the crack opens or its opposite sides become displaced. The kind of mechanical load has to be optimized experimentally. For example, for aircraft structural members [4.49] in which cracks appear at places of stress concentration (near holes), the optimal kind of loading is to screw a bolt into the hole being studied, from the side opposite to the illumination. Figure 4.29 shows how a crack about 1 cm long is made visible in the form of sharp bends of the interference fringes. VEST et al. [4.49] succeeded in detecting cracks less than 0.3 mm long in this way. Cracks in the edges of turbine blades were made visible in a similar way [4.50]. It is evident that one of the main difficulties of this work was the choice of the optimal kind of loading. It was shown that combined loading (bending + twisting) is optimal. It permits cracks as short as 0.2 mm to be made visible. Figure 4.30 shows interferograms of an aircraft-engine turbine blade. They show that an increase of the load makes it possible to detect even a second, smaller crack.

All of the foregoing examples show the possibility of detecting defects according to anomalies of interference patterns. An exception is the inves-

Fig.4.29. Interferogram of the stressed state of an aircraft structural member

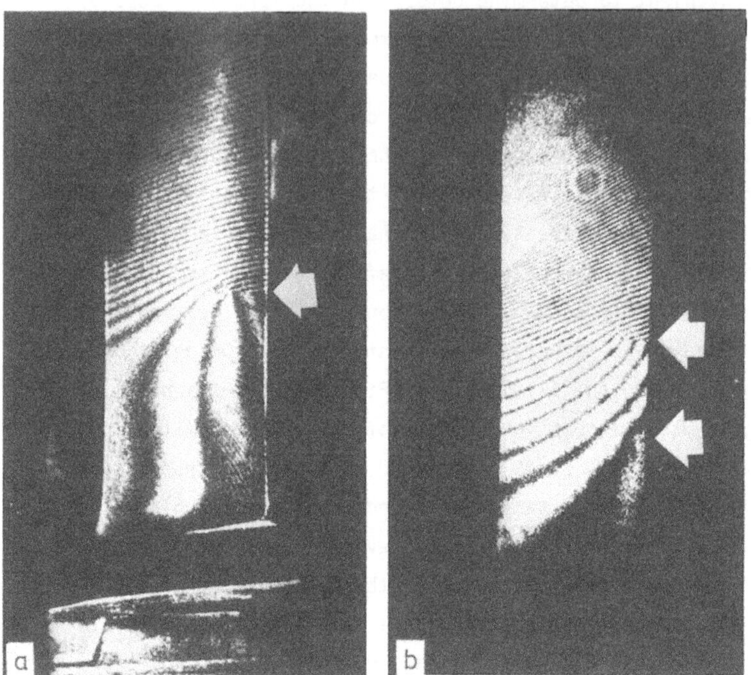

Fig.4.30. Interferogram of a turbine blade with cracks, with two loads applied

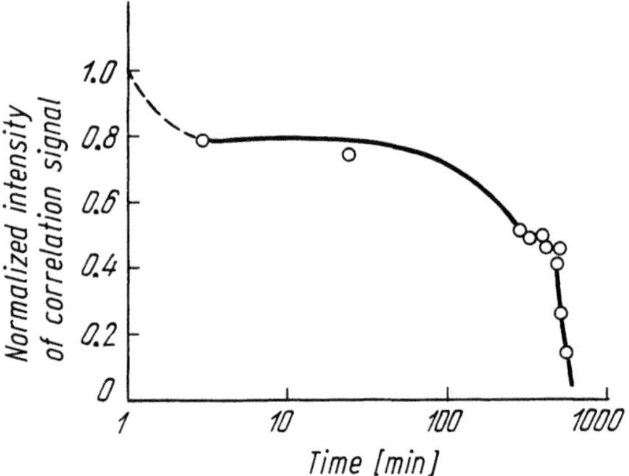

Fig.4.31. Decrease of the intensity of the correlation signal with time, for a vibrating rod

tigation of hollow turbine blades (Fig.4.27) in which the defect was not directly localized. The properties of a specimen were assessed integrally, on the basis of the value of the resonance frequency. Integral assessment, most often made by comparing "interference portraits" of the component being studied and a reference one (or with the initial state of the same component) can also be employed if changes of metallographic composition of the specimen or of the finish of its surface are being tested [4.51]. If the plastic deformation of a specimen results in alteration of the microstructure of its surface then, in accordance with Sect.4.1, the correlation properties of the scattered-light field change. Consequently, be re-

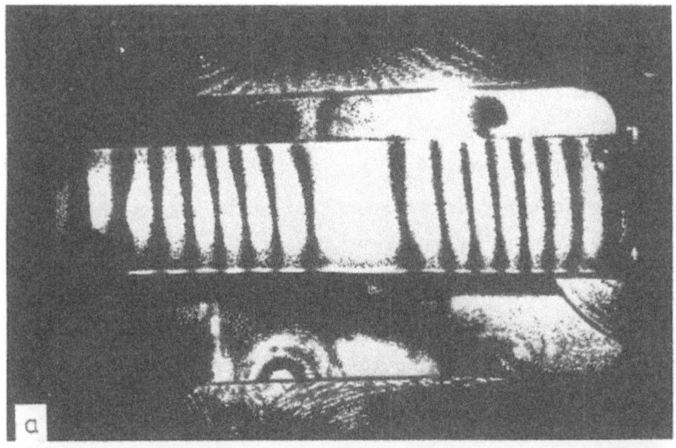

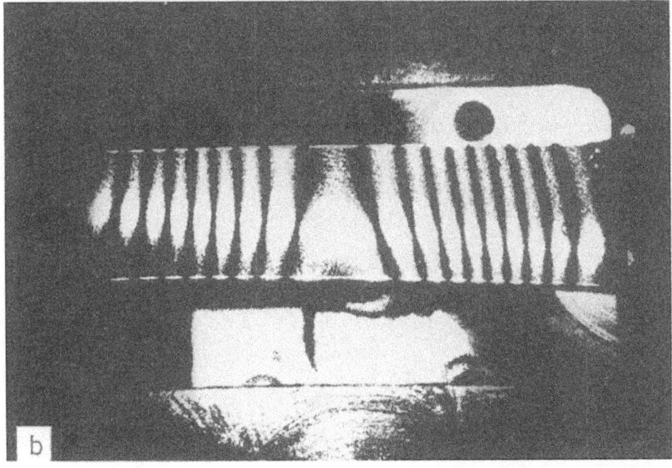

Fig.4.32. Interferograms of standard (a) and weakened (b) turbine blades

cording the wavefront scattered by an object in its initial state on a hologram, we can use this wavefront to reconstruct the initial reference beam from the hologram. When the microstructure of the surface is altered, distortions of the object field result, and the brightness of the reference beam reconstructed from the hologram decreases (Fig.4.31).

Fatigue effects and alterations of the structure of a material can be studied by use of the double-exposure method, as shown in Fig.4.32. We choose the nature and magnitude of the load on the reference specimen so as to obtain several (N) fringes of simple shape on the interferogram (Fig.4.32a). By statistical processing of the interferograms, we can determine the range of values of N that correspond to standard components. Values of N outside of this range indicate decrease of the strength of the component being tested to below the permissible limit. In such a series of experiments, special attention must be given to the design of the specimen-fastening unit and the loading unit so as to provide absolutely identical loading conditions for all articles.

Thus, flaw detection by holographic interferometry can be employed with great success for testing the quality of industrial articles and components of various kinds. Reduction of the time needed to prepare a hologram and the design of specialized small-size installations (see Fig.2.32) should facilitate the widespread industrial introduction of this effective method of nondestructive testing [4.51].

5. Holographic Studies of Vibrations

In 1932, the American physicist OSTERBERG [5.1] used light reflected by a vibrating surface for interferometric analysis of the vibrations of piezoelectric crystals. He showed that the time-averaged distribution of brightness over the surface being studied observed at the interferometer output is described by a function of the kind

$$I = I_0\left[1 + J_0\left(\frac{4\pi A}{\lambda}\right)\right] \quad ,$$

where J_0 is a zero-order Bessel function, A is the amplitude of the vibrations, and λ is the length of the light wave.

OSTERBERG used a Michelson interferometer and replaced the opague mirror in one of its arms with the crystal being studied. A mirror layer was applied to the surface of the crystal.

In 1965, in one of the first papers on holographic interferometry [5.2], POWELL and STETSON proposed a more powerful method of vibration analysis, based on the same principle as that employed by OSTERBERG [5.1]. They used a holographic setup to record the light scattered by a vibrating object. The holographic technique expanded the possibilities of Osterberg's method and provided the means for the interferometry of opaque objects of any shape with any state of their surface - no polishing of the surface being studied nor application of a mirror coating to it is needed.

The idea of the time-average method is very simple. If an object vibrates during an exposure and the duration of the latter exceeds the period of vibrations, then the hologram will record the waves scattered by this object in all of the states through which it consecutively passes.

Owing to the ability of a hologram to record the complex amplitude of a light wave, this amplitude is averaged in the Powell—Stetson method instead of the intensity (as in Osterberg's method). The waves reconstructed by the hologram form an interference pattern that indicates the nature of the motion of different points of the object.

The contributions of the different positions of the object to the total exposure are determined by the velocity with which the object passes through these positions. The greater the velocity, the less is the time spent by the object in a given position and the less is the contribution of this state to the total exposure. Because in its extreme (amplitude) positions the object comes to a full stop, and its velocity near these positions is low, the contribution of these two states of the object to the total exposure are maximum. We can therefore approximately consider the hologram obtained to correspond to the two amplitude positions of the object, i.e., as a hologram produced by the double-exposure method. The reconstructed image of the object will be intersected by interference fringes that connect points that oscillate with the same amplitude. The stationary portions of the object (nodal lines) have maximum brightness – these points of the object were in the same position during the entire exposure. The points for which the path difference of the waves scattered by the object in its amplitude positions equals an odd number of half-waves produce minima of intensity and correspond to the middles of the dark fringes.

The points for which the path difference equals an even number of half-waves will form maxima of the bright fringes. The intensities of these maxima, however, diminish with increasing vibration amplitude because an increase of amplitude reduces the time during which the object is in an extreme position (Fig.5.1).

A more rigorous analysis confirms this qualitative reasoning.

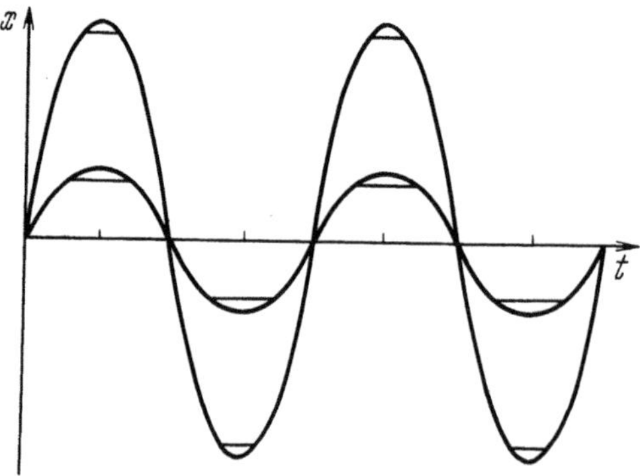

Fig.5.1. Reduction of time during which an object is in its extreme positions for increase of the vibration amplitude

5.1 Influence of Object Displacement on the Brightness of the Reconstructed Image. The Powell–Stetson-Method

Let us consider the general case of displacement of an object. The complex amplitude of the wave scattered by a point O of the stationary object and arriving at the point A of the hologram can be written in the form $a\exp(-i\Psi_0)$, where $\Psi_0 = (2\pi x \sin\theta)/\lambda$. If this point moves with the velocity $\vec{v}$ (Fig.5.2), then the phase of the wave scattered by it to the same point of the hologram will depend on time according to

$$\Psi = \Psi_0 + \frac{2\pi}{\lambda}\int_0^t (\cos\varphi_1 + \cos\varphi_2) v\,dt \quad . \tag{5.1}$$

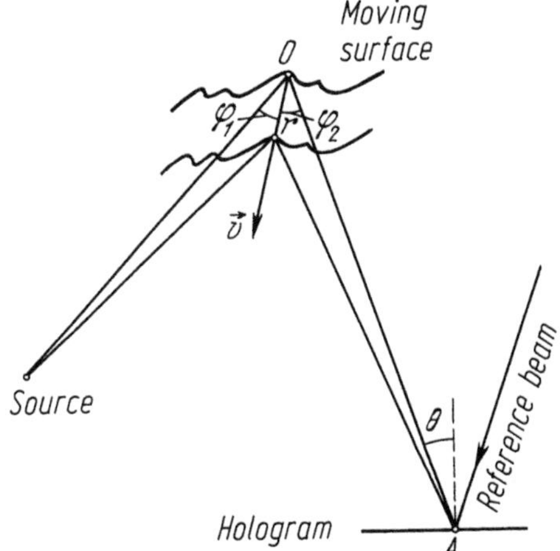

Fig.5.2. Obtaining a hologram of a moving object

The complex amplitude will have the form

$$a\,\exp\left\{-i\left[\Psi_0 + \frac{2\pi}{\lambda}\int_0^t (\cos\varphi_1 + \cos\varphi_2) v\,dt\right]\right\} \quad . \tag{5.2}$$

Its mean value during the exposure τ will determine the brightness of the point O in the reconstructed image of the object[7]

---

[7] Here we are considering the idealized case of linear recording of a hologram.

$$I \sim \left| \frac{a}{\tau} \int_0^\tau \exp\left\{ -i \left[\Psi_0 + \frac{2\pi}{\lambda} \int_0^t (\cos\varphi_1 + \cos\varphi_2) v dt \right] \right\} dt \right|^2 \quad . \tag{5.3}$$

The expression (5.3) is the most general one which covers different cases of motion of the point 0. For a stationary object ($v = 0$), we arrive at the obvious formula $I \propto a^2$.

5.1.1 Motion of an Object with a Constant Velocity

For straight uniform motion ($\vec{v}$ = const), we have

$$I \propto \left| \frac{a}{\tau} \int_0^\tau \exp\{-i[\Psi_0 + \frac{2\pi}{\lambda} (\cos\varphi_1 + \cos\varphi_2) vt]\} dt \right|^2 \quad . \tag{5.4}$$

Integration of this expression gives

$$I \propto a^2 \frac{\sin^2\left[\frac{\pi}{\lambda} (\cos\varphi_1 + \cos\varphi_2) v\tau\right]}{\left[\frac{\pi}{\lambda} (\cos\varphi_1 + \cos\varphi_2) v\tau\right]^2} \quad . \tag{5.5}$$

In the particular case when $\varphi_1 = \varphi_2 = 0$ (the object is illuminated from the side of the hologram and moves in the same direction),

$$I \propto a^2 \frac{\sin^2(2\pi v\tau/\lambda)}{(2\pi v\tau/\lambda)^2} = a^2 \frac{\sin^2(2\pi r/\lambda)}{(2\pi r/\lambda)^2} = a^2 \operatorname{sinc}^2 \frac{2v\tau}{\lambda} \quad . \tag{5.6}$$

A plot of this function is shown in Fig.5.3.

The time-average recording of a hologram of a uniformly moving object was studied by several authors from various viewpoints [5.3-8]. GOODMAN [5.3] formulated the action of a hologram as a temporal filter of the images of a moving object. The light scattered by an object that moves during exposure with the velocity v toward the hologram has the frequency

$$\nu = \nu_0 + \frac{v}{\lambda} \quad ,$$

where ν_0 is the frequency of the light used, and interferes with the reference wave. Owing to the difference between the frequencies of these waves, the fringes of the interference structure run along the hologram. The longer the exposure, the more this effect detracts from the visibility of the structure being recorded. Thus, the action of the hologram as a temporal filter has the result that the intensity of the image of the uniformly moving object reconstructed from the hologram is weakened by the factor $\mathrm{sinc}^2(2v\tau/\lambda)$ times the intensity of the image of a stationary object.

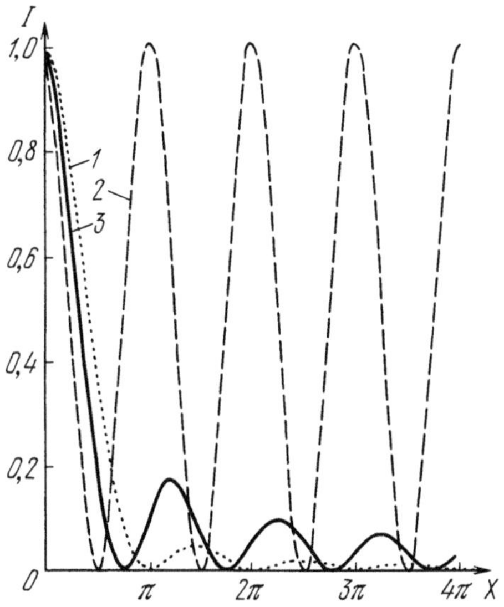

Fig.5.3. Distribution of the intensity in the reconstructed image of an object moving during an exposure: 1 - straight uniform motion $[I \propto (\sin^2 x)/x^2]$; 2 - stepwise motion (double-exposure method, $I \propto (\cos^2 x)$; 3 - harmonic vibration $[I \propto J_0^2(x)]$; the quantity $x = 2\pi r/\lambda$ is shown on the axis of abscissas for curves 1 and 2, and $x = 4\pi A/\lambda$ for the curve 3

5.1.2 Stepwise Motion of an Object

Let us consider another special case of motion of an object — during half of the exposure it is in one position and during the other half it is in another position, displaced by r. This case corresponds to the double-exposure method. The integral (5.3) is divided into two parts that correspond to the first and the second exposures,

$$I \propto \left| \frac{a}{\tau} \left\{ \int_0^{\tau/2} \exp\left\{ -i \frac{2\pi}{\lambda} \int_0^t (\cos\varphi_1 + \cos\varphi_2) v dt \right\} dt + \right.\right.$$

$$\left.\left. + \int_{\tau/2}^{\tau} \exp\left\{ -i \frac{2\pi}{\lambda} \left[\Delta + \int_0^t (\cos\varphi_1 + \cos\varphi_2) v dt \right] \right\} dt \right\} \right|^2 \quad , \qquad (5.7)$$

where $\Delta = r(\cos\varphi_1 + \cos\varphi_2)$ is the path difference of the wave scattered by the point O, caused by the displacement (see Fig.5.2). Because $v = 0$, formula (5.7) becomes

$$I \propto \left| \frac{a}{\tau} \left\{ \frac{\tau}{2} \left[1 + \exp\left(-i \frac{2\pi}{\lambda} \Delta \right) \right] \right\} \right|^2 \quad , \qquad (5.8)$$

whence, after simple transformations, we get

$$I \propto a^2 \cos^2 \left[\frac{\pi}{\lambda} r(\cos\varphi_1 + \cos\varphi_2) \right] \quad . \qquad (5.9)$$

For the particular case we have considered, when the object is illuminated from the side of the hologram and it moves in the same direction, for $\varphi_1 = \varphi_2 = 0$,

$$I \propto a^2 \cos^2 \frac{2\pi r}{\lambda} \quad . \qquad (5.10)$$

A plot of this function is also shown in Fig.5.3. In this case, we get interference fringes with undiminished intensity of the maxima.

5.1.3 Harmonic Vibrations of an Object

Let us finally consider harmonic vibrations of the points of an object, for which

$$\int_0^t (\cos\varphi_1 + \cos\varphi_2) v dt = A(\cos\varphi_1 + \cos\varphi_2) \sin\omega t \quad ,$$

where A is the amplitude of the vibrations. The general expression (5.3) accordingly becomes

$$I \propto \left| \frac{a}{\tau} \int_0^\tau \exp\left\{-i\left[\frac{2\pi A}{\lambda}(\cos\varphi_1 + \cos\varphi_2)\sin\omega t\right]\right\} dt \right|^2 \quad . \tag{5.11}$$

Because

$$\exp\left\{-i\,\frac{2\pi A}{\lambda}(\cos\varphi_1 + \cos\varphi_2)\sin\omega t\right\} = \sum_{n=-\infty}^{\infty} J_n\left[\frac{2\pi A}{\lambda}(\cos\varphi_1 + \cos\varphi_2)\right] e^{in\omega t} \ ,$$

where J_n is an n-th order Bessel function of the first kind, we get

$$I \propto \left| \frac{a}{\tau} \sum_{n=-\infty}^{\infty} \left\{ J_n\left[\frac{2\pi A}{\lambda}(\cos\varphi_1 + \cos\varphi_2)\right] \int_0^\tau e^{in\omega t} dt \right\} \right|^2 \quad . \tag{5.13}$$

After integration,

$$I \propto \left| a \sum_{n=-\infty}^{\infty} \left\{ J_n\left[\frac{2\pi A}{\lambda}(\cos\varphi_1 + \cos\varphi_2)\right] \frac{e^{in\omega\tau} - 1}{in\omega\tau} \right\} \right|^2 \quad . \tag{5.14}$$

If the exposure lasts a whole number of periods of vibration of the object, i.e., $\tau = 2k\pi/\omega$, then all of the addends except that which corresponds to $n = 0$ vanish, and

$$I \propto a^2 J_0^2\left[\frac{2\pi A}{\lambda}(\cos\varphi_1 + \cos\varphi_2)\right] \quad . \tag{5.15}$$

For $\tau >> 2\pi/\omega$, the contribution of the addend with the subscript $n = 0$ to the sum in (5.14) also remains dominant, because the factor $[\exp(in\omega\tau) - 1]/in\omega\tau$ tends to zero when $\omega\tau$ increases, and the distribution of the intensity over the surface of the object is also expressed by (5.15). This can be understood from the physical viewpoint: the transition from the exposure time $\tau_1 = 2k\pi/\omega$ to the time $\tau_2 = 2(k + \delta)\pi/\omega$, where k is a larger integer, and $\delta < 1$ is a small addition to it, introduces practically no changes into the hologram and, consequently, does not change the brightness of the reconstructed image.

For $\varphi_1 = \varphi_2 = 0$, (5.15) will be written in the form

$$I \propto a^2 J_0^2\left(\frac{4\pi A}{\lambda}\right) \quad . \tag{5.16}$$

The function (5.16), like the preceding ones (5.5,10), is an oscillating one. It modulates the brightness of the reconstructed image and forms interference fringes on it. A plot of this function is also shown in Fig. 5.3. It must be noted that the positions of the maxima and minima of the functions (5.10,16) (2 and 3 in Fig.5.3) are displaced insignificantly relative to one another (by less than one-fourth of a fringe). Therefore, calculation of the vibration amplitudes according to a time-average interferogram on the basis of simplified notions of the Powell–Stetson method as a variant of the double-exposure method does not introduce appreciable errors.

The physical reason for reduction of brightness of fringes that correspond to regions of the object with great amplitudes of oscillations is the following. The light scattered by the surface of an oscillating object, apart from the initial frequency of the laser ν, contains components with the frequencies $\nu + n\omega/2\pi$, where n is an integer, and ω is the angular frequency of oscillations of the object. The greater the amplitude of oscillations, the smaller is the fraction of the scattered light that has the initial frequency ν ($n = 0$). This follows from (5.13) because the zero-order Bessel function J_0 diminishes rapidly with increase of the amplitude of vibrations A.

The region on the surface of an object for which $A = 0$ contributes to the object wave only on the initial frequency ν. In other words, the entire wave that is scattered by the stationary portions of the object interferes with the reference wave. The useful contribution to the object wave is considerably less for oscillating portions of the object. For example, if $(2\pi/\lambda)A = 5$, then the useful contribution is only about 2% of intensity.

The instantaneous pattern produced by interference of the n-th harmonic of the object wave with the reference wave having the single frequency ν is displaced during exposure with a velocity determined by the difference frequency $n\omega$, and in accordance with the exposure condition $\tau \gg 2\pi/\omega$, is not registered on the photographic material in the form of a holographic structure. Consequently, the components of the object wave that have frequencies $\nu + n\omega/2\pi$ ($n \neq 0$) are recorded in the form of additional incoherent noise. Hence, when objects that have great vibration amplitudes ($A \geqslant \lambda$) are studied, the optimal exposure of a hologram will be when the intensity of the object wave is equal to or even exceeds that of the reference wave [5.9].

It follows from the foregoing analysis that the Powell—Stetson method consists of recording holograms of a vibrating object with an exposure time that considerably exceeds the period of vibrations, and of analyzing the nature of the vibrations of the object according to the interference pattern obtained. The simplicity of the method, the clear representation of the results, the abundance and value of the data obtained on the nature of the vibrations have all contributed to the appearance of numerous publications on the application and investigation of the Powell—Stetson method [5.9-32]. Part of them (for instance [5.10-17]) are devoted to studying the vibrations of comparatively simple objects — thin membranes, diaphragms, plates, and girders.

Holographic vibrational analysis was used by a number of authors to study the vibrational characteristics of real industrial components of intricate shapes — turbine blades and disks [5.18-27]. Figure 5.4 shows interferograms of an aircraft compressor blade that vibrates at different resonance frequencies [5.26]. Holographic interferometry had an interesting application in [5.28,29], in which the vibration characteristics of the components of a violin body were studied.

The Powell—Stetson method, notwithstanding its simplicity and convenience of use for solving a wide range of questions in vibration analysis noted above, has a number of shortcomings:

1) The brightness of the fringes diminishes rapidly with increase of amplitude — the tenth fringe has only 2% as much brightness as the node line, and the twentieth only 1%. It is therefore practically impossible to study objects whose amplitude of vibrations exceeds 5λ.

2) Although this method makes it possible, in principle, to study vibrations in real time, it gives only the node lines.

3) It is impossible to obtain objective information on the relative phases of vibrations of different points on the surface of the object.

4) It is impossible to analyze vibrations with very small amplitude ($A < \lambda/4$).

The first three shortcomings can be eliminated by use of the stroboholographic method of vibration analysis. The fourth shortcoming can be elimi-

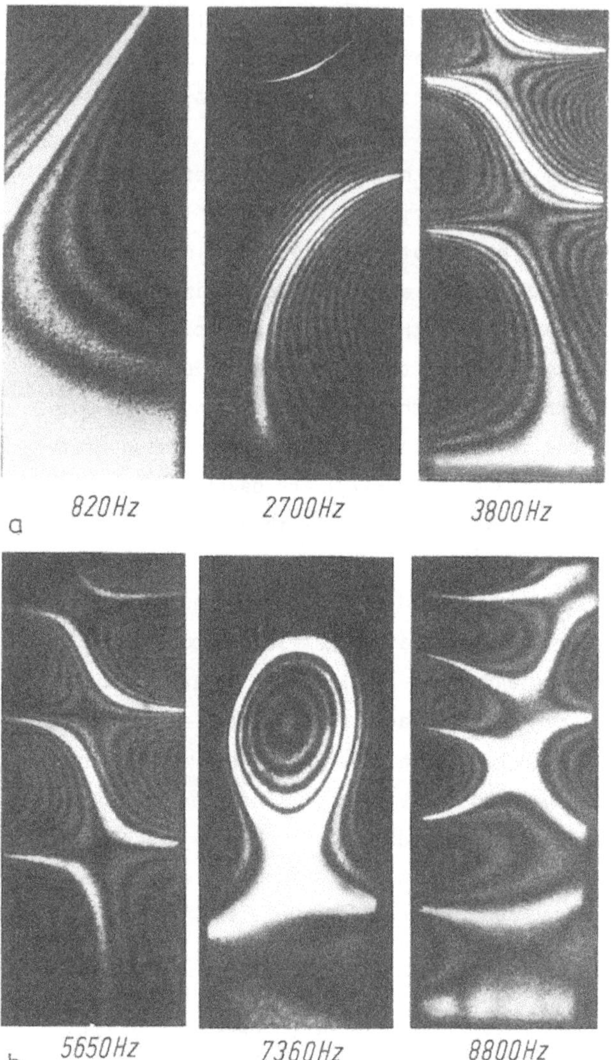

Fig.5.4. Interferograms of a compressor blade vibrating at different frequencies, obtained by the Powell–Stetson method (without a shutter)

nated by phase modulation of the reference beam and the method of "holographic subtraction", which will be treated later.

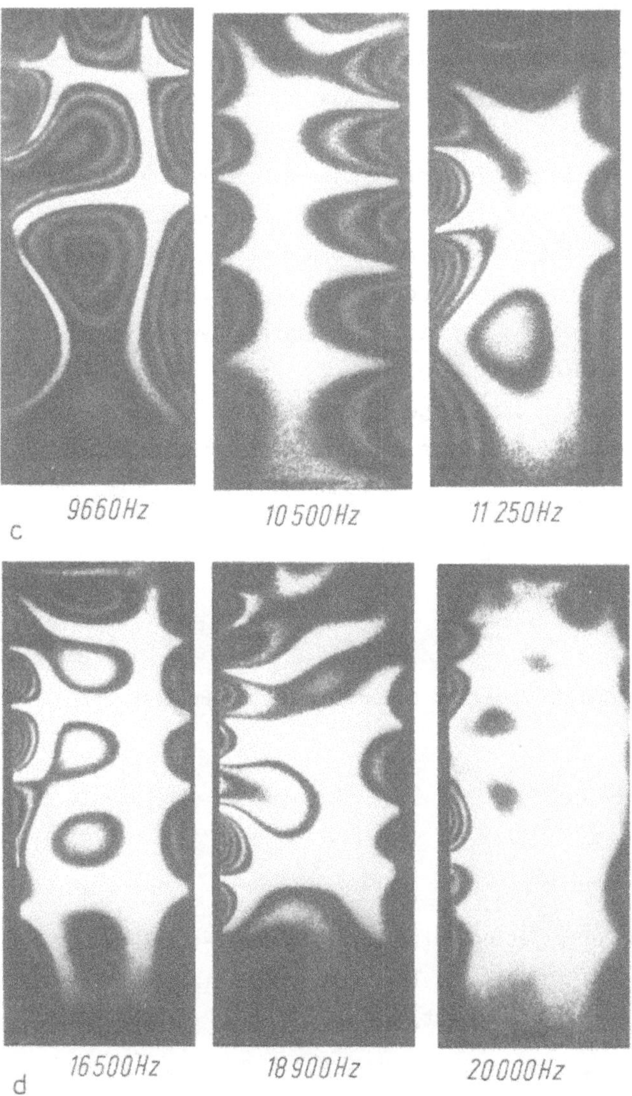

Fig.5.4. (Continued)

5.2 The Strobohologrephic Method

We shall use the term stroboholography (stroboscopic holography) to denote holographic methods of studying repeating processes in which the holograms are exposed in the light of repeated light pulses that are synchronized with a definite phase of the process. This method was first proposed in 1967 for study of a gas-discharge plasma fed by an alternating current [5.33]. The idea of stroboholography, however, was found to be most fruitful for vibration analysis [5.34-37].

Let us consider the basic theoretical prerequisites of the stroboholographic method. We shall limit ourselves to the case when the object is illuminated from the side of the hologram and vibrates along the same direction ($\varphi_1 = \varphi_2 = 0$). The complex amplitude of the object wave that falls on the hologram from a point of the object O that vibrates with the amplitude A equals $a \exp\{i[\Psi_0 + (4\pi A/\lambda) \sin\omega t]\}$.

The intensity of the reconstructed image is proportional to the square of the magnitude of the mean value of this quantity during the exposure time τ, i.e.,

$$I = \left| \frac{ae^{-i\Psi_0}}{\tau} \int_0^\tau \exp\left(i \frac{4\pi A}{\lambda} \sin\omega t\right) dt \right|^2 , \tag{5.17}$$

or, expanding the integrand into a series according to the Bessel functions,

$$I = \left| \frac{a}{\tau} \sum_{n=-\infty}^{\infty} J_n\left(\frac{4\pi A}{\lambda}\right) \int_0^\tau e^{in\omega t} dt \right|^2 . \tag{5.18}$$

After setting the duration of a gating pulse equal to one-k-th of the period of vibrations T, i.e., $\tau = 2\pi/\omega k = T/k$, and synchronizing it with one of the extreme positions of the object, we get

$$I = a^2 \left| \frac{k}{T} \sum_{n=-\infty}^{\infty} J_n\left(\frac{4\pi A}{\lambda}\right) \int_{\frac{T}{4}-\frac{T}{2k}}^{\frac{T}{4}+\frac{T}{2k}} e^{in\frac{2\pi t}{T}} dt \right|^2 . \tag{5.19}$$

The integral inside the summation sign equals $\frac{T}{k} \frac{\sin(n\pi/k)}{n\pi/k} \times \exp(in\pi/2$, and we have

$$I = a^2 \left| \sum_{n=-\infty}^{\infty} J_n\left(\frac{4\pi A}{\lambda}\right) \frac{\sin(n\pi/k)}{n\pi/k} e^{in\pi/2} \right|^2 . \tag{5.20}$$

Because $J_{-n} = (-1)^n J_n$, and $\exp(in\pi/2) = i^n$, we have $J_n\exp(in\pi/2) = $ $= J_{-n}\exp(-in\pi/2)$, and

$$I = a^2 \left| J_0\left(\frac{4\pi A}{\lambda}\right) + 2\sum_{n=1}^{\infty} J_n\left(\frac{4\pi A}{\lambda}\right) \frac{\sin(n\pi/k)}{n\pi/k}\, i^n \right|^2 . \tag{5.21}$$

Similarly, for illumination of the image reconstructed from a hologram exposed to pulses of the same duration T/k, but synchronized with the other extreme position of the vibrating object, we get

$$I = a^2 \left| \frac{k}{2T} \sum_{n=-\infty}^{\infty} J_n\left(\frac{4\pi A}{\lambda}\right) \int_{\frac{3T}{4}-\frac{T}{2k}}^{\frac{3T}{4}+\frac{T}{2k}} e^{in\frac{2\pi t}{T}} dt \right|^2 =$$

$$= a^2 \left| J_0\left(\frac{4\pi A}{\lambda}\right) + 2\sum_{n=1}^{\infty} J_n\left(\frac{4\pi A}{\lambda}\right) \frac{\sin(n\pi/k)}{n\pi/k} (-i)^n \right|^2 . \tag{5.22}$$

If the hologram is exposed both at the former and the latter moments, then both images (5.21,22) will be reconstructed, and we shall observe the interference pattern formed by their superposition, which will have the intensity distribution

$$I = a^2 \left| \frac{k}{T} \sum_{n=-\infty}^{\infty} J_n\left(\frac{4\pi A}{\lambda}\right) \left\{ \int_{\frac{T}{4}-\frac{T}{2k}}^{\frac{T}{4}+\frac{T}{2k}} e^{in\frac{2\pi t}{T}} dt + \int_{\frac{3T}{4}-\frac{T}{2k}}^{\frac{3T}{4}+\frac{T}{2k}} e^{in\frac{2\pi t}{T}} dt \right\} \right|^2 =$$

$$= a^2 \left| J_0\left(\frac{4\pi A}{\lambda}\right) + \sum_{n=1}^{\infty} J_n\left(\frac{4\pi A}{\lambda}\right) \frac{\sin(n\pi/k)}{n\pi/k} [i^n + (-i)^n] \right|^2 . \tag{5.23}$$

The quantity in brackets inside the summation sign equals $2i^n$ for even n's and zero for odd n's. Accordingly, all of the addends with odd subscripts vanish from the sum, and the interference pattern has the intensity distribution

$$I = a^2 \left| J_0\left(\frac{4\pi A}{\lambda}\right) + 2\sum_{p=1}^{\infty} J_{2p}\left(\frac{4\pi A}{\lambda}\right) \frac{\sin(2p\pi/k)}{2p\pi/k} (-1)^p \right|^2 . \quad (5.24)$$

It is not difficult to see that, in the particular case k = 2, which corresponds to continuous exposure of the object, $\sin(2p\pi/k) = 0$ for all p's, and we arrive at the Powell–Stetson formula

$$I = a^2 J_0^2\left(\frac{4\pi A}{\lambda}\right) . \quad (5.25)$$

The double-exposure method corresponds to the limit $k \to \infty$. Indeed, substituting $\cos p\pi$ for $(-1)^p$ in (5.24) and assuming that $[\sin(2p\pi/k)/(2p\pi/k)] = 1$, we have

$$I = a^2 \left| J_0\left(\frac{4\pi A}{\lambda}\right) + 2\sum_{p=1}^{\infty} J_{2p}\left(\frac{4\pi A}{\lambda}\right) \cos 2p\frac{\pi}{2} \right|^2 , \quad (5.26)$$

or, in accordance with the known property of Bessel functions that have even subscripts [5.38]

$$I = a^2 \cos^2\left(\frac{4\pi A}{\lambda}\right) , \quad (5.27)$$

which is exactly the intensity distribution in the double-exposure method (5.10).

Figure 5.5 shows the dependence of the intensities of the interference-fringe maxima on the amplitude of the vibrations calculated by use of (5.24,25,27) [(5.24) for different k's]. It follows from Fig.5.5 that the intensities of the fringe maxima diminish much more slowly in the strobo-holographic method than in the Powell–Stetson method. Even with small duty cycles, i.e., $k \approx 10$-20, vibrations can be studied whose amplitudes are scores of wavelengths.

Now let us examine the case when the strobe pulse is synchronized not with an extreme position of the object, but with another arbitrary position of it. It is evident that the requirements on the duration of the pulse are least strict for the case considered above when a hologram of the object is recorded near the position of a full stop. If the amplitude of vibrations of the object is so great, however, that the number of fringes in the reconstructed image of the object becomes too large and their resolution is no longer possible, it pays to shift the timing of the strobe pulse to reduce the sensitivity of the method. Such experiments have been done [5.34].

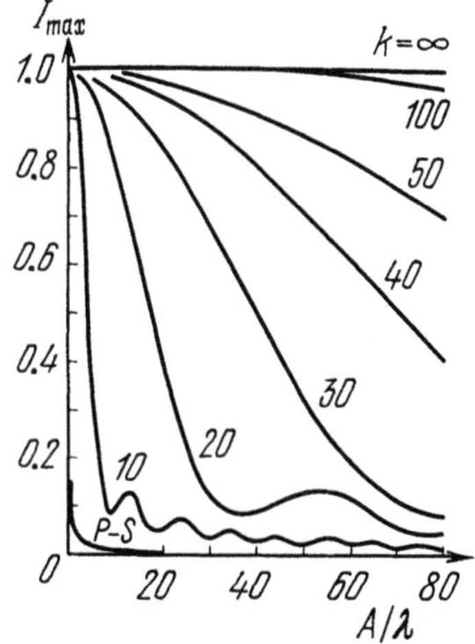

Fig.5.5. Intensity of interference-fringe maxima vs the vibration amplitude, for the Powell–Stetson method (P-S) and for the stroboholographic method with different duty cycles k between the strobe pulses

Let us assume, as previously, that the duration of a strobe pulse is T/k, but that its middle is at the arbitrary moment αT (and $\alpha T + T/2$) instead of at the moment $T/4$ (and $3T/4$). This case should reduce to the one already considered if we assume that $\alpha = 1/4$. Thus, our task is to find the function (5.19) for the new integration limits from $\alpha T - T/2k$ to $\alpha T + T/2k$, i.e.,

$$I = a^2 \left| \frac{k}{T} \sum_{n=-\infty}^{\infty} J_n\left(\frac{4\pi}{\lambda} A\right) \int_{\alpha T - \frac{T}{2k}}^{\alpha T + \frac{T}{2k}} e^{in\frac{2\pi t}{T}} dt \right|^2 \quad . \tag{5.28}$$

Integration yields

$$I = a^2 \left| \sum_{n=-\infty}^{\infty} J_n\left(\frac{4\pi A}{\lambda}\right) \frac{\sin(n\pi/k)}{n\pi/k} e^{in2\pi\alpha} \right|^2 \quad . \tag{5.29}$$

A plot of the function (5.29) for $k = 100$ and different values of α is given in Fig.5.6. Inspection of the figure shows that shifting of the moment of strobing by only a small fraction of the period of vibrations leads to a sharp reduction of the brightness of the fringes, especially of the high orders. If the strobe duty cycle is sufficiently great, however, then the shift of the strobe pulse may be useful for analyzing vibrations.

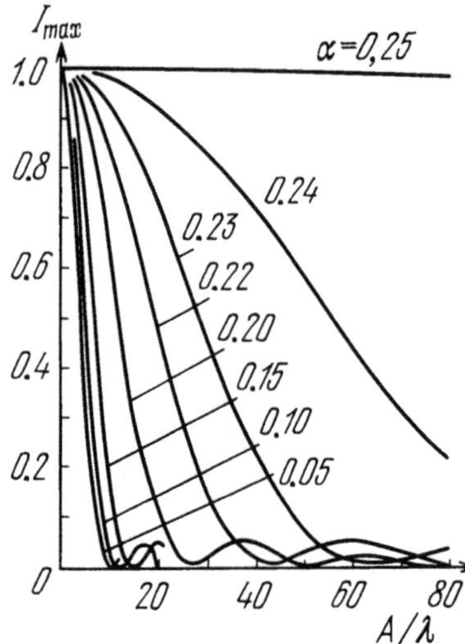

Fig.5.6. Intensity of interference-fringe maxima vs the vibration amplitude when the strobe pulse is shifted by αT from the moment when the vibrating body passes through the unshifted position, for various values of α. The value $\alpha = 0.25$ corresponds to the extreme (amplitude) position of the object. The strobe duty cycle is $k = 100$

Obtaining light pulses strictly synchronized with the vibrations of the object is the cornerstone in the carrying out of stroboholographic analysis. Many different mechanical interrupters have been used for this purpose [5.34-36, 39-41]. A typical setup of this kind, which makes it possible to study vibrations at frequencies up to 12 kHz is shown in Fig.5.7.

In this setup the principal element that controls the vibrations of the object and the interruptions of the laser pulse is a disk with several rows of holes (Fig.5.8). One row interrupts the laser beam, and the other interrupts the light flux that passes from the incandescent lamp onto the photodiode. The fundamental harmonic of the current is amplified at the output of the photodiode and produces vibrations of the membrane being studied..The simplicity of this setup provides considerable conveniences. The disk is rotated by a motor with a speed up to 200 s^{-1}, which when a row with sixty holes is used corresponds to a maximum frequency of 12 kHz. The speed of the motor is controlled by a rheostat.

If the laser beam is interrupted by a row of twice as many holes as in that which produces the vibrations (thus obtaining two strobe pulses during

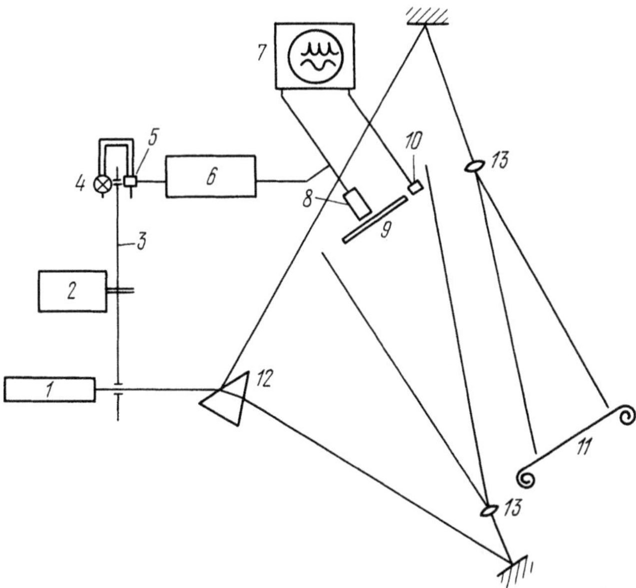

Fig.5.7. Arrangement for stroboholographic studying of vibrations:
1 - laser; 2 - motor; 3 - disk; 4 - lamp; 5, 10 - photodiode; 6 - amplifier; 7 - oscillograph; 8 - vibrator; 9 - membrane; 11 - hologram; 12 - beamsplitting wedge; 13 - beam expander

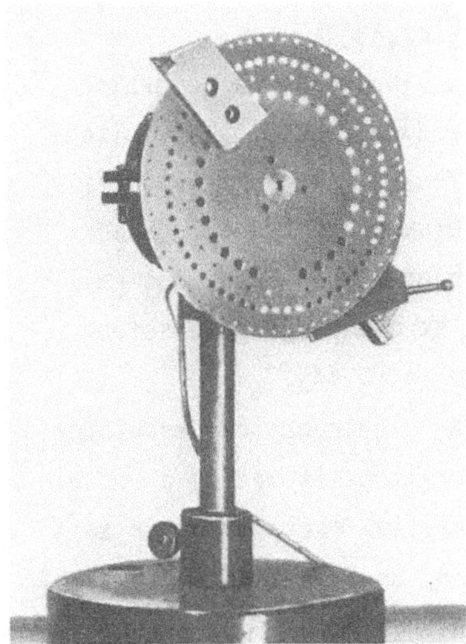

Fig.5.8. Rotating disk

one period of vibration of the membrane), then the recorded hologram will be equivalent to a doubly exposed hologram.

This arrangement also makes it possible to observe vibrations in real time. For this purpose, a hologram of the stationary membrane is exposed with the disk stopped, and then the vibrating membrane is observed with the disk rotating through the hologram developed "on the spot". Thus, we can watch the dynamics of development of the vibrations by smoothly increasing the speed of the motor. The laser beam was interrupted by using a row with the same number of holes as that which produced the current for exciting the vibrations of the membrane (one strobe pulse per period of vibration). The timing of the strobe phase could be smoothly changed by turning the lamp and photodiode system about the axis of the disk.

The main shortcomings of the system described, as well as of other mechanical systems, are the difficulty of setting a definite disk speed and its instability, the limited range of frequencies, and also the impossibility of smoothly controlling the durations of the strobe pulses.

The stroboholographic method that employed a disk stroboscope was also used to study the vibrations of round membranes [5.40] and of a cantilever beam [5.41].

The use of a stroboscope in the form of an electrooptical modulator of light on a Pockels cell was first mentioned in [5.37]. [Detailed information on electrooptical modulators is given in [5.42,43]].

A thin round plate fixed at its periphery was vibrated by a generator of sinusoidal oscillations. A pulse generator produced rectangular voltage pulses for starting the Pockels cell with a half-wave voltage of 1300 V. The possibility was provided for changing the duration of the pulses and synchronizing them with the vibrations of the object. When open, the cell passed about 90% of the light from the laser. A stroboholographic setup with an electrooptical shutter was also employed in [5.44].

A block diagram of a holographic setup with an electrooptical modulator [5.26] is shown in Fig.5.9. The frequency of vibrations is set by a sound generator that vibrates the object with a piezocrystal oscillator, or an electromagnetic element. The same sound generator, through a phase shifter, starts a generator of shifted rectangular pulses that produce pulses of controlled duration with an amplitude of 100 V. An electrooptical light modulator of type МЛ-3 with a KDP Pockels cell was used as a stroboscope.

Because the half-wave pulse by the voltage modulator is about 600 V, the pulses from the modulator are first amplified to this level.

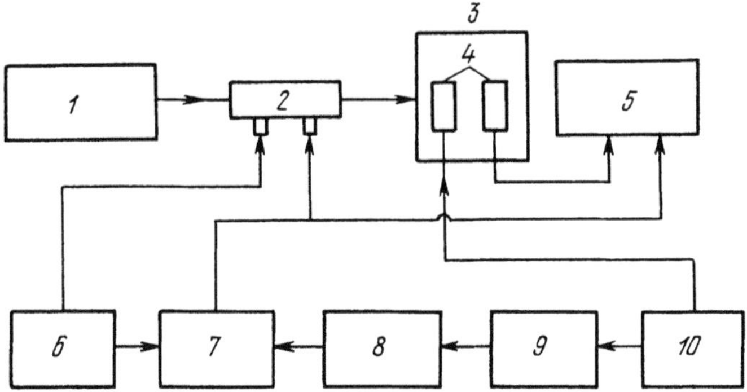

Fig.5.9. Block diagram of a stroboholographic setup with an electrooptical shutter: 1 - helium-neon laser; 2 - modulator; 3 - specimen; 4 - piezocrystals; 5 - double oscillograph; 6 - power source; 7 - amplifier; 8 - pulse generator; 9 - phase shifter; 10 - sound generator

The synchronization of the strobe pulses with different moments in the period of object vibrations was controlled by observation of the screen of a double oscillograph. The pulse going to the modulator was fed to one channel of the oscillograph, while the signal from a piezotransmitter glued onto the object was fed to the second channel. Synchronization of the strobe pulse with any phase of object vibration could be achieved by using a phase shifter and a delay of the shifted pulse generator.

The ratio of the intensities of the reference and object beams was from three to five. An exposure of 10 seconds was used in this case to obtain holograms of a good quality ФПГВ film in the time-average method. When using strobing, the duration of a half exposure (with a vibrating object) was 3 min 40 s at $k = 20$, and 7 min 20 s at $k = 40$. An optical diagram of the setup is shown in Fig.5.10. Figure 5.11 shows interferograms of an aircraft turbocompressor blade obtained with this setup.

A pulsed laser whose generation moment can be synchronized with any phase of vibration of a component being studied also makes it possible to produce holograms of a vibrating component similar to the ones obtained by the stroboholographic method [5.45]. In this case, however, the fringe visibility is not affected by frequency or amplitude instabilities of the vibrations because the recording process continues only a small fraction of the vibration period (the duration of a pulse is usually about 5×10^{-8} s).

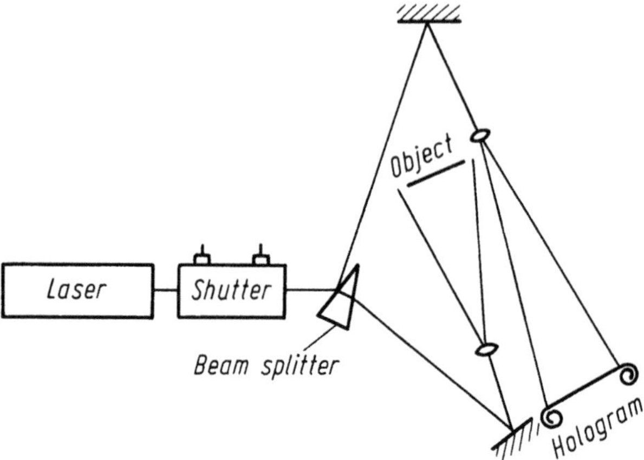

Fig.5.10. Optical diagram of arrangement with an electrooptical shutter

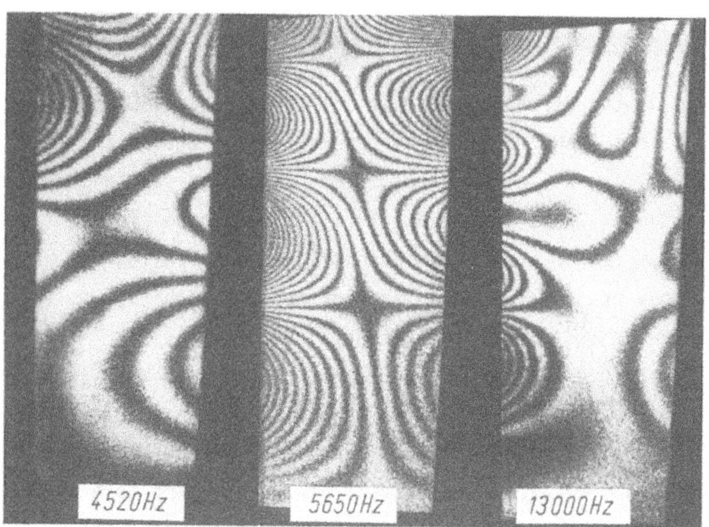

Fig.5.11. Interferograms of a turbocompressor blade obtained by the strobo-holographic method

The stroboholographic technique makes it possible to use multiple wave methods of holographic vibrometry. Let us consider some ways of realizing these methods.

One of them is repeated exposure of a hologram of a vibrating object during the period of vibrations. Definite moments must be chosen for the exposures, so as to ensure equal changes of the phase of the object wave during the time interval between any two consecutive exposures. Figure 5.12 shows that this results in different intervals between the strobe pulses. In the case shown in Fig.5.12, the number of strobe pulses during the period of vibrations is 16, i.e., it is eight times as great as in conventional stroboholography when the illuminating pulses are made to coincide with both amplitude positions of the object. The sensitivity of the method is accordingly reduced to one eighth. Because the object is exposed in nine different positions, the profile of the interference fringes corresponds to nine-wave interferometry, i.e., the fringes are only 1/4.5 = 0.22 times as wide as ordinary two-wave fringes.

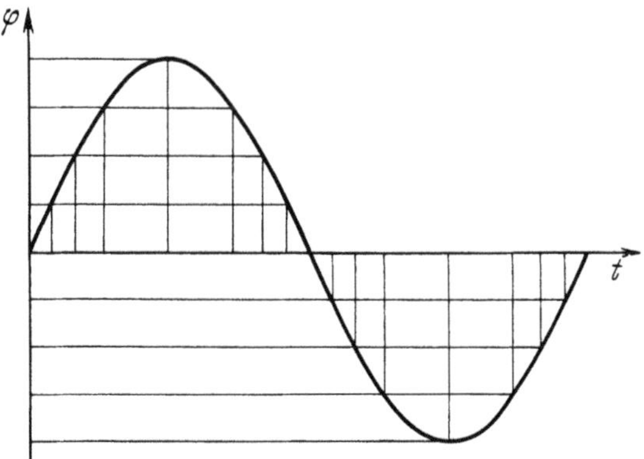

Fig.5.12. Principle of multiple-wave stroboholographic vibrometry

5.3 Phase Modulation of the Reference Beam

A number of authors [5.46-50] proposed to introduce phase modulation of the reference beam when recording a hologram, in order to broaden the possibilities of the Powell–Stetson method. Let us consider analytically what is gained by this. In the time-average method, the light wave scattered by a point of the ojbect O in harmonic oscillation is modulated in phase with a depth of modulation proportional to the amplitude of vibrations of this point.

The complex amplitude of the wave that arrives at the hologram from the vibrating object will, in accordance with (5.2), have the form

$$a \exp\left\{-i\left[\Psi_0 + \frac{2\pi}{\lambda}(\cos\varphi_1 + \cos\varphi_2)A \sin\omega t\right]\right\} , \tag{5.30}$$

where A is the amplitude of vibrations of the object, which determines the depth of phase modulation of the object wave, and Ψ_0 is the initial phase of the object wave.

If we introduce phase modulation of the object wave (for example by displacing its source) with a frequency equal to that of the vibrations of the object, then its complex amplitude can be written in the form

$$a_r \exp\left\{-i\left[\theta_r + \frac{2\pi}{\lambda} M \sin\omega t\right]\right\} , \tag{5.31}$$

where M is the amplitude of displacement of the reference source wich determines the depth of modulation of the reference-wave phase, and θ_r is the initial phase of the reference wave.

Hence, similar to (5.3), the brightness of the point 0 in the reconstructed image of the object will be determined by the formula

$$I = \left|\frac{a}{T}\int_0^T \exp\left\{-i\left\{\frac{2\pi}{\lambda}[(\cos\varphi_1 + \cos\varphi_2)A - M]\sin\omega t + (\Psi_0 - \theta_r)\right\}\right\}dt\right|^2 . \tag{5.32}$$

For $\varphi_1 = \varphi_2 = 0$, assuming that $\Psi_0 = \theta_r$, we have

$$I = \left|\frac{a}{T}\int_0^T \exp\left\{-i\left[\frac{4\pi}{\lambda}(A - B)\sin\omega t\right]\right\}dt\right|^2 , \tag{5.33}$$

where we have introduced the symbol B = M/2 for convenience in writing the formula.

Expanding the exponential function inside the integral into a series similar to (5.12), and in accordance with (5.13,14), we have

$$I = a^2 J_0^2 \left[\frac{4\pi}{\lambda}(A - B)\right] . \tag{5.34}$$

The expression obtained is similar to (5.15) for the Powell–Stetson method. The only difference is that the argument of the zero-order Bessel function depends on the depth of modulation B of the reference-beam phase.

This circumstance opens up the following possibilities for the method of phase modulation of the reference beam.

5.3.1 Determining Large Amplitudes of Vibrations

Examination of (5.34) shows that, by selecting the depth of phase modulation of the reference beam, we can compensate the corresponding part of the amplitude of object vibrations. Hence, the interference fringes for which $A = B$ have the maximum brightness in the reconstructed image instead of those for which $A = 0$ as in the Powell–Stetson method. Each fringe will be formed of points that have the same difference $A - B$.

This result was first experimentally illustrated by ALEKSOFF [5.46] in observing the mode structure of vibrations of an ADP crystal ($NH_4H_2PO_4$) in which standing waves were generated electrically.

Using phase modulation of the reference beam, ALEKSOFF [5.47] successfully investigated the vibrations of a loudspeaker with an amplitude exceeding 6λ. It has been shown [5.48] that modulation of the reference beam makes it possible to study inhomogeneities of vibrations of objects when these vibrations contain a large reciprocating (piston) component. These authors analyzed the vibrations of the surface of a piezoceramic electromechanical transducer with an amplitude that exceeded 10λ.

MOTTIER [5.49] proposed a method of automatic modulation of the reference wave phase with the aid of a small mirror fastened on the moving object. MOTTIER's method was shown [5.51] to be a particular case of holography with a local reference beam [5.52,53]. This name combines all the methods in which a reference beam is used that is formed by a part of the object beam. Any small movement of the object illuminated with light from a laser is attended by a phase change in the optical field of the reference beam. MOTTIER obtained a hologram of a creeping snail to whose shell a small mirror was glued. It formed the reference beam.

MOTTIER later proposed [5.50] to achieve synchronized phase modulation of the reference beam with the aid of a mirror fastened on a separate piezoelectric vibrator. A second piezoelectric vibrator was fastened to the cantilever beam whose vibrations were being studied. This method of phase modulation of the reference beam was also used by BELOGORODSKY et al. [5.48].

5.3.2 Determining Small Amplitudes of Vibrating

If in (5.34) we select the depth of modulation B for which the function $J_0^2(B)$ has its first root, then the possibility appears of observing the regions of small amplitudes of vibrations of the object in the form of light spots on the general dark background of the object's image [5.47,48].

The function $J_0[(4\pi/\lambda)(A - B)]$ from formula (5.34) at small values of A can be expanded into a Taylor series,

$$J_0\left[\frac{4\pi}{\lambda}(A - B)\right] = J_0\left(\frac{4\pi}{\lambda}B\right) + \frac{4\pi}{\lambda}AJ_0'\left(\frac{4\pi}{\lambda}B\right) + \left(\frac{4\pi}{\lambda}\right)^2\frac{A^2}{2}J_0''\left(\frac{4\pi}{\lambda}B\right) + \ldots \quad (5.35)$$

Limiting ourselves to the first two terms of the expansion (5.35) and taking into account that the first term of the expansion vanishes at the value of B we have selected, and because $J_0'(z) = -J_1(z)$, we can write for the distribution of the intensity in the reconstructed image of the object,

$$I = a^2\left(\frac{4\pi}{\lambda}A\right)^2 J_1^2\left(\frac{4\pi}{\lambda}B\right) \quad . \quad (5.36)$$

In other words, the distribution of the intensity in the reconstructed image is proportional to the square of the amplitude A instead of to the square of the zero-order Bessel function. The "dark field" method makes it possible to determine vibrations with an amplitude of about 0.01λ. Thus, for recording very small amplitudes of vibrations, the sensitivity of the time-average method can be increased by means of phase modulation of the reference beam during an exposure. The technical realization of this method, however, requires a quite complicated experimental arrangement.

The method of phase modulation of the reference beam can be replaced for determining small vibration amplitudes with the relatively simple method of "holographic subtraction" proposed in [5.54]. In this method, two holograms of an object are consecutively recorded on one photographic plate, but with a phase shift of 180° of one of the beams betwenn exposures. The regions of the object that do not change between the exposures vanish in reconstruction. Only those regions are seen that change between exposures or during one of them [5.55].

Superposition of a hologram of an object in the quiet state on a hologram of the vibrating object recorded by the Powell—Stetson method has been proposed, to obtain a 180° phase shift of the object beam between the exposures [5.56]. A similar procedure for observing the fringe pattern in real time, was proposed by BIEDERMANN and MOLIN [5.57]. A phase shift of 180° between exposures was achieved in the reference beam by use of a half-wave plate [5.58,59].

5.4 Determining the Phases of Vibrations of an Object

Of great interest in vibration analysis in various fields of engineering is information on the relative phases of vibrations of different points on an object. That phase modulation of the reference beam makes it possible to determine the phase of vibrations of an object was shown in [5.46,47, 50].

A way was proposed of determining the vibration phases of various points of an object when the time-average method or its real-time variant are used, with phase modulation of the reference beam in which a mirror is placed that vibrates with the same frequency as the vibrations of the object [5.60]. The interference pattern obtained depends on the amplitude and phase of the vibrations of the mirror with respect to the vibrations of the object. By adjusting the amplitude and phase of oscillations of the mirror, we can displace the node lines to any point on the surface of the object. The phases of vibrations of points on the object are determined by the phase of the vibrating mirror. Its value is read on the scale of the generator of sinusoidal vibrations that vibrate the mirror.

A proposal was made to determine the vibration phases of an object by turning the hologram through a small angle between exposures in stroboscopic illumination [5.61]. If the object does not vibrate, a system of equidistant parallel fringes is formed on its image. If the object vibrates during one of the exposures, then in the image reconstructed from the hologram the pattern of parallel fringes is distorted in accordance with the displacement of the object. The system of fringes comprises lines of equal displacement of the object relative to the hologram; the sign of the phase change is known with reference to the direction of turning of the hologram. Therefore, the local distortions of the resultant pattern are used to determine the sign of the phases of vibrations of various portions of the object. A similar procedure has been used with the real-time method [5.40].

The simple and convenient wedge method was used to create a system of parallel equidistant fringes [5.26]. This made it possible to interpret unambiguously the signs of the phases of different points of the vibrating object, according to the bending of the fringe pattern in the reconstructed image. Two exposures were made. During the first, the object (a turbine blade) was stationary; the second exposure was made with the blade vibrating and with stroboscopic illumination of the hologram. The object wave was inclined before the second exposure by turning the wedge (Fig. 5.13).

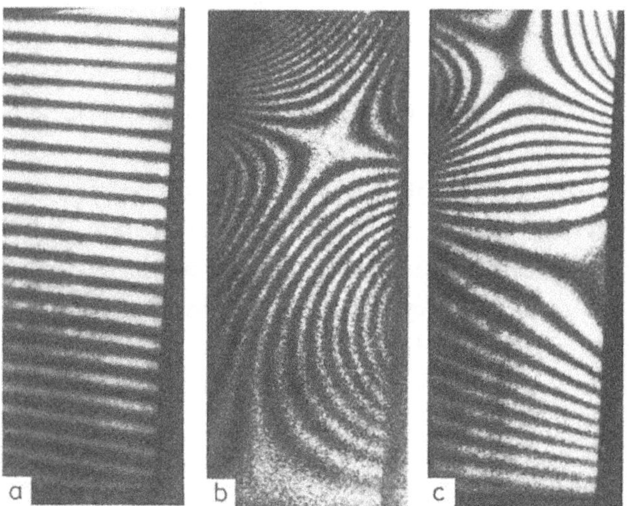

Fig.5.13. Hologram strobe interferogram obtained with a wedge: (a) interferogram of a fixed blade, obtained by the double-exposure method; the wedge is rotated before the second exposure. (b) interferogram of a vibrating blade, obtained by the hologram strobe double-exposure method; before the second exposure, the strobe phase is changed by 180°, and the wedge remains fixed. (c) interferogram of the vibrating blade, obtained as in case b, but with the wedge rotated before the second exposure; the vibrational frequency is 2720 Hz

References

Most of the cited Soviet journals are published in English by the American Institute of Physics, New York, under the following titles:

Opt. i spektr. - Optics and Spectroscopy

Zh. eksp. teor. fiziki - Soviet Physics - JETP

DAN SSSR - Soviet Physics - Doklady

Usp. fiz. nauk - Soviet Physics - Usp.

Zh. tekh. fiz. - Soviet Physics - Technical Physics

1.1 I. Newton: Optics: or a Treatise of the Reflections, Refractions, Inflections and Colours of Light, 3rd corrected ed. (William & John Innys, London 1721)

1.2 Th. Young: "On the Theory of Light and Colours", in Philos. Trans. Roy. Soc. Lond. 91, part 1 (1802)

1.3 Oeuvre complètes d'Augustin Fresnel (publices par M.M. Henri de Senarmon, Émile Verdet et Léonor Fresnel, Paris 1866)

1.4 J.R. Klauder, E. Sudarshan: Fundamentals of Quantum Optics (Benjamin, New York 1968)

1.5 G.A. Korn, T.M. Korn: Mathematical Handbook for Scientists and Engineers (McGraw-Hill, New York 1968)

1.6 Yu.N. Denisyuk: Zh. Tekh. Fiz. 44, 131 (1974)

1.7 M. Born, E. Wolf: Principles of Optics, 3rd ed. (Pergamon Press, Oxford 1965)

1.8 M. Françon, S. Slansky: Cohérence en optique (Ed. centre Nat. rech. sci., Paris 1965)

1.9 A. Angot: Compléments de mathématiques, A l'usage des ingénieurs de l'élektrotechnique et des télécommunications (Paris Revue d'optique 1957)

1.10 A.A. Michelson: Studies in Optics (Univ. of Chicago Press 1927)

1.11 A.N. Zakharevsky: Interferometry (Interferometers) (Oborongiz, Moscow 1952)

1.12 M. Françon: Optical Interferometry (Academic Press, New york, London 1966)

1.13 G.V. Dreiden, Yu.I. Ostrovsky, E.N. Shedova: Opt. Spektrosk. 32, 367 (1972)

1.14 D.S. Rozhdestvensky: Raboty po anomal'noi dispersii v parakh metallov (Works on the Anomalous Dispersion in Metal Vapours)(Izd. AN SSSR, Moscow 1951)

1.15 A.N. Zaidel, G.V. Ostrovskaya, Yu.I. Ostrovsky: Tekhnika i praktika spektroskopii (Techniques and Practice of Spectroscopy), 2nd ed. (Nauka, Moscow, 1976)

1.16 N.R. Batarchukova: Novoe opredelenie metra (The New Definition of the Meter) (Gos. komitet standartov, mer i izmer. priborov SSSR, Moscow 1964)

1.17 H. Fizeau: Ann. Chem. Phys. 66, 429 (1862)

1.18 P. Jacquinot: Rep. Prog. Phys. 23, 267 (1960)

1.19 Yu.A. Tolmachev: Novye spekral'nye pribory (New Spectral Instruments) (Leningrad University Press 1976)

1.20 G.V. Rozenberg: Optika tonkosloinykh pokrytii (The Optics of Thin Coatings) (Fizmatgiz, Moscow 1958)

1.21 I.V. Skokov: Mnogoluchevye interferometry (Multiple Wave Interferometers) (Machinostroenie, Moscow 1969)

1.22 G.G. Dolgov-Savelev, M.I. Pergament, M.M. Stepanenko, A.I. Yaroslavsky: in Diagnostika plazmy (Diagnostics of Plasma), vyp. II (Atomizdat, Moscow 1968) p. 13

1.23 S. Tolansky: Revolution in Optics (Penguin, London 1968)

1.24 I.M. Nagibina: Interferentsiya i difraksiya sveta (Interference and Diffraction of Light (Mashinostroenie, Leningrad 1974)

1.25 F.A. Korolev: Spektroskopiya vysokoi razreshayushchei sily (High Resolution Spectroscopy) (Gostekhizdat, Moscow 1953)

1.26 S. Tolansky: High Resolution Spectroscopy (Methuen, London 1947)

1.27 D. Gabor: Nature 161, 777 (1948)

1.28 D. Gabor: Proc. Roy. Soc. (London) A197, 454 (1949)

1.29 D. Gabor: Proc. Phys. Soc. B64, 449 (1951)

1.30 E. Leith, J. Upatnieks: J. Opt. Soc. Am. 53, 1377 (1963); 54, 1295 (1964)

1.31 E. Leith, J. Upatnieks: J. Opt. Soc. Am. 51, 1469 (1961); 52, 1123 (1962)

1.32 Yu.N. Denisyuk: DAN SSSR 144, 1275 (1962)

1.33 Yu.N. Denisyuk: Opt. Spektrosk. 15, 522 (1963)

1.34 G.W. Stroke: An Introduction to Coherent Optics and Holography, 2nd ed. (Academic Press, New York 1969)

1.35 J.W. Goodman: Introduction to Fourier Optics (McGraw-Hill, New York 1968)

1.36 R.J. Collier, C.B. Burckhardt, L.H. Lin: Optical Holography (Academic Press, New York 1971)

1.37 M. Francon: Holographie (Masson, Paris 1969)

1.38 Yu.I. Ostrovsky: Holography and its Application, Trans. by G. Leib (Mir Publishers, Moscow 1977)

1.39 L.M. Soroko: Osnovy golografii i kogerentnoi optiki (Fundamentals of Holography and Coherent Optics) (Nauka, Moscow 1971)

1.40 J.N. Butters: Holography and Its Technology (Peter Peregrinus, London 1971)

1.41 J.Ch. Viénot, P. Smigelski, H. Royer: Holographie Optique (Dunod, Paris 1971)

1.42 H.M. Smith: Principles of Holography (Wiley/Interscience), New York 1969)

1.43 V.M. Ginzburg, B.M. Stepanov (eds.): Golografiya. Metody i apparatura (Holography. Methods and Apparatus) (Sov. radio, Moscow 1974); Opticheskaya golografiya, prakticheskie primeneniya (Optical Holography, Practical Applications) (Sov. radio, Moscow 1978)

1.44 S. Lowenthal, Y. Belvaux: Rev. d'Optique 1, 1 (1967)

1.45 H. Kiemle, D. Röss: Einführung in die Technik der Holographie (Akad. Verlagsges., Frankfurt 1969)

1.46 K. Švarcs, A. Ozols: Holografija revolucija optika (Zinatne, Riga 1975)

1.47 M. Miler: Holografie (SNTL-Nakladatelstvy Technicke Literatury, Prague 1974)

1.48 R.W. Meier: J. Opt. Soc. Am. 56, 219 (1966)

1.49 B.P. Konstantinov, A.N. Zaidel, V.B. Konstantinov, Yu.I. Ostrovsky: Zh. Tekh. Fiz. 36, 1718 (1966)

1.50 O. Bryngdahl, A. Lohmann: J. Opt. Soc. Am. 58, 1325 (1968)

1.51 B.P. Konstantinov: Usp. Fiz. Nauk 100, 185 (1970)

1.52 D. Gabor: Proc. R. Inst. Gr. Brit. 43, 200 (1969)

1.53 V.G. Komar, G.A. Sobolev: in Sovremennoe sostoyanie i perspektivy razvitiya golografii (The Present State and Prospects of Development of Holography) (Nauka, Leningrad 1974) p. 120

1.54 S.B. Gurevich, G.A. Gavrilov, D.F. Cherhykh: In Ref. 1.53, p. 142

1.55 A.F. Metherell, H.M.A. El-Sum, L. Larmore (eds.): Acoustical Holography, Vols. 1-3 (Plenum Press, New York 1969-1971)

1.56 G.S. Safronov, A.P. Safronova: Vvedenie v radiogolografiyu (Introduction to Radioholography) (Sov. radio, Moscow 1973)

1.57 L.D. Bakhrakh, A.P. Kurochkin: Golografiya v mikrovolnovoi tekhnike (Holography in Microwave Techniques) (Sov. radio, Moscow 1979)

1.58 W.T. Cathey: Optical Information Processing and Holography (Wiley, New York 1974)

1.59 R.E. Brooks, L.O. Heflinger, R.F. Wuerker: Appl. Phys. Lett. 7, 248 (1965)

1.60 L.O. Heflinger, R.F. Wuerker, R.E. Brooks: J. Appl. Phys. 37, 642 (1966)

1.61 J. Burch: Prod. Eng. 44, 431 (1965)

1.62 M.H. Horman: J. Opt. Soc. Am. 55, 615 (1965)

1.63 R.I. Collier, E.T. Doherty, K.S. Pennington: Appl. Phys. Lett. 7, 223 (1965)

1.64 R.L. Powell, K.A. Stetson: J. Opt. Soc. Am. 55, 1593 (1965)

1.65 K.A. Haines, B.P. Hildebrand: Phys. Lett. 19, 10 (1965)

1.66 G.W. Stroke, A.E. Labeyrie: Appl. Phys. Lett. 8, 42 (1966)

1.67 A. Kakos, G.V. Ostrovskaya, Yu.I. Ostrovsky, A.N. Zaidel: Phys. Lett. 23, 81 (1966)

1.68 A.G. Smirnov, in Opticheskaya golografiya i ee primenenie (Optical Holography and Its Application) (Izd. LDNTP, Leningrad 1974) p. 45

2.1 M. Born, E. Wolf: Principles of Optics, 3rd ed. (Pergamon Press, Oxford 1965)

2.2 R.J. Collier, C.B. Burckhardt, L.H. Lin: Optical Holography (Academic Press, New York 1971)

2.3 G. Froehly, A. Lacourt, J.C. Viênot: Nouv. Rev. Opt. 4, 183 (1973)

2.4 M. Françon, S. Slansky: Cohêrence en optique (Ed. centre Nat. rech. sci., Paris 1965)

2.5 J.W.C. Gates: Proc. Electro-Opt. Syst. Design Conf., Brighton, England (1972)

2.6 A.K. Levine, A. DeMaria (eds.): Lasers, Vol. 1, 2 and 3 (Marcel Dekker, New York 1966, 1968 and 1971)

2.7 M. Young: Optics and Lasers, Springer Series in Optical Sciences, Vol.5 (Springer, Berlin, Heidelberg, New York 1977)

2.8 M. Ross (ed.): Laser Applications, Vol. 1 (Academic Press, New York 1971)

2.9 W. Koechner: Solid-State Laser Engineering, Springer Series in Optical Sciences, Vol.1 (Springer, Berlin, Heidelberg, New York 1976)

2.10 F.T. Arecchi, E.O. Schulz-Du Bois: Laser Handbook (North-Holland, Amsterdam, Elsevier, New York 1972)

2.11 D.A. Ansley: Appl. Opt. 9, 815 (1970)

2.12 V.A. Burtsev, V.N. Litunovski, A.G. Smirnov, V.G. Smirnov: Preprint K-0291, NIIEFA (1976)

2.13 R.G. Hall, J.W. Gates, I.N. Ross: J. Phys. E3, 789 (1970)

2.14 J.C. Binder: Rev. d'Optique Appl. 4, 5 (1971)

2.15 L.S. Ushakov, Yu.I. Filenko, in Trudy VI Vsesoyuznoi shkoly po golografii (Proc. 6th All-Union School on Holography) (Izd. LIYaF, Leningrad 1974) p. 253

2.16 F.P. Schäffer (ed.): Dye Lasers, Topics in Applied Physics, Vol. I, 2nd ed. (Springer, Berlin, Heidelberg, New York 1977)

2.17 H.M. Smith (ed.): Holographic Recording Materials, Topics in Applied Physics, Vol. 20 (Springer, Berlin, Heidelberg, New York 1977)

2.18 O. Bryngdahl: J. Opt. Soc. Am. 59, 1171 (1969)

2.19 O. Bryngdahl, A.W. Lohmann: J. Opt. Soc. Am. 58, 141 (1968)

2.20 K.S. Mustafin, V.A. Seleznev, E.I. Shtyrkov: USSR Inventor's Certificate No. 272.602 (1968); Byull. izobr. No. 19 (1970); Opt. Spektrosk. 28, 1186 (1970)

2.21 K. Matsumoto, M. Takashima: J. Opt. Soc. Am. 60, 30 (1970)

2.22 C.H.F. Velzel: Opt. Commun. 2, 289 (1970)

2.23 G.V. Ostrovskaya, Yu.I. Ostrovsky: Zh. Tekh. Fiz. 40, 2419 (1970)

2.24 Yu.N. Solodkin: Avtometriya 5, 64 (1973)

2.25 R.L. van Renesse, F.A.T. Bouts: Optik 38, 156 (1973)

2.26 M.M. Butusov, S.N. Gulyaev, V.V. Sudarikov: Kvantovaya Elektron. 2, 13 (1975)

2.27 W.T. Cathey: J. Opt. Soc. Am. 55, 457 (1965)

2.28 J.E. Sollid, J.B. Swint: Appl. Opt. 9, 2717 (1970)

2.29 L.K. Vagin, V.A. Vanin, L.G. Nazarova: In Problemy golografii (Problems of Holography), vyp. III (Izd. MIREA, Moscow 1973) p. 84

2.30 N.I. Kirillov, V.A. Barachevsky (eds.): Registriruyushchie sredy dlya golografii (Recording Media for Holography) (Nauka, Leningrad 1975)

2.31 N.G. Nakhodkin, N.G. Kuvshinsky, I.M. Pochernyaev: In Sposoby zapisi informatsii na besserebryanykh nositelyakh (Ways of Recording Information on Non-Silver Carriers), vyp. 5 (Visshaya shkola, Kiev 1974) p. 3

2.32 J. Urbach, R.W. Meier: Appl. Opt. 8, 2269 (1969)

2.33 J. Bordogna, S.A. Keneman, J.J. Amodei: In Materialy V Vsesoyuznoi shkoly po golografii (Materials of the 5th All-Union School on Holography) (Izd. LIYaF, Leningrad 1973) p. 567

2.34 A.A. Bugaev, B.P. Zakharchenya, F.A. Chudnovsky: FTIROS - novy registriruyuschii material dlya golografii (FTIROS - a New Recording Material for Holography) (Izd. LDNTP 1976)

2.35 B.P. Zakharchenya, I.K. Meshkovsky, E.I. Terukov, F.A. Chudnovsky: Pis'ma (Letters) Zh. Tekh. Fiz. 1, 8 (1975)

2.36 A.A. Bugaev, B.P. Zakharchenya, I.K. Meshkovsky, V.M. Ovchinnikov, F.A. Chudnovsky: Pis'ma Zh. Tekh. Fiz. 1, 209 (1975)

2.37 L.H. Tanner: J. Sci. Instr. 43, 81, 346, 353 (1966)

2.38 H.J. Caulfield: Laser Focus 21, 23 (1967)

2.39 G.L. Rogers: J. Opt. Soc. Am. 61, 784 (1971)

2.40 M.M. Butusov (ed.): Primenenie golograficheskoi interferometrii i lazernoi tekhniki dlya kontrolya promyshlennykh izdelii (Application of Holographic Interferometry and Laser Techniques for Controlling Industrial Articles) (Izd. VNIINMASh, Gorky 1976)

2.41 M.M. Butusov, Yu.G. Turkevich: Prib. Tekh. Eksp. 6, 179 (1969)

2.42 B.G. Turukhano, N. Turukhano: Zh. Tekh. Fiz. 38, 757 (1968)

2.43 M.M. Butusov, Yu.G. Turkevich: Prib. Tekh. Eksp. 6, 179 (1970)

2.44 R. Roy: Amer. J. Phys. 37, 748 (1969)

2.45 R.M. Brown: Appl. Opt. 9, 1726 (1970)

2.46 L. Levi: Applied Optics (Wiley, New York 1968)

2.47 O.B. Gusev, V.B. Konstantinov: Zh. Tekh. Fiz. 39, 354 (1969)

2.48 L.P. Boivin: Appl. Opt. 11, 1782 (1972)

2.49 M.M. Butusov, in Golograficheskie izmeritel'nye sistemy (Holographic Measuring Systems), ed. by A.G. Kozachok (Izd. NETI, Novisibirsk 1976) p. 100

2.50 M.M. Butusov, Yu.G. Turkevich: Zh. Nauchn. Prikl. Fotogr. Kinematogr. 16, 303 (1971)

2.51 A. Hioki, T. Suzuki: Jpn. J. Appl. Phys. 4, 817 (1965)

3.1 R.W. Ladenburg (ed.): Physical Measurements in Gas Dynamics and Combustion (Oxford University Press, London 1955)

3.2 L.A. Vasilev: Tenevye metody (Shadow Methods) (Nauka, Moscow 1968)

3.3 V.M. Ginsburg, B.M. Stepanov (eds.): Golografiya. Metody i apparatura (Holography. Methods and Apparatus) (Sov. radio, Moscow 1974)

3.4 F.D. Bennett, D.D. Shear, H.S. Burden: J. Opt. Soc. Am. 50, 212 (1960)

3.5 F.C. Jahoda, E.M. Little, W.E. Quinn, F.L. Ribe, G.A. Sawyer: J. Opt. Soc. Am. 35, 2351 (1964)

3.6 W. Hauf, U. Grigull: Optical Methods in Heat Transfer (Academic Press, New York 1970)

3.7 A.G.Smirnov, V.G. Smirnov, D.I. Staselko: Preprint NIIEFA T-0195 (1974). A.G. Smirnov, in Opticheskaya golografiya i ee primenenie (Optical Holography and Its Application) (Izd. LDNTP, Leningrad 1974), p. 45

3.8 A.F. Belozerov, A.N. Berezkin, A.I. Razumovskaya, N.M. Spornik: Zh. Tekh. Fiz. 43, 777 (1973)

3.9 A.F. Belozerov, A.N. Berezkin, A.I. Razumovskaya, N.M. Spornik: Proc. 10th Congress of High Speed Photography, Nice, France (1974) p. 401

3.10 V.K. Demkin, V.A. Nikashin, V.K. Sakharov, V.K. Tarasov: Zh. Tekh. Fiz. 40, 1424 (1970)

3.11 M.M. Butusov: In Materialy III Vsesoyuznoi shkoly po golografii (Materials of the 3rd All-Union School on Holography) (Izd. LIYaF, Leningrad 1972) p. 195

3.12 R.E. Brooks, L.O. Heflinger, R.G. Wuerker: IEEE J. QE-2, 275 (1966)

3.13 S. Lowenthal, Y. Belvaux: C.R. Acad. Sci. Paris 263, 904 (1966)

3.14 S. Lowenthal, Y. Belvaux: Revue d'Optique 1, 1 (1967)

3.15 F.C. Jahoda: Holographic Plasma Diagnostics. Preprint of Los Alamos Sci. Lab. of the Univ. of Calif. LA-3968 (1968)

3.16 C.M. Vest, D.W. Sweeney: Appl. Opt. 9, 2321 (1970)

3.17 A. Kakos, G.V. Ostrovskaya, Yu.I. Ostrovsky, A.N. Zaidel: Phys. Lett. 23, 81 (1966)

3.18 F.C. Jahoda, R.A. Jeffries, G.A. Sawyer: Appl. Opt. 6, 1407 (1967)

3.19 A.N. Zaidel, Yu.I. Ostrovsky: Proc. 8th Int. Congress of High Speed Photography, Stockholm (1968) p. 309

3.20 E.B. Aleksandrov, A.M. Bonch-Bruevich: Zh. Tekh. Fiz. 37, 360 (1967)

3.21 T. Tsuruta, N. Shiotake, Y. Itoh: Opt. Acta 16, 723 (1969)

3.22 R.J. Collier, C.B. Burckhardt, L.H. Lin: Optical Holography (Academic Press, New York 1971)

3.23 I.S. Zeilikovich, V.A. Komissaruk: Opt. Spektrosk. 39, 985 (1975)

3.24 B.J. Thompson, J.H. Ward, W. Zinky: Appl. Opt. 6- 519 (1967)

3.25 W.T. Welford: Appl. Opt. 5, 872 (1966)

3.26 B.G. Turukhano: Zh. Tekh. Fiz. 40, 181 (1970)

3.27 D.I. Staselko, V.A. Kosnikovsky: Opt. Spektrosk. 33, 365 (1973)

3.28 W.D. Pearce: In Conference on Extremely High Temperatures, ed. by H. Fischer, L.C. Mansur (Wiley, New York 1958)

3.29 K. Bokasten: J. Opt. Soc. Am. 51, 943 (1961)

3.30 O.H. Nestor, H.N. Olsen: SIAM Rev. 2, 200 (1960)

3.31 W.L. Bohn, M.U. Beth, G. Nedder: J. Quant. Spectrosc. Radiat. Transfer 7, 661 (1967)

3.32 M.P. Freeman, S. Katz: J. Opt. Soc. Am. 50, 826 (1960)

3.33 M. Kock, J. Richter: Ann. Phys. 52, 885 (1962)

3.34 L.T. Larkina: In Primenenie plazmotrona v spektroskopii (The Use of a Plasmatron in Spectroscopy) (Ilim, Frunze 1970)

3.35 V.A. Gribkov, V.Ya. Nikulin, G.V. Sklizkov: Kvantovaya elektronika (Quantum Electronics), ed. by N.G. Basov, No. 6, 60 (1971)

3.36 I.D. Kulagin, L.M. Sorokin, E.A. Dubrovskaya: Opt. Spektrosk. 32, 865 (1972)

3.37 V.V. Pikalov, N.G. Preobrazhensky: Fiz. Goreniya Vzryva 6, 923 (1974)

3.38 E.I. Kuznetsov, D. Shcheglov: Diagnostika vysokotemperaturnoi plazmy (Diagnostic of High-Temperature Plasma) (Atomizdat. Moscow 1974)

3.39 V.F. Turchin, V.P. Kozlov, M.Ts. Malkevich: Usp. Fiz. Nauk 102, 345 (1970)

3.40 M.M. Lavrentev: Uslovno-korrektnye zadachi dlya differentsial'nykh uravnenii (Conditionally Correct Problems for Differential Equations (Novosibirsk State Univ. 1973)

3.41 G.V. Ostrovskaya: Zh. Tekh. Fiz. 46, 2529 (1976)

3.42 T.V. Bazhenova, Z.S. Leonteva, V.S. Pushkin: In Gazodinamika i fizika goreniya (Gas Dynamics and Physics of Combustion) (Izd. AN SSSR, Moscow 1950)

3.43 S.M. Belotserkovsky, V.S. Sukhorukikh, V.S. Tatarenchik: Zh. Prikl. Mekh. Tekh. Fiz. 3, 95 (1964)

3.44 P.D. Rowley: J. Opt. Soc. Am. 59, 1496 (1969)

3.45 R.D. Matulka, D.J. Collins: J. Appl. Phys. 42, 1109 (1971)

3.46 K. Iwato, R. Nagata: J. Opt. Soc. Am. 60, 133 (1970)

3.47 D.W. Sweeney, C.M. Vest: Appl. Opt. 11, 205 (1972)

3.48 D.W. Sweeney, C.M. Vest: Appl. Opt. 12, 2649 (1973)

3.49 A.N. Zaidel, N.I. Kaliteevsky, L.V. Lipis, M.P. Chaika: Emissionny spektral'ny analiz atomnykh materialov (The Emission Spectral Analysis of Atomic Materials) (Fizmatgiz, Moscow 1960)

3.50 F. Weigl. O.M. Friedrich, Jr., A.A. Dougal: IEEE J. QE-6, 41 (1970)

3.51 A.F. Belozerov, A.N. Berezkin, L.T. Mustafina, A.I. Razumovskaya: in Fizicheskie metody issledovaniya prozrachnykh neodnorodnostei (Physical Methods of Studying Transparent Non-Homogeneities) (Izd. MDNTP, Moscow 1977)

3.52 T. Tsuruta, Y. Itoh: Appl. Opt. 8, 2033 (1969)

3.53 M. De, L. Sevigny: Appl. Opt. 6, 1665 (1967)

3.54 K.S. Mustafin, V.A. Seleznev: Opt. Spektrosk. 30, 154 (1971)

3.55 K.A. Haines, B.P. Hildebrand: J. Opt. Soc. Am. 57, 155 (1967)

3.56 F. Weigl, O.M. Friedrich, Jr., A.A. Dougal: IEEE Catalog, No. 69, C-16, SWIECO (1969)
F. Weigl.: Appl. Opt. 10, 187 (1971)

3.57 F. Weigl.: Appl. Opt. 10, 1083 (1971)

3.58 E. Leith, J. Upatnieks: J. Opt. Soc. Am. 57, 975 (1967)

3.59 J.M.J. Tokarski: Appl. Opt. 7, 989 (1968)

3.60 O. Bryngdahl, A.W. Lohmann: J. Opt. Soc. Am. 58, 141 (1968)

3.61 K.S. Mustafin, V.A. Seleznev, E.I. Shtyrkov: USSR Inventor's Certificate No. 272, 602 (1968); Byull. izobr. No. 19 (1970); Opt. Spektrosk. 28, 1186 (1970)

3.62 K. Matsumoto, M. Takashima: J. Opt. Soc. Am. 60, 30 (1970)

3.63 C.H.F. Velzel: Opt. Commun. 2, 289 (1970)

3.64 G.V. Ostrovskaya, Yu.I. Ostrovsky: Zh. Tekh. Fiz. 40, 2419 (1970)

3.65 A.B. Ignatov, I.I. Komissarova, G.V. Ostrovskaya, L.L. Shapiro: Zh. Tekh. Fiz. 41, 417 (1971)

3.66 3.66 I.I. Komissarova, G.V. Ostrovskaya: In Problemy golografii (Problems of Holography), vyp. 3 (Izd. MIREA, Moscow (1973) p. 50

3.67 Yu.N. Denisyuk, G.B. Semenov, N.A. Savostyanenko: Opt. Spektrosk. 29, 994 (1970)

3.68 A.V. Alekseev-Popov, I.I. Komissarova, G.V. Ostrovskaya: Opt. Spektrosk. 37, 1143 (1974)

3.69 A. Unsöld: Physik der Sternatmosphären mit besonderer Berücksichtigung der Sonne, 2nd ed. (Springer, Berlin 1955)

3.70 D.A. Lenard, J.C. Keck: ARS J. 32, 142 (1962)

3.71 Yu.I. Ostrovsky: USSR Inventor's Certificate No. 268, 732 (1961); Byull. Izobr. No. 14 (1970)

3.72 A.F. Belozerov, K.S. Mustafin, A.I. Sadykova, V.S. Fedoseev, E.I. Shtyrkov, V.A. Yakovlev, V.I. Yanichkin: Opt. Spektrosk. 29, 384 (1970)

3.73 Yu.I. Ostrovsky, L.V. Tanin: Zh. Tekh. Fiz. 45, 1756 (1975)

3.74 G.V. Ostrovskaya, N.A. Pobedonostseva: Zh. Tekh. Fiz. 45, 1562 (1975)

3.75 G.V. Dreiden, G.V. Ostrovskaya, N.A. Pobedonostseva, V.N. Filippov: Pis'ma Zh. Tekh. Fiz. 1, 106 (1975)

3.76 G.V. Dreiden, Yu.I. Ostrovsky, E.N. Shedova, A.N. Zaidel: Opt. Commun. 4, 209 (1971)

3.77 G.V. Dreiden, A.N. Zaidel, Yu.I. Ostrovsky, E.N. Shedova: Zh. Tekh. Fiz. 43, 1537 (1973)

3.78 K.M. Measures: Appl. Opt. 9, 737 (1970)

3.79 G.V. Dreiden, A.N. Zaidel, G.V. Ostrovskaya, Yu.I. Ostrovsky, N.A. Pobedonostseva, L.V. Tanin, V.N. Filippov, E.N. Shedova: Fiz. Plazmy 1, 462 (1975)

3.80 G.V. Ostrovskaya, Yu.I. Ostrovsky: Pis'ma Zh. Eksp. Teor. Fiz. 4, 121 (1966)

3.81 F.C. Jahoda: Proc. 8th Int. Conf. Phen. Ioniz. Gases, Vienna (1967) p. 509

3.82 A.N. Zaidel, G.V. Ostrovskaya, Yu.I. Ostrovsky: Zh. Tekh. Fiz. 38, 1406 (1968)

3.83 G.V. Ostrovskaya: In Materialy III Vsesoyuznoi shkoly po golografii (Materials of the 3rd All-Union School on Holography) (Leningrad 1972) p. 257

3.84 A.N. Zaidel, G.V. Ostrovskaya, Yu.I. Ostrovsky: Trudy GOI 42, 20 (1975)

3.86 A.N. Zaidel, G.V. Ostrovskaya, Yu.I. Ostrovsky, T.Ya. Chelidze: Zh. Tekh. Fiz. 36, 2208 (1966)

3.86 J.C. Buges, A. Plet, A. Terneaud: C.R. Acad. Sci. Paris 267, 1271 (1968)

3.87 J.L. Bobin, J.C. Buges, P. Rouzaud, A. Terneaud: Proc. 9th Int. Conf. Phen. Ioniz. Gas., Bucharest (1969) p. 638

3.88 I.I. Komissarova, G.V. Ostrovskaya, L.L. Shapiro: Zh. Tekh. Fiz. 38, 1369 (1968)

3.89 I.I. Komissarova, G.V. Ostrovskaya, L.L. Shapiro, A.N. Zaidel: Phys. Lett. 29A, 262 (1969)

3.90 I.I. Komissarova, G.V. Ostrovskaya, L.L. Shapiro: Zh. Tekh. Fiz. 40, 1072 (1970)

3.91 A.B. Ignatov, I.I. Komissarova, G.V. Ostrovskaya, L.L. Shapiro: Zh. Tekh. Fiz. 41, 701 (1971)

3.92 I.I. Komissarova, G.V. Ostrovskaya, L.L. Shapiro, A.N. Zaidel: Proc. 9th Int. Conf. Phen. Ioniz. Gas., Bucharest (1969) p. 641

3.93 A.B. Ignatov, I.I. Komissarova, G.V. Ostrovskaya, L.L. Shapiro: In Diagnostika plazmy (Diagnostic of Plasma), vyp. 3 (Atomizdat, Moscow 1973) p. 162

3.94 A.H. Guentner, W.K. Pendleton, C. Smith, C.H. Skeen, S. Zivi: J. Opt. Soc. Am. 61, 688 (1971)

3.95 A.H. Guenther, W.K. Pendleton, C. Smith, C.H. Skeen, S. Zivi: Opt. Laser Technol. 2, 20 (1973)

3.96 R. Sigel: Phys. Lett. 30a, 103 (1969)

3.97 R. Sigel: Z. Naturforsch. 25A, 488 (1970)

3.98 I.I. Ashmarin, Yu.A. Bykovsky, N.N. Degtyarenko, V.F. Elesin, A.I. Larkin, I.P. Sipailo: Zh. Tekh. Fiz. 41, 2369 (1971)

3.99 R. Belland, C. de Michelis, M. Mattioli: Opt. Commun. 3, 7 (1971)

3.100 Yu.V. Ascheulov, A.D. Dymnikov, Yu.I. Ostrovsky, A.N. Zaidel: Phys. Lett. 24A, 61 (1967)

3.101 V.A. Nikashin, G.I. Rukman, V.K. Sakharov, V.K. Tarasov: Teplofiz. Vys. Temp. 7, 1198 (1969)

3.102 V.M. Ginzburg, B.M. Stepanov, Yu.I. Filenko: Radiotekh. Electron. 17, 2219 (1972)

3.103 A.P. Burmakov, G.V. Ostrovskaya: Zh. Tekh. Fiz. 40, 660 (1970)

3.104 A.P. Burmakov, V.B. Avramenko, A.A. Labuda, L.Ya. Minko: In Problemy golografii (Problems of Holography), vyp. 3 (Izd. MIREA, Moscow 1973) p. 43

3.105 A.P. Burmakov, A.A. Labuda, V.M. Lutkovsky: Inzh. Fiz. Zh. 29, 499 (1975)

3.106 A.P. Burmakov, V.A. Zaikov, G.M. Novik: In Teoreticheskaya fizika. Fizika plazmy (Theoretical Physics. Physics of Plasma) (Nauka i tekhnika, Minsk 1975) p. 75

3.107 T.D. Butler, J. Henins, F.C. Jahoda, J. Marshall, R.L. Morze: Phys. Fluids 12, 1904 (1969)

3.108 F.C. Jahoda: Appl. Phys. Lett. 14, 341 (1969)

3.109 M.V. Stabnikov, M.Sh. Tombak: Zh. Tekh. Fiz. 41, 1310 (1971)

3.110 K.S. Mustafin, V.I. Protasevich, V.N. Rzhevsky: Opt. Spektrosk. 30, 406 (1971)

3.111 R.A. Jeffries: Phys. Fluids 13, 210 (1970)

3.112 R.F. Gribble, W.E. Quinn, R.E. Siemon: Phys. Fluids 14, 2042 (1971)

3.113 K.S. Thomas, C.R. Harder, W.E. Quinn, R.E. Siemon: Phys. Fluids 15, 1658

3.114 L.V. Dubovoi, A.G. Smirnov, V.G. Smirnov, D.I. Staselko: In Opticheskaya golographiya i ee primenenie (Optical Holography and Its Application) (Izd. LDNTP, Leningrad 1974) p. 47

3.115 V.G. Smirnov, A.G. Smirnov: Preprint T-0194 NIIFA (Leningrad 1974)

3.116 G.V. Dreiden, A.N. Zaidel, V.S. Markov, A.M. Mirzabekov, G.V. Ostrovskaya, Yu.I. Ostrovsky, N.P. Tokarevskaya, A.G. Frank, A.Z. Khodzhaev, E.N. Shedova: Pis'ma Zh. Tekh. Fiz. 1, 141 (1975); Fiz. plazmy 3, 45 (1977)

3.117 G.V. Ostrovskaya, A.N. Zaidel: Phys. Lett. 26A, 393 (1968)

3.118 R. Huddlestone, S. Leonard (Eds.): Plasma Diagnostic Techniques (Academic Press, New York 1965)

3.119 S.E. Frish: Opticheskie spektry atomov (Optical Spectra of Atoms) (Fizmatgiz, Moscow 1963)

3.120 M. Born, E. Wolf: Principles of Optics, 3rd ed. (Pergamon Press, Oxford 1965)

3.121 G.G. Dolvog, G.D. Petrov: Materialy X Vsesoyuznogo soveshchaniya po spektroskopii (Materials of the 10th All-Union Conference on Spectroscopy), Vol. 2 (Lvov University Press, 1958) p. 68

3.122 R.A. Alpher, D.R. White: Phys. Fluids 1, 452 (1958); 2, 153, 162 (1959)

3.123 W. Braun: Phys. Lett. A47, 144 (1974); Z. Phys. B20, 195 (1975)

3.124 P.R. Forman, S. Humphries, Jr., R.W. Peterson: Appl. Phys. Lett. 22, 537 (1937)

3.125 R. Kristal: Appl. Opt. 14, 628 (1975)

3.126 R.G. Hall, J.W. Gates, I.N. Rose: J. Phys. E3, 789 (1970)

3.127 G.V. Dreiden, Yu.I. Ostrovsky, E.N. Shedova: Opt. Spektrosk. 32, 367 (1972)

3.128 R.E. Brooks, L.O. Heflinger, R.F. Wuerker: IEEE J. QE-2, 275 (1966)

3.129 E.A. Kuznetsova, L.N. Prokhorova, V.Ya. Tsarfin: In Golograficheskie metody i apparatura, primenyaemaya v fizicheskikh issledovaniyakh (Holographic Methods and Apparatus Used in Physical Research) (Izd. VNIIOFI, Moscow 1974) p. 38

3.130 A.G. Rozanov: Opt. Spektrosk. 33, 1188 (1972)

3.131 V.Z. Bryskin, A.G. Smirnov: In Opticheskaya golografiya i ee primenenie (Optical Holography and Its Application) (Izd. LDNTP, Leningrad 1974) p. 26

3.132 L. Heflinger, R. Wuerker, R. Brooks: J. Appl. Phys. 37, 642 (1966)

3.133 A.N. Berezkin, Yu.E. Kamach, E.N. Kozlovsky, V.M. Ovchinnikov, A.I. Razumovskaya: Zh. Tekh. Fiz. 42, 219 (1972)

3.134 A.F. Belozerov, A.N. Berezkin, N.P. Mudrevskaya, L.T. Mustafina, A.I. Razumovskaya: Proc. 11th Int. Congr. High Speed Photogr., London (1974) p. 301

3.135 E.P. Kazandzhan, V.S. Sukhorukikh: USSR Inventor's Certificate No. 280, 920, Byull, izobr. No. 28 (1970)

3.136 E.P. Kazandzhan, V.S. Sukhorukikh: In Materialy II Vsesoyuznoi shkoly po golografii (Materials of the 2nd All-Union School on Holography) (Leningrad 1971) p. 166

3.137 C.J. Reinheimer, C.E. Wiswall, R.A. Schmiege, R.J. Harris, J.E. Dueker: Appl. Opt. 9, 2059 (1970)

3.138 L.H. Tanner: J. Sci. Instrum. 43, 81 (1966)

3.139 L.H. Tanner: J. Sci. Instrum. 43, 353 (1966)

4.1 A.A. Michelson: Philos. Mag. 13, 236 (1882)
A.A. Michelson: Studies in Optics (University of Chicago Press, 1927)

4.2 M. Born, E. Wolf: Principles of Optics, 3rd ed. (Pergamon Press, Oxford 1965)

4.3 N.G. Vlasov: Zh. Tekh. Fiz. 40, 1656 (1970)

4.4 V.A. Kizel: Otrazhenie sveta (Light Reflection) (Nauka, Moscow 1973)

4.5 F.G. Bass, I.M. Fuks: Rasseyanie voln na statisticheski nerovnoi poverkhnosti (Wave Scattering on a Statistically Rough Surface) (Nauka, Moscow 1972)
A.P. Khusu, Yu.D. Vitenberg, V.A. Palmov: Sherokhovatost' poverkhnostei - teoretiko-veroyatnostny podkhod (Surface Roughness - A Theoretical-Probabilistic Approach) (Nauka, Moscow 1975)

4.6 R.J. Collier, C.B. Burckhardt, L.H. Lin: Optical Holography (Academic Press, New York 1971)

4.7 J.C. Dainty (ed.): Laser Speckle and Related Phenomena, Topics in Applied Physics, Vol. 9 (Springer, Berlin, Heidelberg, New York 1975)

4.8 R. Dandliker, E. Marom, F.M. Mottier: Opt. Commun. 6, 368 (1972)

4.9 Lord Rayleigh: Theory of Sound (Dover, New York 1945)

4.10 A. Angot: Compléments de Mathématique. A l'Usage des ingénieurs de l'élektrotechnique et des télécommunications (Paris 1957)

4.11 N.E. Molin, K.A. Stetson: Optik 31, 3 (1970)

4.12 E.B. Aleksandrov, M.A. Bonch-Bruevich: Zh. Tekh. Fiz. 37, 360 (1967)

4.13 J.Ch. Viénot, P. Smigelski, H. Royer: Holographie Optique (Dunod, Paris 1971)

4.14 S. Walles: Ark. Fys. 40, 26 (1969)

4.15 J. Tsuruta, N. Shiotake, Y. Itoh: Jpn. J. Appl. Phys. 7, 1092 (1968)

4.16 J.E. Sollid: Opt. Commun. 2, 282 (1970)

4.17 A.E. Ennos: J. Sci. Instrum. Ser. 2, 1, 731 (1968)

4.18 J.E. Sollid: Appl. Opt. 8, 1587 (1969)

4.19 N. Abramson: Appl. Opt. 8, 1235 (1969)

4.20 N. Abramson: Appl. Opt. 9, 97 (1970)

4.21 N. Abramson: Appl. Opt. 9, 2311 (1970)

4.22 N. Abramson: Appl. Opt. 10, 2155 (1971)

4.23 N. Abramson: Appl. Opt. 11, 1143 (1972)

4.24 N. Abramson: Appl. Opt. 11, 2562 (1972)

4.25 N. Abramson: Nature 231, 2065 (1971)

4.26 Yu.I. Ostrovsky: Holography and Its Application, Transl. by G. Leib (Mir Publishers, Moscow 1977)

4.27 K. Shibajama, H. Uchijama: Appl. Opt. 10, 2150 (1971)

4.28 Yu.N. Solodkin: Avtometriya 5, 64 (1973)

4.29 C.H.F. Velzel: J. Opt. Soc. Am. 60, 419 (1970)

4.30 P.M. Boone, L.C. de Backer: Optik 37, 61 (1973)

4.31 I.S. Klimenko, E.G. Matinian: Opt. Spektrosk. 27, 367 (1969)

4.32 I.S. Klimenko, G.V. Skrotsky: Usp. Fiz. Nauk 109, 269 (1973)

4.33 E. Archbold, J.M. Burch, A.E. Ennos: Opt. Acta 17, 883 (1970)

4.34 E. Archbold, A.E. Ennos: Opt. Acta 19, 253 (1972)

4.35 L. Hode, K. Biedermann: Abstr. Swed. Phys. Congr. Lund, Sweden (1972)

4.36 B.P. Hildebrand, K.A. Haines: J. Opt. Soc. Am. 57, 155 (1967)

4.37 J.S. Zelenka, R.R. Varner: Appl. Opt. 8, 1431 (1969)

4.38 S.T. De, A.G. Kozachok, A.V. Loginov: In Opticheskaya golografiya i ee primenenie (Optical Holography and Its Application) (Izd. LDNTP, Leningrad 1974) p. 12

4.39 W. Schmidt, A. Vogel, D. Preussler: Appl. Phys. 1, 103 (1973)

4.40 L.O. Heflinger, R.F. Wuerker: Appl. Phys. Lett. 15, 28 (1969)

4.41 N. Shiotake, T. Tsuruta, Y. Itoh: Jpn. J. Appl. Phys. 7, 904 (1968)

4.42 E.S. Marrone, W.B. Ribbens: Appl. Opt. 14, 23 (1975)

4.43 D.M. Bavelsky, V.S. Listovets, Yu.I. Ostrovsky, V.F. Sidorenko, V.V. Trofimovsky, V.A. Chubarov, M.I. Etinberg: Energomashinostroenie No. 8, 21 (1976)

4.44 I.M. Nagibina, V.L. Kazak, T.A. Ilyinskaya, A.S. Dorofeyuk: In Opticheskaya golografiya i ee primenenie (Optical Holography and Its Application) (Izd. LDNTP, Leningrad 1974) p. 3

4.45 V.M. Ginzburg, B.M. Stepanov (eds.): Golografiya. Metody i apparatura (Holography. Methods and Apparatus) (Sov. radio, Moscow 1974)

4.46 R.M. Grant, G.M. Brown: Mater. Eval. 27, 79 (1969)

4.47 V.S. Obraztsov, D.N. Sitnik: In Opticheskaya golografiya i ee primenenie (Optical Holography and Its Application) ed. by Yu.N. Denisyuk, Yu.I. Ostrovsky (Nauka, Leningrad 1977)

4.48 M.M. Butusov, B.A. Belogordsky: Golograficheskii analiz elektromekhanicheskikh preobrazovatelei (Holographic Analysis of Electromechanical Transducers) (Izd. LDNTP, Leningrad 1974)

4.49 C.M. Vest, E.L. McKague, A.A. Friesem: J. Basic Eng. 6, 237 (1971)

4.50 M.A. Zarutsky: Izv. AN Latv. SSR, Ser. Fiz. Tekh. Nauk 4, 79 (1976)

4.51 A.E. Ennos: In Research Techniques in Nondestructive Testing, ed. by R.S. Sharpe (Academic Press, New York 1970)

5.1 H. Osterberg: J. Opt. Soc. Am. 22, 19 (1932)

5.2 R.L. Powell, K.A. Stetson: J. Opt. Soc. Am. 55, 1953, 1694 (1965)

5.3 J.W. Goodman: Appl. Opt. 6, 857 (1967)

5.4 J.D. Redman: J. Sci. Instrum. 44, 1032 (1967)

5.5 M. Lurie: J. Opt. Soc. Am. 57, 573A (1967)

5.6 M. Lurie: J. Opt. Soc. Am. 58, 614 (1968)

5.7 D.B. Neumann: J. Opt. Soc. Am. 58, 1003 (1968)

5.8 A.S. Bogomolov, N.G. Vlasov, E.G. Solovev: Opt. Spektrosk. 31, 481 (1971)

5.9 M.M. Butusov: Opt. Spektrosk. 37, 532 (1974)

5.10 M.R. Wall: Symp. Engineering Uses of Holography, Glasgow, Sept. 1968 (Cambridge, 1970)

5.11 M. Lurie, M. Zambuto: Appl. Opt. 7, 2323 (1968)

5.12 M. Zambuto, M. Lurie: Appl. Opt. 9, 2066 (1970)

5.13 M.O. Fein, E.L. Green: Appl. Opt. 7, 1864 (1968)

5.14 N.E. Molin, K.A. Stetson: J. Sci. Instrum. 2, 609 (1969)

5.15 A.D. Wilson, D.H. Strope: J. Opt. Soc. Am. 60, 1162 (1970)

5.16 R. Aprahamian, D.A. Evensen: J. Appl. Mech., ser. E, 37, Nor. 2, June, 287 (1970)

5.17 R. Aprahamian, D.A. Evensen: J. Appl. Mech., ser. E, 37, No. 4, December, 1083 (1970)

5.18 P. Waddell, W. Kennedy: New Scientist, June, 633 (1968)

5.19 P. Waddell, W. Kennedy: Symp. Engineering Uses of Holography, Glasgow, Sept. 1968 (Cambridge, 1970)

5.20 W.G. Alwang, R. Burr, L.A. Cowanaugh: SAE Preprint No, 265, 1 (1969)

5.21 P. Spicer, J.W. Langworthy, D. Holt: Flight No. 3126, 253 (1969)

5.22 B.S. Hockley, R.S. Hill: Aircraft Eng. 41, No. 8, 6 (1969)

5.23 B.S. Hockley, J.N. Butters: J. Photogr. Sci. 18, 16 (1970)

5.24 B.S. Hockley, J.N. Butters: J. Mech. Eng. Sci. 12, No. 1 (1970)

5.25 W.F. Fagan, P. Waddell, W. McCracken: Opt. Laser Technol. Ref. 5.25, 167 (1972)

5.26 V.S. Listovets, Yu.I. Ostrovsky: Zh. Tekh. Fiz. 44, 1345 (1974)

5.27 D.M. Bavelsky, V.S. Listovets, Yu.I. Ostrovsky, V.F. Sidorenko, V.V. Trofimovsky, V.A. Chubarov, M.I. Etinberg: Energomashinostroenie No. 8, 21 (1976)

5.28 E. Jansson, N.E. Molin, H. Sundin: Phys. Scr. 2, 243 (1970)

5.29 C.-H. Ågren, K.A. Stetson: J. Acoust. Soc. Am. 51, No. 6 (Part 2), 1971 (1972)

5.30 B.A. Belogorodsky, M.M. Butusov, Yu.G. Turkevich: Akust. Zh. 17, 455 (1971)

5.31 M. Lashkari, V.I. Veingarten: Exper. Mech. March, 120 (1973)

5.32 B.A. Belogordsky, E.P. Smirnov, Yu.G. Turkevich, E.I. Kheifets: Akust. Zh. 18, 303 (1971)

5.33 A.N. Zaidel, Yu.I. Ostrovsky: Proc. VIII Int. Conf. Phen. in Ioniz. Gases, p. 508 Vienna (1967)

5.34 A.N. Zaidel, L.G. Malkhasian, G.V. Markova, Yu.I. Ostrovsky: Zh. Tekh. Fiz. 37, 1825 (1968)

5.35 B. Watrasiewicz, P. Spicer: Nature 217, 1142 (1968)

5.36 E. Archbold, A.E. Ennos: Nature 217, 942 (1968)

5.37 P. Shaenko, C.D. Johnson: Appl. Phys. Lett. 13, 44 (1968)

5.38 For example, G.A. Korn, T.A. Korn: Mathematical Handbook for Scientists and Engineers (McGraw-Hill, New York 1968)

5.39 V.I. Gerasev, Yu.I. Ostrovsky: Tez. dokl. i soobshen. Vsesoyuzn. konf. po avtomatizatsii izmerenii (Abstracts of Reports to All-Union Conf. on Measurement Automation) (Novosibirsk 1970) p. 37

5.40 C.R. Hazell, C.D. Liem: Proc. Int. Symp. Appl. of Holography. Besançon, France (1970)

5.41 C.T. Moffatt, B.W. Watrasiewicz: Proc. Int. Symp. Appl. of Holography. Besançon, France (1970)

5.42 D.F. Nelson: Sci. Am. 218, 17 (1968)

5.43 F.S. Chen: Proc. IEEE 58, 1440 (1970)

5.44 A.J. Waddell, W. Kennedy, P. Waddell: Proc. Int. Symp. Appl. of Holography. Besancon, France (1970)

5.45 M.M. Butusov, Yu.G. Turkevich, V.Ya. Demchenko: Prib. Tekh. Eksper. No. 2, 203 (1971)

5.46 C.C. Aleksoff: Appl. Phys. Lett. 14, 23 (1969)

5.47 C.C. Aleksoff: Appl. Opt. 10, 1329 (1971)

5.48 B.A. Belogorodsky, M.M. Butusov, Yu.G. Turkevich: Avtometrya, 1, 47 (1972)

5.49 F.M. Mottier: Appl. Phys. Lett. 15, 44 (1969)

5.50 F.M. Mottier: Appl. Phys. Lett. 15, 285 (1969)

5.51 H.S. Caulfield: Appl. Phys. Lett. 16, 234 (1970)

5.52 H.S. Caulfied, J.L. Harris, H.W. Hemstreet, Jr., J.C. Gobb: Proc. IEEE 55, 1758 (1967)

5.53 W.T. Cathey: U.S. Patent 3, 415, 587 (1968)

5.54 D. Gabor, G.W. Stroke, R. Restrick, A. Funkhouser, D. Brumm: Phys. Lett. 18, 116 (1965)

5.55 L.F. Collins: Appl. Opt. 7, 203 (1968)

5.56 M.R. Wall: Opt. Technol. 1, 266 (1969)

5.57 K. Biedermann, N.E. Molin: J. Phys. 3, 669 (1970)

5.58 P. Hariharan, B.S. Ramprasad: J. Phys. 5, 976 (1972)

5.59 P. Hariharan: Appl. Opt. 12, 143 (1973)

5.60 D.B. Neumann, C.F. Jacobson, G.M. Brown: Appl. Opt. 9, 1357 (1970)

5.61 G.M. Mayer: J. Appl. Phys. 40, 2863 (1969)

Subject Index

www.ingramcontent.com/pod-product-compliance
Ingram Content Group UK Ltd.
Pitfield, Milton Keynes, MK11 3LW, UK
UKHW021835190726
13853UKWH00003B/1301
* 9 7 8 3 6 6 2 1 3 4 8 8 7 *